D1480373

THE BLACK DOUGLASES

Also by Michael Brown:
James I

THE BLACK DOUGLASES

War and Lordship in Late Medieval Scotland
1300–1455

Michael Brown

TUCKWELL PRESS

First published in Great Britain in 1998 by
Tuckwell Press Ltd
Phantassie
East Linton
East Lothian EH40 3DG
Scotland

Reprinted 1999

ISBN 1 86232 036 5 (cased)

The Publishers acknowledge subsidy from
The Scottish Arts Council towards the publication of
this volume

British Library Cataloguing-in-Publication Data
A catalogue record for this book is available
from the British Library

Typeset by Carnegie Publishing, Chatsworth Road, Lancaster
Printed and bound by Biddles Limited, Guildford and King's Lynn

Contents

ACKNOWLEDGEMENTS

As befits a study of a family, like the Douglases, with international ambitions and connections, this book, first conceived in Wales, has been written and researched in Scotland, England, France and Ireland. In the course of these travels I have had the pleasure of meeting and working with numerous colleagues who have given me enormous help and encouragement for which I am grateful. In particular, I would like to thank the Institute of Scottish Historical Studies at the University of Strathclyde and Professor Tom Devine for awarding me the research fellowship which allowed me to initiate and do much of the groundwork for this project. I am also grateful to the University of Aberdeen for giving me a second research fellowship, during which I was able to devote the time necessary to complete the writing of *The Black Douglases*. I would like to thank University College Dublin for funding my research in France.

I also owe many personal debts of gratitude. Dr Sandy Grant of the University of Lancaster kindly made available to me his collection of the acts of the fourth earl of Douglas and has been generous with his insights into the Douglas family and his own work on late medieval Scotland. Dr Steve Boardman of the University of Edinburgh has also been free with his own research on fourteenth-century Scotland and has been prepared to listen to my own ramblings. The Tuckwells have once more proved a pleasure to work with, enthusiastic with the project and patient with its recalcitrant author. I owe a special debt to Dr Norman Macdougall of the University of St Andrews. His careful reading of, and comments on, the text were above and beyond the call of duty and his support and encouragement of my research have been a major factor in the completion of this book.

I would like to take this opportunity to thank my family and especially my uncles and aunts, Eric and Evelyn Ireland and Bob and Ella Tollervey, for their hospitality to their itinerant nephew during research trips to Edinburgh in the course of writing this book. Finally, my wife, Margaret, has been a constant source of support and encouragement in the writing of this book, even if I suspect her interest in the complexities of Scottish political society may have remained strictly limited. For this I forgive her; for the rest I am deeply grateful.

The photographs in the plate section, of Château Gaillard and of Loches, are my own. The artefacts found at Threave are reproduced by permission of the Public Record Office, London. All the remaining photographs are reproduced with the permission of Historic Scotland.

ILLUSTRATIONS

MAPS

GENEALOGICAL TABLES

ABBREVIATIONS

A.B. Ill.	*Illustrations of the Topography and Antiquities of the Shires of Aberdeen and Banff*
A.P.S.	*The Acts of the Parliaments of Scotland*
Cal. Docs. Scot.	*Calendar of Documents Relating to Scotland*
Calendar of Papal Letters	*Calendar of Entries in the Papal Registers in the Papal Registers relating to Great Britain and Ireland: Papal Letters*
Calendar of Papal Petitions	*Calendar of Entries in the Papal Registers relating to Great Britain and Ireland: Petitions to the Pope*
C.S.S.R.	*Calendar of Scottish Supplications to Rome*
Copiale	*Copiale Prioratus Sanctiandree*
Dumfriesshire Trans.	*Transactions of the Dumfriesshire and Galloway Natural History and Antiquarian Society*
E.R.	*The Exchequer Rolls of Scotland*
Foedera	*Foedera, Conventiones, Litterae et Cuiscunque Generis Acta Publica*
H.M.C.	*Reports of the Royal Commission on Historical Manuscripts*
N.L.S.	National Library of Scotland
N.L.S., Adv. MSS.	National Library of Scotland, Manuscripts of the Advocates' Library Collection
N.R.A.S.	National Register of Archives (Scotland)
P.P.C.	*Proceedings of the Privy Council*
R.M.S.	*Registrum Magni Sigilli Regum Scotorum*
R.R.S.	*Regesta Regum Scotorum*
Rot. Scot.	*Rotuli Scotiae in Turri Londiniensi et in Domo Capitulari Westmonasteriensi Asservati*
R.C.A.H.M.S.	*Reports of the Royal Commission on the Ancient and Historical Monuments of Scotland*
S.H.R.	*Scottish Historical Review*
S.H.S. Miscellany	*The Miscellany of the Scottish History Society*
S.R.O.	Scottish Record Office
S.P.	*The Scots Peerage*

Introduction: 'A Large and Attractive Book'

The Black Douglases were amongst the most powerful and certainly the most notorious of the great aristocratic families of late medieval Scotland. Their name and reputation creates an image of warfare on the borders with England, of the defence of king and kingdom by the Good Sir James Douglas, the first of the family to achieve widespread fame, and his heirs. The family was also associated with darker deeds. The great castles built or held by Douglas lords, Tantallon, Threave, Bothwell, the Hermitage and others, were seen to symbolise the power and arrogance of these great magnates within Scotland. This power rested on the rule, or misrule, of the family over its tenants and neighbours, a dominance maintained by fear and force and only ended by a bloody conflict with their own lord, the king of Scots. The climax of this conflict, the stabbing of the Douglas earl by King James II himself, was a fitting culmination to a history punctuated by similar acts of violence by lords of the Douglas name. This two-sided reputation, as patriotic heroes and as overmighty subjects defying their king, was born, not simply from the actions of Douglas lords, but from the changing perceptions and preoccupations of Scottish historians from the fourteenth to the twentieth centuries. Whatever their view, though, these writers all regarded the rise and fall of the Douglas dynasty as central to the development of Scotland in the later middle ages. This importance is not just because the relationships of such great lords with the local communities they ran, and with the crown which, in theory, ran them, has consistently been identified as 'the dominant theme' of all European political societies in the later middle ages.[1] It is also because the history of the Douglases as great magnates is bound up with the emergence of Scotland as an independent European kingdom in the later middle ages. For a nation lacking such status in the modern world, the place of great nobles in the independent kingdom, as defenders of its liberties and existence or as a check on the development of the Scottish state, assumes a special importance. As patriots or robber barons no magnates have greater significance for the history of Scotland than the Douglas earls.

To their contemporaries, the fame of the Douglases sprang overwhelmingly from their feats in warfare against England. While to these English enemies the Douglases were dangerous foes, whose successes could be attributed to pacts with the devil, and to continental observers they were a dynasty of chivalrous and cultured knights, in Scotland they were defenders of kingdom and community. It was on the efforts of Douglas lords in the cause of Robert Bruce and his heirs, regarded by the historians of late medieval Scotland as a patriotic struggle, that their own compatriots concentrated. In the 1370s John Barbour portrayed the first great lord of Douglas, the Good Sir James, as the foremost of King Robert's paladins in his epic poem about the war for Scotland's freedom, *The Bruce*. Seventy years later in his history of the Scottish people, *Scotichronicon*, Walter Bower wrote of just one Douglas lord's struggle with the English 'that if anybody could retrace the story, he would by himself have produced a large and attractive book'.[2]

These contemporaries also recognised that, linked to the role of the family as leaders in war, the Douglases were great lords, possessing 'great conquests', giving 'fair judgement' and leading 'the greatest company of knights and men at arms'. Their lordship, measured in lands and in followers, also marked the house of Douglas out as the principal source of authority in many parts of the kingdom, especially the south. Though Bower and his contemporaries could criticise the Douglases and the style of lordship they exercised, comparing such magnates to wolves ravening the flock, these authors also believed that the family's power had been achieved in the service of the kingdom. The generations of historians and writers who followed Bower were more reluctant to recognise the positive elements of Douglas lordship. Instead of a desire to record and praise the role of the Douglases in war, which was present in the works of men who saw victory in the conflict with England as the keystone of Scotland's survival, later writers with different perceptions were far less sympathetic.[3] From 1500 and especially after the Reformation, the war against the English seemed far less central to those concerned with the past and future of the Scottish realm. Instead, historians and jurists dwelt most heavily on the development of the state. Their focus was on the exercise and extension of central authority, and their attitude overwhelmingly in favour of the crown and its powers. Though the period produced the earliest history of the family in the shape of David Hume of Godscroft's *History of the House and Race of Douglas and Angus*, a study in which, from mythical origins to his own day, Hume sought to praise his subjects, in this royalist and centrist atmosphere the place of magnates like the Douglases was generally seen as obstructive to the crown and destructive to the kingdom. For example,

Robert Lindsay of Pitscottie, in his 'Historie and Cronicles of Scotland', dating from the 1570s, spoke of the Black Douglas earls as greedy, arrogant, lawless and violent, retaining bands of thieves as their followers and fomenting factional conflict across the kingdom. His attitude was echoed in more general statements from other authors over the next two centuries. The Scottish nobility of the late middle ages were seen as 'mutinously proud', and their maintenance of armed retinues meant that 'violence and rapine prevailed over law and justice'. The aristocracy were selfish, law-breaking and obstructive, a block on the efforts of successive kings to create a centralised and, by implication, better-run and more civilised Scotland. Royal attacks on such men, even James II's murder of William earl of Douglas, a protected guest at the king's court, were justified by Pitscottie, and, before the Reformation, by Hector Boece, as the removal of a local tyrant who terrorised his tenants and defied the crown's authority.[4]

This identification of aristocratic lordship as a negative force remained the prevailing tone of historical writing on late medieval Scotland well into this century. Though not unquestioning 'king's friends', historians such as E. W. M. Balfour-Melville and A. I. Dunlop looked to royal government for anything approaching effective authority. Balfour-Melville, in his *James I, King of Scots*, spoke of 'the wreck of the king's high purpose on the rock of magnate self-interest', while Dunlop described the Black Douglases as a 'menace' and 'steeped in treason', contrasting sharply with the loyal support given by her hero, Bishop James Kennedy, to his royal lord. In their approach to late medieval politics, Dunlop and Balfour-Melville held a viewpoint not altogether different from their predecessors since the sixteenth century. The Black Douglases were the worst of a bad breed, holding pride of place in the rogues' gallery of great nobles, their private power a check on public justice and government and a recipe for anarchy. Efforts to temper this hostile reputation proved uncomfortable. In *The Douglas Books*, written in the 1880s, Sir William Fraser produced an indispensable edition of family papers alongside biographical portraits which sought to clear the Douglas magnates of the worst charges against them. Fraser stressed the loyalty of Douglas lords to the crown and presented major conflicts with the government as the product of misunderstandings on both sides. Writing for the descendants of the men under examination, Fraser almost presents the Douglas earls as Victorian men of affairs, dignified, respectable but with the occasional skeleton in the family closet to add a touch of colour. As the works of Dunlop, Balfour-Melville and others showed, later historians were not convinced by Fraser's defence of the Douglases. The family's origins as patriotic supporters of the Bruce were remembered but were overshadowed by what was seen as the abuse of royal generosity.[5]

Over the last three decades, though, there has been a reappraisal of political society in late medieval Scotland. This revision presents the kingdom, not as a cultural and political backwater, but as a confident and secure political unit whose government, though unsophisticated, met the needs of the Scots. At the heart of such views stands a far more developed understanding of the role and interests of the Scottish nobility. The works of Drs A. Grant and J. Wormald, in particular, have shown the magnates to be willing and vital partners of the crown in the running of the kingdom. Rather than a disruptive force, the local dominance of these lords is shown to have been a force for stability. The bonds between magnates and their retinues, identified earlier as leading to local feuding and disorder, are associated with the maintenance of peace through arbitration by great men between their retainers. Scottish government is regarded as functioning in self-regulating regional blocs dominated by magnates who acted as intermediaries between crown and local communities. The result was 'low-key politics' in which conflict was rare and resolved by compromise. As part of this revision, the role of the Black Douglases in Scottish political life has been viewed in a more positive light. The Douglas earls are presented as responsible guarantors of local order, capable of regulating their relations with other great men. For example, the alliance between the earl of Douglas and the duke of Albany in 1409 is seen, not as the product of sinister 'political ambition' nor as 'the flouting of order', but as an agreement to keep the peace between the two magnates and their followers.

However, the history of the Black Douglases does not always sit comfortably with such assertions. The fall of the family after a prolonged and violent conflict with the crown hardly suggests stable co-operation, and has, as a result, been regarded as an isolated exception to these general rules of Scottish political behaviour. Dr Grant suggests that the conflict represented a break in the Douglases' 'remarkably well-maintained record of loyalty and service' to the crown, Dr Wormald that the family 'threatened the ideal' of crown-magnate co-operation and that the Douglas earls represented 'too great a concentration of power in the hands of one family', a return to the old idea of such great magnates as overmighty subjects of the king. Uncertainty clearly persists about the character of Douglas lordship in late medieval Scotland and the extent to which they conformed to, or contradicted, the general characteristics of the Scottish nobility as identified in these recent works.[6]

However, the importance of this research lies in its recognition that the predominance of great aristocratic houses over wide areas of Scotland was, in general, an essential and legitimate element in the government and political society of a kingdom where, for geographical, cultural and economic reasons, power was decentralised. There are, though, problems with the conclusions drawn from this recognition, as the difficulty of fitting the Douglases into patterns of crown-magnate co-operation shows. There must be suspicions that, in the presentation of Scotland as a more stable and better-run society than had previously been recognised, the place of competition and conflict in the political life of both the kingdom as a whole and, perhaps more importantly, in regional communities has been played down. In his survey of the British Isles in the middle ages, Robin Frame adds a note of caution to the revised view of Scottish regional society: 'Unfortunately the fourteenth century evidence rarely allows us to glimpse the realities of regional life; it would be naive to imagine that monopolistic aristocratic power was wholly benign'. Owing an acknowledged debt to K. B. McFarlane, whose work on the nobility revolutionised perceptions of late medieval England, recent studies of the role of the Scottish aristocracy have looked to English political society as their yardstick. Though the differences between Scotland and England have been stressed, the identification of the English polity as working on comparable lines has contributed to the emphasis on 'low-key politics'. As an alternative to the centralised, legalistic and professionalised government and political society of England, late medieval Ireland, a land far more similar to Scotland in size, wealth and in its highly regionalised structure, provides a closer and more revealing parallel. Anglo-Irish magnates were not a 'dark force', but 'necessary and positive' to the maintenance of the English lordship, but the power of such men was also militarised and coercive. They gave leadership and protection, but demanded service and payments in war and politics. Ireland, 'a land of war', had an English-speaking nobility which saw self-regulation in terms of physical and forceful competition as much as agreement and compromise. Scotland, like Ireland a 'patchwork of lordships', may have shared many characteristics of this model.[7]

The Black Douglases is a study of 'the realities of regional life' in late medieval Scotland through the accumulation and exercise of power by its greatest non-royal dynasty during the century and a half of its power. Its focus is the house of Douglas from its emergence into the front rank of the Scottish nobility in the years of major war with England from 1296, through to the loss of the family's predominance in the borders and their principal estates with the downfall of the Black Douglas earls in the 1450s. By dealing with a single magnate house, rather than studying an individual reign or the nobility as a group, this study examines both the continuity and instability of aristocratic power over a long period, and looks at the specific characteristics and relationships at the heart of one dynasty's exercise of lordship. To regard late medieval Scotland from an aristocratic perspective is not to take a narrow view of the workings of political society but to examine it from what was, for most of the century and a half from 1300, its most effective level. Analysis of the role played by the Douglases in the survival and development of the Scottish realm, of the relationships between Douglas lords and both kings and local communities, and of the family's view of its own place in the kingdom, as servants of kings or a special dynasty with its own rights, raises issues fundamental to an understanding of the structure and character of late medieval Scotland, its identity, culture and society.

The rise and fall of the Douglases is treated in terms of the dual elements associated with the Douglas name, leadership in war against England and the power of the family as great and independent lords. The link between the family's role in war and their political predominance in those regions which were in the front line of Anglo-Scottish conflict is the principal theme of the book. The effects of this relationship were felt far beyond the marches, leaving their mark not just on Scottish politics but on warfare and diplomacy across western Europe. From Moray and Galloway to the Loire, *The Black Douglases* displays the power and ambition of a late medieval noble house at its greatest extent.

NOTES

1. A. Tuck, *Crown and Nobility* (London, 1985), 9–12.
2. Walter Bower, *Scotichronicon*, ed. D. E. R. Watt, 9 vols (Aberdeen, 1987-) vii, 108–109; *Barbour's Bruce*, ed. M. P. McDiarmid and J. A. C. Stevenson, Scottish Text Society (Edinburgh, 1980–84); *The Bruce*, ed. W. M. Mackenzie (London, 1909).
3. *Scotichronicon*, viii, 34–35, 63, 216–19, 292–93.
4. David Hume of Godscroft, *History of the House and Race of Douglas and Angus*, 2 vols (Edinburgh, 1748); Robert Lindsay of Pitscottie, *The Historie and Cronicles of Scotland*, Scottish Text Society (Edinburgh, 1899–1911), i, 40, 47, 88–92, 126; Hector Boece, *Scotorum Historia* (Paris, 1526), 372–73; 'The History of the Lives and Reigns of the Five James's, King of Scotland', in *The Works of William Drummond of Hawthornden* (Edinburgh, 1711); Andrew MacDouall, lord Bankton, *An Institute of the Laws of Scotland* (Edinburgh, 1751). The latter two sources both quoted from J. M. Wormald, *Lords and Men in Scotland: Bonds of Manrent 1442–1603* (Edinburgh, 1985), 1–13, which contains an excellent discussion of attitudes to the Scottish nobility from the sixteenth to eighteenth centuries.
5. E. W. M. Balfour-Melville, *James I, King of Scots* (London, 1936); A. I. Dunlop, *The Life and Times of James Kennedy, Bishop of St. Andrews* (Edinburgh, 1950), 24, 142–45, 151–52, 208–209; W. Fraser, *The Douglas Books*, 4 vols (Edinburgh, 1885), i, especially 465–72, 482–96.
6. A. Grant, *Independence and Nationhood* (Edinburgh, 1984), 120–99; J. M. Brown, 'The Exercise of Power', in J. M. Brown (ed.), *Scottish Society in the Fifteenth Century* (London, 1977), 33–65; J. M. Wormald, 'Taming the Magnates?', in K. J. Stringer (ed.), *Essays on the Nobility of Medieval Scotland* (Edinburgh, 1985), 270–80; A. Grant, 'Crown and Nobility in Late Medieval Britain', in R. A. Mason (ed.), *Scotland and England 1286–1815* (Edinburgh, 1987), 34–59. Recent accounts of crown-magnate relations which are less convinced of the inherent stability of Scottish politics include R. Nicholson, *Scotland,*

The Later Middle Ages (Edinburgh, 1974), C. A. McGladdery, *James II* (Edinburgh, 1990); M. Brown, *James I* (Edinburgh, 1994); S. Boardman, *The Early Stewart Kings, Robert II and Robert III* (East Linton, 1996). For a discussion of these recent works see M. Brown, 'Scotland Tamed: Kings and Magnates in Late Medieval Scotland: a review of recent work', in *Innes Review*, 45 (1994), 120–46.

7. K. B. McFarlane, *The English Nobility in the Later Middle Ages* (London, 1973); M. A. Hicks, 'Bastard Feudalism: Society and Politics in Fifteenth Century England', in M. A. Hicks, *Richard III and his Rivals* (London, 1991), 1–40; R. Frame, *The Political Development of the British Isles 1100–1400* (Oxford, 1990), 191; R. Frame, 'Power and Society in the Lordship of Ireland, 1272–1377', in *Past and Present*, no. 76 (1977), 3–33; R. Frame, *English Lordship in Ireland* (Oxford, 1982), 46–51.

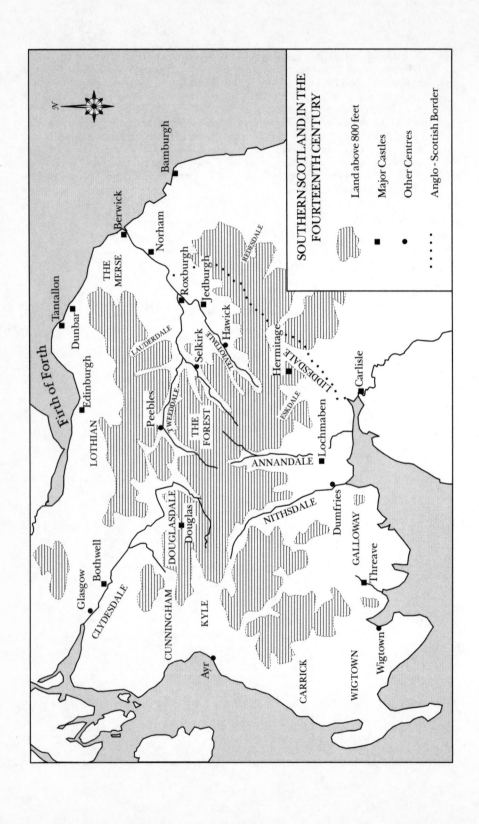

SOUTHERN SCOTLAND IN THE
FOURTEENTH CENTURY

Land above 800 feet

■ Major Castles

● Other Centres

⋯⋯ Anglo - Scottish Border

CHAPTER ONE

The Black Douglas

THE GREEN TREE

In 1450 the power and prestige of the house of Black Douglas seemed to stand at its height. In lands, titles and followers the family was without a rival amongst the nobility of Scotland. They were the greatest of the great lords, the magnates of the kingdom. Only the young king of Scots, James II, possessed wider estates and higher status than the Black Douglases. At the head of this powerful family were William, the earl of Douglas, eighth man to bear that title, and his four brothers. Between them the Douglas brothers held five earldoms in the Scottish kingdom and a collection of lordships which reached from Galloway in the south-west to the Black Isle in the north. The great castles at the heart of these lordships, Threave in Galloway, Bothwell on the Clyde, Abercorn by the Forth and Balvenie and Darnaway on the Moray coast, all stood as physical proof of the power and wealth of the Black Douglases. For almost a century and a half the name of the family had been linked with the fortunes of the independent kingdom of Scotland. In the years from the great wars against England in the early fourteenth century, the Black Douglases' accumulation of lands and influence shaped the Scottish realm. The fame which they had won in their rise made the Douglas name known far beyond Scotland. As crusaders in Spain and the Baltic and as soldiers, allies and vassals in the service of the kings of France, the Black Douglas magnates carved out a great reputation among the noble houses of Europe. In 1450 this fame was still at its height. On his pilgrimage to Rome in the autumn of that year, Earl William of Douglas was fêted by the rulers of western Europe, the kings of England and France and the duke of Burgundy. The arrival of the earl in the Holy City for the Papal celebrations of the half-century was that of a great prince, exciting the Romans and winning honour from the Pontiff.[1] In their reputation, their wealth and their power, the Black Douglases were recognised as lords of European status, men whose support was sought by kings and rulers across the west.

The lands and powers of Earl William and his brothers and the fame of their house represented the achievements of their forefathers as earls and lords of Douglas. The heirs of a great magnate dynasty, the Black Douglas brothers shared the conscious pride in inherited rights and duties which was at the heart of the medieval nobility's view of the world and their place in it. In Scotland and beyond, aristocratic families showed keen interest in their real or imagined past in the commissioning of genealogies, romances and memorials which glorified their history.[2] The satisfaction of such dynastic ambitions was a major element in the activities of the great houses of late medieval Europe. The Douglases held just such an understanding of the role and the rise of their dynasty. It was in the years just before 1450, when the earl of Douglas and his brothers seemed to dominate the Scottish kingdom, that the strongest statement of the Black Douglases' self-image was made. *The Buke of the Howlat* (the book of the owl), a poetic fable of nearly a thousand lines, was produced in the household of Earl William's younger brother, Archibald, earl of Moray.[3] Within its allegorical tale of a parliament of birds, the author, Moray's secretary, Richard Holland, introduced twenty verses which expressed the inherited status and honour of his master's house, the Black Douglas family. As his guide to this inheritance, Holland used 'the armes of Dowglas ... knawin throw all Cristindome'. The arms of Douglas lords were a visible measure of family achievement. The heraldic rules of late Medieval Europe were not just part of a taste for decoration but were about the obsession of noble society with rank, kinship and status. The lion of Galloway, the stars of the Murrays and the red heart of the Bruce in the arms of the earl of Douglas were a pictorial display of the Douglas family's rise in the century and a half before Holland described them.[4]

The heraldic bearings carried by the members of the Black Douglas family on their shields stood in *The Howlat* as a visual display of lands, honourably won and defended, and of the services of the Douglas lineage to the crown and community of Scotland. The poem gave the family arms a position 'next the soverane signe', the 'lyon ... rampand' of the king, and the Douglases were 'ever servable', always ready to serve their royal master. There was no challenge to the power of their king but Holland placed his masters as the next in rank after the crown, as great lords with their own inherited rights and duties. Above all, honour and fame were the rewards for their defence of the kingdom. 'Of Scotland, the werwall ... our fais force to defend ... Baith barmekyn and bar to Scottis blud'. In warfare the Douglases had been a wall, a bastion and barrier in resisting the foes of the kingdom, roles which gave them, in Holland's view, a heroic status in the realm. To those 'of a trewe Scottis hart' the Douglases were 'our lois and our lyking', our glory and delight, and a family with a special place within the kingdom.[5]

Holland's description of his master's family in *The Buke of the Howlat* was in part propaganda, justifying Douglas dominance in the years up to 1450. However, there can be little doubt that Richard Holland, a servant of the Black Douglases, also presented his masters' views of their inheritance. The lands which are represented by Holland on the arms of the Black Douglases include the family's great southern Scottish lordships, Galloway, 'wyn ... on weir', conquered from its people who 'rebellit the croune', and 'the forest of Ettrick' and the 'landis of Lawder' taken 'with dynt of his derf swerd ... fra the sonnis of Saxonis'.[6] Holland described a dynasty of war-lords. From the opening of the fourteenth century to the middle of the fifteenth, in an age of war for Scotland and the British Isles, the Black Douglases had carved out a role and a private empire for themselves as the war leaders of the kingdom in the new marcher zone against the English enemy. The key to the power and standing of the house as the greatest magnate dynasty in late medieval Scotland rested on their ability to wage war in conflicts which combined national, regional and private goals. From the 1320s to the 1440s, Douglas dominance in all or part of the border region was a fact of Scottish life. The Douglases began and remained a noble house whose rights were won and maintained in war, the most spectacular product of the militarised world of the Anglo-Scottish border.

For Richard Holland, looking back from the late 1440s, the pride of place in the family's 'armes of ancestry' belonged to the shield of one man, the 'douchty Dowglas', Sir James Douglas, 'quilk oft blythit the Bruse in his distres'. It was to James Douglas, the right arm of Robert Bruce, already a hero in Scottish chronicles and poetry by the time Holland wrote, that most space is given in *The Howlat*. James's arms of blue set with three stars over a silver field represented heavenly constancy and courage, but it was the 'bludy hert' set on the silver on which Holland's focused.[7] The heart stood as the principal badge of the Douglas line, a reminder of the service beyond death of James to 'his singular soverane', King Robert Bruce. Douglas's death on crusade bearing his dead master's heart was unshakeable proof of the affection borne by 'Good King Robert' for the lord of Douglas and his line. Holland depicted James Douglas as valiant in war, a marytred crusader and as a steadfast and loyal champion of the hero-king of medieval Scotland. The achievements of James were part of the glorious inheritance of the Black Douglas dynasty. The arms of his successors bore the 'bludy heart' and marked Sir James Douglas as the progenitor of the 'grene tre, gudly and gay' from whose branches were hung the shields and helms of the Earl of Douglas and his brothers in Holland's fantasy.[8] Pride in the Black Douglas lineage, the green tree, and in the founder of its growth, James Douglas, justified the ambitions and emphasised the honour of his descendants.

In the eyes of the Black Douglas lords and their servants in the mid-fifteenth century, the roots of their house's fame and fortune could be traced
back to James lord of Douglas, the 'Good Sir James' of Scottish historical
tradition. Such a view is easy to understand. It was from James's career
that the status and fame of the Douglas line grew, and within fifty years
of his death he was being presented as a model of knighthood and aristocratic values. Yet James was not the founder of the Douglas family as a
part of the Scottish nobility. Instead he was probably the sixth in the line
of minor nobles who dwelt in, and took their name from, the lordship of
Douglas and Douglasdale in the upper Clyde valley. These men were not
great lords like their descendants and their ancestry was unclear. As early
as 1400, when Scottish chroniclers sought to explain the rise of the family
beyond these limited beginnings, there was doubt about the origins of the
Douglases. 'Sindry men spekis syndrely' as the chronicle of Andrew Wyntoun described it. However, it seems likely that the dynasty which held the
lands and name of Douglas arrived in Scotland in the mid-twelfth century
in the person of a knight from the plains of Flanders, William son of
Erkenbald.[9]

William's arrival was a small part of a process of settlement and change
which brought Flemings and Frenchmen to all regions of the British Isles.
At the invitation of the Canmore kings of Scots, small bands of Flemings
arrived to settle in Clydesdale, on the edge of the powerful sub-kingdom
of Galloway.[10] The new lords of Douglas came in the wake of this group
but the ignorance of later writers about the origins of the family is an
indication that distinctions between native and incomer in the Scottish
aristocracy had lost significance more quickly than elsewhere in the British
Isles. However, if he knew nothing of 'thare begynnyng', the chronicler,
Andrew Wyntoun, retained knowledge of one element of their ancestry.
The Douglases were kinsmen of a second noble house, that of Murray,
which took its name from the great northern province which the '*de Moravia*'
family had tamed for the crown. Once again the arms of the two families
showed their kinship. Wyntoun and Holland describe the three silver stars
on the shields of both Douglas and Murray and saw the alliance of the
two houses in the fourteenth century as the re-unification of an ancient
kindred.[11]

While the Murrays were great lords in the north, their cousins, the lords
of Douglas were, by the 1280s, well-connected but minor barons. Though
the family acquired minor lands in Lothian and elsewhere, the heart of
their estates remained the lordship of Douglas. Even after the acquisition
of far greater lordships, the Douglases would retain close personal
bonds with the estate, taking their title as earls from the valley in 1358.[12]

Attachment to their lands and tenants was an element in the process of assimilation which saw the Douglases connected by marriage to a range of Scottish houses, both native and incomer, in the century from their settlement. The Douglases were part of a nobility which, in its relationships and ties to the Canmore kings, saw its loyalties in broadly similar terms. However, this was not a Scottish nobility in any sense of exclusive identification with Scotland or its rulers. The lords of Douglas were normal in possessing lands and connections beyond Scotland, holding estates in Northumberland and Essex by 1296. The conflict of loyalties between English and Scottish rulers which this suggests was rarely a problem in the thirteenth century and such divided loyalties were an unexceptional part of aristocratic society. The Anglo-Scottish border was no bar to landholding. It simply divided the fiefs of the Scottish king from those of his own lord, the king of England.[13] The stability of British landholding did not automatically translate to local peace. The career of James Douglas's father was one of feud and dispute. William Douglas, nicknamed *le Hardi*, the bold, managed to antagonise both English and Scottish kings in the early 1290s, suffering imprisonment and temporary dispossession at the hands of Edward I of England. His neighbours and overlords found William equally hard to deal with.[14] Unaffected by prison, William's boldness looks more like a readiness to use violence, a characteristic which he passed on to many of his descendants in the next two centuries.

In general, though, for the south of Scotland, where the bulk of the Douglases' estates lay, the period from 1230 to 1296 was one of stability unmatched for many centuries. The sub-kingdom of Galloway was largely absorbed into the structures of the kingdom after 1236, its native lords replaced by the Balliols from northern England. The wide interests of the Balliols and the other great magnates of southern Scotland, the Bruce lords of Annandale, the de Soulis lords of Liddesdale and the Anglo-Saxon earls of Dunbar, all reflected the ease of links with England and the stability of relations between Scottish crown and nobility in the region. The great border lordships, which after 1300 were to form the front line of Douglas territorial power in a militarised frontier region, were, before 1296, the focuses of stable local government based on geographically distinct and ancient divisions of land. The south, from Galloway to Berwickshire, was a secure heartland of the later thirteenth-century Scottish kingdom whose military frontier lay far to the north and west. The Anglo-Scottish border with its royal forests and great religious houses was, by contrast, a land of peace by medieval standards. It was in this regional society that the minor, but well-established and connected, house of Douglas operated, a small and unexceptional family of Lanarkshire barons.[15]

THE GOOD SIR JAMES

After a century and a half of this primarily local significance, the fortunes of the Douglases were transformed in a single generation. As later Scots were well aware, it was in the lifetime and through the career of James lord of Douglas that the family rose in rank and reputation. His exploits did more than bring fame to the Douglas name. The place of his heirs as war leaders, as lords who held power in the south of Scotland and as men who followed a tradition of adherence to the Scottish crown and community, all had their beginnings in the deeds of James, the 'Black Douglas'. The house of Douglas emerged as winners from a period of political and military crisis between 1286 and 1330 which left its mark on the whole Scottish nobility and on the structure and status of the kingdom itself. The failure of the Canmore dynasty, which more than any other force had forged the kingdom of Scots and its noble class, created a crisis which threatened to engulf both. The sudden death of Alexander III in 1286 was, over the next ten years, exploited by Edward I of England to extend his rights as superior lord of the Scottish king and realm. The demands of Edward's kingship were unacceptable to the nobility of Scotland who, whether landholders in England or not, saw their 'custom' and status in the northern kingdom as being infringed. The weight of Edward's overlordship provoked the defiance of the bulk of the Scottish nobility. Edward's response was war.[16]

The war, which began in 1296, was to last for over thirty years before a negotiated peace. Its continuation was to shape Scotland's internal structure and place in the world for three centuries to come. It was in this long war between the two realms of late medieval Britain that the role of the Douglas family was established. The support which James Douglas gave to Robert Bruce identified his house with what Scots writers saw, by 1400, as a patriotic cause. At the time, however, the issue was far less clear-cut. Douglas's adherence to Bruce was clouded by the experience of a decade of war and its effects which cut across the networks of allegiance which had existed before 1296. Service to the king of Scots was now incompatible with ties of lordship to the English crown. Friendship to one meant the emnity of the rival ruler. The cutting of cross-border ties was to be a long-term reality which shaped the political society of those lands closest to the new frontier. In the short term, choices of allegiance were forced on the whole Scottish nobility. Such choices could not be simply determined by identification with a national community of either kingdom. Obligations of kinship and personal lordship as well as the strong desire to survive with lands and status intact created competing calls on noble loyalties. Lords who had defended 'the community of the realm of Scotland' against Edward in negotiation were not prepared to take up arms against their feudal superior.[17] William lord of Douglas, James's father, had no such qualms. Personal hostility towards the English king, who had earlier imprisoned him, fuelled William's opposition to Edward in 1296 and, despite capture in the Scottish defeat of that year, again in 1297. Forgiven twice by the king, William's third captivity was to be permanent. Edward allowed such an unrepentant enemy to rot in prison. 'Savage and enraged' at his fate, Douglas was dead by 1299, according to later sources, a 'martyr' to English tyranny.[18]

William's young and disinherited heir was left with an uncertain future, sheltered in the households of his kin and connections amongst the Scottish nobility. Men like his uncle, James Stewart, Bishop Lamberton of St. Andrews, and Robert Keith, the marischal, were political survivors who, unlike William Douglas, had managed to win Edward's peace and even trust despite earlier opposition. In the years of Scottish defeat from 1302 the young James Douglas was brought into English allegiance in the service of these lords. Whatever his doubts, he was to remain in the English camp for the next five years.[19] James Douglas must have considered his loyalties against the experiences of this period. The fate of his father showed the price of defeat. James had been left landless. The lordship of Douglas had been granted to the English magnate Robert Clifford. It was the recovery of Douglasdale and the status of his family which, above all else, determined James's allegiance.[20] The only fourteenth-century account of Douglas's motives comes from the epic poem *The Bruce* by John Barbour. Written in the 1370s and a work which glorified both Robert Bruce and James Douglas, the poem showed the young James identifying from the first with Robert's seizure of the Scottish throne in 1306. For all its inaccuracies, *The Bruce* captured the key to James's decision. Douglas joined the king because '"throw hym I trow my land to wyn, magre the Clyffurd and his kyn"'. The recovery of landed title was the key to the restoration of status. Barbour described Douglas's earlier attempt to obtain them from Edward I which he saw, not as self-seeking, but as a quest for denied rights. Douglas was made to identify with Bruce as another man deprived of his rightful estate, the kingdom of Scotland. The bond between king and nobleman in the recovery of their rights is a running theme of the poem. The meeting of Bruce with Douglas at Ericstane near the heads of Tweed, Clyde and Annan was the meeting of the men whose search for 'fredome' was also expressed as a desire for their own status.[21]

Barbour's description of the impetuous Douglas joining Bruce at the beginning of his exploits hides a more cautious reality. In the summer of

1306, when Robert Bruce was fighting to maintain the throne he had seized, James Douglas remained in English allegiance.[22] He joined Bruce only after the king returned from exile in February 1307. His decision was based on a specific desire to reclaim his lands, and as Edward I had refused to restore them, Douglas turned to the alternative source of power in the kingdom. Bruce, based in his south-western earldom of Carrick, was for the first time in a position to provide support. In the spring Douglas and the king attacked Douglasdale, driving out the English garrison, but, within a month, James was treating with the English for a return to the peace.[23] Having shown he could disrupt their hold on his lands, Douglas was not prepared to risk capture by the advancing English. In these circumstances, his loyalty to Bruce was hardly in the heroic mould of Barbour.

In the event there was no return to English allegiance. The retreat of English forces after Bruce's victory at Loudoun Hill in May 1307 caused James to end efforts to make his peace.[24] For the next twenty-three years, Douglas's career was one of constancy to the king as the poets claimed. Given the traitors' deaths inflicted on Bruce's followers in 1306, Douglas's hesitation is understandable. However, in both 1307 and thereafter the key to winning and keeping the adherence of men like Douglas was Bruce's ability to use the powers of the Scottish crown. Service to a successful king of Scots held out the prospect of greater rewards than anything that could be won by a Scot in Edward's administration. The promise of good lordship in rewarding men for service was a fundamental principle of medieval society and, as Barbour stressed, the bond between Bruce and Douglas rested on the king's favour to a loyal follower. From 1307, Douglas was a rising power in Bruce's regime. He had no need to look elsewhere for lordship. The success of both king and Douglas strengthened the personal bond of allegiance. James's loyalty was never strained by the threat of defeat which faced previous Scottish leaders. To later Scots, Douglas's adherence to the king forged his family's reputation and fortunes as leaders of the patriotic causes in the marches, a region where the realities of war would continue to compel changes of allegiance.[25] His entry into Bruce's service was a crucial landmark in the rise of the house of Douglas.

The Douglas claim to be the defenders of the Scottish kingdom was a task and a boast which rested, not just on loyalty, but on the deeds of James in support of Bruce. The kingship of Robert I itself depended overwhelmingly on his success in war. Decisive victories against his Scottish opponents and the expulsion of all but a handful of English garrisons from the kingdom between 1307 and 1313 were the basis of his political strength. In 1309 it was claimed by the king's own propaganda that success in winning and holding the kingdom was the proof and test of his right to rule.[26] It was certainly the cause of his transformation from usurper to hero-king in the eyes of many Scots. War was the principal task of Robert I's kingship, and the principal service performed by his adherents amongst the nobility was as military subordinates. Whatever ties of affection grew up between the king and Douglas, the Bruce's friendship and the rewards which it brought rested on Douglas's abilities as a leader in the warfare which secured King Robert his kingdom.

Barbour described James Douglas as the king's close companion and, from the outset, as a man whose exploits in war won him fame and were vital to his lord. Yet, as Barbour himself admitted, when James joined the king he was 'bot littill of mycht', an untried and dispossessed nineteen-year-old.[27] His value to Bruce in 1307 related to the sudden significance of south-western Scotland in the war. Bruce had made the area his first base on his return to the kingdom, and the local influence of even the disinherited lord of Douglas was useful to the king. As will be clear, James was able to raise support from the region of his family's estates and he used it to harry local English forces during the summer.[28] When Bruce countered the new king of England, Edward II's, campaign in Nithsdale by devastating Galloway, his force probably included the followings of local lords like James Douglas and two Ayrshire landowners, Robert Boyd and Alexander Lindsay, who were active on Bruce's behalf in the south-west. The same local nobles, Boyd, Lindsay and Douglas, were the leading men in another raid against the Galwegians in June 1308 under Edward Bruce. Galloway, loyal to its lord, John Balliol, the exiled king of Scots, was to remain a source of opposition to the Bruce cause for the next fifty years and the king, who had gone to deal with his enemies further north in late 1307, had left his brother, Edward Bruce, to recover it.[29] In the raids of 1307 and 1308 Douglas provided support for the Bruces' warfare in the south. Like Boyd, Lindsay and the Clydesdale neighbours of the Douglases, Simon Lockhart of the Lee and Thomas Somerville of Carnwath, James Douglas was a figure of purely local significance in the following of King Robert.[30]

Between 1307 and 1315, Douglas was to emerge from this role to become one of the king's closest and most valuable supporters. By 1315 James had outstripped the other south-western adherents of Bruce to become the leading figure in the war on, and beyond, the borders. This rise proved to be a key stage in the growth of Douglas power, but from 1307 the career of James in the south is shrouded in darkness. *The Bruce* sheds the only light on these early years. It shows Douglas as firmly based in the south and concentrates on James's private campaign to win Douglasdale from the English garrison. In a war of raid and ambush, James forced the abandonment of Douglas Castle, destroying it to prevent Clifford 'pesabilly holding

his lands'.[31] Though the recovery of Douglas drew Barbour's focus, it was
not in Clydesdale that James established his base but to the south and east.

> In all this tym James of Douglas
> In the Forest travelland was
> And it throu hardiment and slicht
> Occupyit, magre all the mycht
> Of his feill fayis . . .

Against many foes, James Douglas had occupied and brought 'to the kingis
pes' by his 'travale' the Forest of the south.[32]

The 'great forest of Selkirk' was taken to include a vast swathe of upland
dale, moor and woodland stretching from Tweeddale and Annandale to
the dales of Esk and Teviot. Before 1296 the hunting reserves of kings and
nobles, it had become a strategically vital area of the south in the years of
war. The Forest divided the eastern borders from the west and Clydesdale,
and the nature of the ground made its control by English field forces and
by the garrisons at Roxburgh, Jedburgh, and Selkirk a constant problem.
The Scots appreciated this value as a refuge and used the Forest as a base
for attacks on English-held regions of the south and across the border and
as a recruiting ground and source of supply for their armies in the years
before 1304.[33] In 1307, Douglas was well placed to take advantage of this
natural stronghold. The lands of his family lay near the western edge of
the upland region and Robert Keith, in whose household Douglas had
served, had been the warden of the Forest in 1299. There was moreover
no alternative leader in the region in 1307. Keith and the lord of nearby
Liddesdale, William de Soulis, remained in English allegiance and the main
local lord, Simon Fraser of Oliver Castle, had been executed for his support
of Bruce a year earlier. With his fall the Forest had been brought under
English control by its new lord, Edward II's lieutenant, Aymer de Valence,
earl of Pembroke, who had laid waste Fraser's lands.[34]

By December 1307 de Valence's control had gone. Men and tenants of
Selkirk Forest and in Selkirk- and Peebles-shires had joined Bruce's party
and seized local lands. Their change of loyalty may be linked to the arrival
in the Forest of James Douglas. After 1314 James Douglas was the king's
chief agent in the region, residing in the Forest at 'Etybredshiels' near Selkirk
and holding official powers in both Selkirk and Jedburgh Forests.[35] This
position, won in the years after 1307, was to become one of the key centres
of Douglas power. James's control of the Forest had been established by
local warfare. English campaigns in 1308 and 1311 and raids by English
garrisons meant continuous fighting up to 1314, but the winning of the region
by war, one of the boasts of *The Buke of the Howlat*, marked the arrival of the

Douglases as a military force in the marcher zone they would come to dominate.[36] The Forest gave Douglas, a young and relatively minor noble, a place in the forefront of the war with England and in the ranks of his king's supporters. The value of the Forest as a base for Bruce's recovery of the south was first shown on the night of Shrove Tuesday, 19 February, 1314. In a sudden assault in the dark, James Douglas, with a party of men clad in black surcoats to disguise their armour, scaled the walls to capture Roxburgh Castle. Roxburgh, the principal fortress of the central borders, with a garrison of a hundred men, had been the bedrock of English control in the surrounding country. With its fall, Barbour reports the submission of Teviotdale to 'the Kingis pes ... outane Jedworth'. Even in Jedburgh, where the castle was supplied from England, the English monks of the abbey fled across the border on the day after Roxburgh's fall, aware of the local implications of Douglas's success.[37] Until its final capture and destruction by the Scots in 1460, Roxburgh provided the key to the allegiance of the men of Teviotdale. The castle's garrison, English or Scots, exercised control of the lands about and, as such, Roxburgh was to be a long-lasting factor affecting Douglas influence in Teviotdale. Its capture in 1314 brought this area of the south into the orbit of James Douglas.

Roxburgh's fall marked James Douglas's emergence as one of the principal adherents of the king. With Thomas Randolph, the king's nephew, who had captured Edinburgh Castle a month after Douglas took Roxburgh, he was to emerge as a key figure in Bruce's following after 1314.[38] Douglas showed himself to have mastered the type of warfare practised by the king in the local fighting which won the Forest and in the surprise assault on Roxburgh. His effectiveness was used to the full by Bruce and it was in the fifteen years from 1314 that the reputation of James Douglas was established. In Scottish eyes, this reputation was as the 'Good Sir James', a constant and victorious servant of the king and a model of chivalry. To the English, though, his black hair, sallow skin and the reign of terror he inflicted on the counties between Tweed and Humber earned him a different name. 'The blak Dowglas' was a dark force 'mair fell than wes ony devill in hell' who made pacts with his fellow-demons to achieve his evil goals. By adopting the name of Black Douglas, first applied to them by the enemy, James's son and grandsons were consciously binding themselves to the fearsome reputation of their forebear.[39] Along with King Robert and Thomas Randolph, from 1315 Douglas's partner in war, James was responsible for the military ascendancy of the Scots in a war of destruction and attrition in the marches between the kingdoms. To his English victims, Douglas, even more than his royal master, was the symbol of this ascendancy.

The first mark of Douglas's rising star was his knighting by King Robert on the eve of the great victory of the Bruce cause at Bannockburn in June 1314. Success in the pitched battle confirmed the achievements of the previous seven years. Its effects were felt even in the borders where Jedburgh Castle surrendered on news of the English defeat.[40] However, the war against the claims of the English crown remained to be won. Even before Bannockburn, a growing strand in the Scottish strategy was the devastation of northern England. From 1315, the tactical implementation of the king's strategy was placed in the hands of Douglas above all. He was to be the first of the breed of magnates who built their power in the war-zone of northern Britain. Within two months of Bannockburn, James Douglas was to be raiding across the border in a force led by the king's brother, Edward Bruce, and from 1314 to 1319, when the battered English agreed to a truce, Douglas would raid the northern shires at least once a year.[41] His attacks reached as far as the Humber and Pontefract and their aim was not simply the devastation of enemy territory. Sustained raiding also brought the Scots king and his leading captains cash, paid in return for local truces by defenceless English communities, and displayed the inability of Edward II to protect his people from Bruce and Douglas.[42] James Douglas was the most active exponent of this border warfare. He seems to have taken the role of Edward Bruce as the king's principal southern lieutenant when Edward departed to seek his ill-fated kingdom in Ireland. When King Robert himself went to support the Irish venture in 1317, he left Douglas as his joint warden in the south, a position which must have included command of the war.[43] Douglas effectively remained in this role as joint leader of the war in the marches until the peace of 1328. His partner was Thomas Randolph, earl of Moray. As an earl and the king's nephew, Randolph held formal authority, but, to the English and later Scots writers, Douglas was the driving force in the campaigns against northern England.[44] This reputation was not just posthumous glorification. It reflected the genuine ability of James Douglas in mobile warfare which went beyond just destructive forays. He harried invading English armies in the borders at Lintalee in 1317 and Melrose in 1322, he supplied the siege of Carlisle by plundering the surrounding land in 1315 and, with Moray, raised the English attempt to recover the newly taken town of Berwick in 1319 by launching a devastating diversionary raid into Yorkshire.[45] Twice Douglas's efforts sought an even greater impact on the war. In the 1319 raid into Yorkshire and then again in 1322, James deliberately attempted the capture of Queen Isabella of England and her husband, Edward II, respectively. Though unsuccessful, his efforts added to the diabolical reputation of the Black Douglas.[46]

From 1314 to 1327 Douglas led or participated in all the chief acts in the war on the borders. In 1318 he was one of the leaders in the Scots' capture of Berwick, a town which James reportedly said he would rather enter than the kingdom of heaven. His and Moray's rout of the clergy and levies of York at the 'chapter of Myton' in 1319 and the night attack on the camp of the young Edward III in 1327 were exploits which further built James's fame and displayed his talents. By 1327, when Douglas and Randolph outmanoeuvred the host of Edward III in Weardale, James's reputation and ability were at their height. The Scots' war-cry of 'Douglas, Douglas' recorded by an eye-witness of the night assault on Edward III's army bore testament to the potency of his name in Anglo-Scottish warfare.[47] The 'general war between England and Scotland' created both new demands and new opportunities for the Scottish nobility. Far more than his predecessors, James Douglas's power and purpose revolved around the waging of war. The war, in which he was involved from 1307, was of greater intensity than any experienced by the Scots in recent centuries and Douglas, his reputation and his legacy were a product of this war. His methods and style as a leader of the 'Scottis fay' reflected the developing character of national conflict between England and Scotland. The English description of Douglas as a man who waged war with brutality, who, for example, mutilated any of the battle-winning English archers he captured, is the picture of a professional soldier cold-bloodedly damaging the military potential of his enemy.[48] His attitude was the product of a warfare which, like the subsequent Hundred Years' War between England and France, combined chivalric interludes with the atrocities and fervour of a prolonged clash between national communities.

The power of James Douglas and of his family was built on the front line of conflict between rival kingdoms and communities. The role of the Douglases as leaders of the Scots community in war depended on their 'folk', the men who looked to James and his successors for command and reward. In fourteenth-century Scotland such ties of lordship were bound up with the waging of war. The 'cumpanny' of the Douglas lords were armed bands, prepared to campaign with their lords. In this light, Barbour's romanticised view of the roots of Douglas military lordship rings true in its essentials. He describes James Douglas's earliest support as coming from the family heartlands of Douglasdale where the influence of Thom Dicson, 'that wes of frendis richt mychty', was crucial in winning local support. Thomas was probably a freeholder of substance, a group vital in the local successes of the Bruce party. He had served William Douglas before 1297 and his support of James earned him and his heirs the nearby barony of Symington. Such bonds of lordship could last generations. As neighbours, followers and hereditary keepers of Douglas Castle, the lords of Symington remained servants of the Douglases for the next century and a half.[49]

While James could draw on the traditional influence of his family in Clydesdale, the years of war, dispossession and shifting loyalties cut across the structures of lordship elsewhere. Douglas was able to exploit this in extending his own connection and following across the south. He won the Forest for Bruce by 'inciting or compelling the inhabitants to rebel', the methods ascribed to his royal lord.[50] The success was achieved, not just by driving out the English, but by gaining the personal submissions of local men to his authority. As with his successors in the marches, the power of James to protect and punish drew local lords into his 'cumpany'. Barbour describes how Laurence Abernethy, an 'Inglis man', a Lothian landowner in English allegiance,

> quhen that he herd how it wes,
> He left the Inglis mennys pes,
> And till the lord Douglas richt thar
> For till be leill and trew he swar;

Abernethy's personal oath to Douglas was combined with his entry into Scottish allegiance. He was not just switching loyalties between rival realms but, with greater immediate impact, was entering the following of a great lord. With him he brought his own 'cumpany' of 80 men into Douglas's service.[51] In the first half of the fourteenth century, power in the south was won by such ties of personal loyalty between nobles and followers. The Douglas lords exploited the opportunities of war to build their own armed retinues. As Douglas's significance in war increased from 1314, so did the scale and range of his military lordship. In February 1316 he led a force drawn from across the central borders which cut up a raid of the Berwick garrison into the Merse. Douglas's lead was followed by the major territorial lord of the region, William Soulis of Liddesdale, and by the sheriff of Roxburgh, Henry Balliol.[52] His prominent role in the war gave James a significance in leading men which did not rest on conventional structures of landed status and resources. Until the 1320s his lands identified him as a minor Lanarkshire lord, but he was acknowledged as one of the principal leaders of the kingdom. Along with Randolph he led forces in war which were drawn from across the south, his authority accepted by men of greater or equal status.[53]

This leadership in war translated into more peaceful expressions of lordship. Men like Laurence Abernethy, who had entered Douglas's military orbit, looked to James's influence in politics and landholding as well. This process was fostered by the growing signs of royal trust in Douglas and the landed rewards he received from 1318 to 1320. The tenants and neighbours of these lands were natural adherents of James. For example, Douglas's patronage and intervention on behalf of Roger Murray, a neighbour of James from Lothian, shows the creation of new connections with lesser men.[54] His growing landed and political status made Douglas a natural focus of protection and support in Lothian and across the south. Even lords who were significant landowners and royal councillors, like Robert Keith, Henry Sinclair, Alexander Seton and Robert Lauder, were frequent associates of Douglas in both war and the regulation of the regional community. These men, of whom Keith and Sinclair had existing connections with the Douglases, seem to have been comfortable with the rise of James in Lothian and beyond.[55]

However, although the power and status of Douglas had their origins in his abilities as a war leader and protector, they also reflected the patronage and trust of the king. As *The Bruce* makes clear, the rise of Douglas rested on the rewards bestowed by the 'worthy, wicht and wys' king. From 1307 Bruce maintained the personal link with his increasingly powerful southern supporter and Douglas spent at least one sustained spell in the king's immediate retinue, in 1309.[56] After 1314, in parallel with his rising importance to the king, Douglas occupied a place in the royal circle with increasing regularity. Such contact was a vital part of the effective royal lordship which Bruce established in the kingdom. If the king looked to Douglas as a key adherent in the south as a result of the successes which he had won by his own efforts, the powers James received were as a royal lieutenant. Even in the Forest, Douglas's power was as the king's 'official', the local agent of the crown. The office of joint warden during the king's absence in 1317 probably gave James vice-regal powers in war and justice but only in the specific period of the king's absence.[57]

Such delegation of royal power to trusted supporters was a necessity of medieval government. Medieval kingship was personal. It functioned best in the presence of a ruler, but poor communications, the needs of government and, above all, a state of permanent war, made it impossible for Robert Bruce to provide continuous royal leadership across Scotland. He needed local deputies for the tasks of kingship; men with the standing and ability to perform their role but who could be trusted not to usurp royal authority. For the next century, the needs of permanent local leadership, especially in the new war-zone in the borders, made the handing of wide powers to magnates by the crown an unavoidable fact of life. Bruce acknowledged this by leaving Douglas and Randolph to prosecute the war. Royal rights of war leadership were delegated but Robert's supremacy was maintained by his frequent presence in the south and resumption of command in warfare.

A king who had achieved such success faced no challenge from within his own following. Bruce's links in the south with, not just Douglas and Randolph, but also the rank of influential barons like Robert Keith, Henry Sinclair and Alexander Seton who provided their principal backing in war and government, indicates the strength of royal influence in the region. Moreover, although Randolph and Douglas were men with their own regional ambitions, they were prepared to co-operate effectively in the south as lieutenants of the king. Their obedience was a testament to Robert's authority but was also because the king had essentially given the two men their status as great lords. From 1319, Douglas's service was rewarded with more than just royal offices. In the last ten years of Robert I's reign, James received the rights and titles which would form the core of the Douglas dynasty's landholding in the marches. The key estates in this network of lordships were the Forests of Jedburgh and Selkirk, where Douglas's powers were made into hereditary fiefs, the constabulary of Lauder, the lands of Stablegordon and Westerkirk in Eskdale and Buittle in Galloway. The king's patronage gave James lands across the south from Berwickshire through Lothian and the central borders to Galloway and Douglas itself in Lanarkshire.[58]

Reward for services rendered was only part of the equation. In his generosity to Douglas, King Robert was serving his own needs and what he saw as the needs of his dynasty. The timing of this sudden flow of patronage hints at these motives. Douglas received his lands long after the king had rewarded his kinsmen, Edward Bruce and Thomas Randolph. By late 1312 Bruce was lord of Galloway and Randolph, earl of Moray and lord of Annandale.[59] Just as Douglas's role as the leader of the war on the borders stemmed from Edward Bruce's departure for Ireland, so his promotion as a landowner may have been a reaction to the death of Edward in October 1318. Only two months later, Douglas received his first grant of new lands in conjunction with being named as regent after Randolph in any minority following King Robert's death.[60] The removal of Edward created the need for a new major magnate in the south, and the forfeiture of William Soulis of Liddesdale for conspiracy against the king in 1320 further cleared the south of established landowners. James received lands taken from William Soulis in Liddesdale and in 1325 was granted Buittle, the traditional centre of the lordship of Galloway.[61] The grant of Buittle, in particular, was designed to give the Douglases a base from which to control the potentially rebellious province of Galloway, replacing Edward Bruce's role in the province. Beyond Galloway, the king was attempting to create the Douglases as a powerful landed dynasty in the south which would provide leadership in war and a focus of allegiance for lesser men.

The role played by Douglas in the region as an arbitrator and guardian of religious houses was a further product of his local power. In the middle marches between Moffat and Lauder, Bruce was prepared to support these functions by granting wide estates for James Douglas from the lands and rights of the crown. The royal forests of the south were given into private hands and in 1324 the king made over powers of justice and freedom from royal exactions on goods and revenue to Douglas in all his lands.[62] This was not just an alienation of rights but the recognition of the new needs of lordship in the marches. The judicial powers granted to Douglas were the basis for the potent local lordship necessary to run a military frontier. The lands and rights given to James created the legal basis for the power of the Douglases in the marches.

Bruce was not seeking to create Douglas as dominant lord in the borders. Other grants, of Annandale to Thomas Randolph, of Liddesdale to Bruce's bastard son, Robert, and of lesser estates in the marches to Walter and Robert Stewart, the king's son-in-law and grandson, created a network of royal kin and supporters in control of those areas most vulnerable to English pressure.[63] These men were bound to the survival of the king's regime. Their title to these lands rested on the Bruce dynasty's rights to the crown. The fall of this dynasty would bring the return of the 'disinherited', the heirs of Bruce's Scottish enemies, the Balliols, Soulis and Comyns, and English recipients of lands like the Cliffords and Percies, who cherished rival claims in the marches.[64] For the next 130 years the leadership provided by Douglas lords for the Scottish cause, which at its heart was the claim of Bruce and his heirs to the kingship free of English influence, was not simply the recreation of the Good Sir James's loyalty to his lord. Instead, the defence of Scotland was the defence of the status and power of the house of Douglas against the challenges of lords in English allegiance. The rewards of King Robert to James Douglas cemented his heirs to Scottish allegiance in the years of doubt and crisis to come after 1330. The replacement of the old lords of the region by this new Bruce 'establishment' marked a dramatic change in the structures of lordship in the south of Scotland. Only the Dunbars with their earldom in the eastern borders provided continuity of tenure with the thirteenth century. Even there the change of the Dunbars' title to earl of March reflected the new reality of political society in the region.[65] The lordships from Annandale to Dunbar were now in the front line of warfare. The place of magnates and lesser men in the south was to be determined by this reality over the next half century.

If the rewards and regional influence of Douglas served the king in this way, this does not deny the new status and power of James. Nor does it indicate that Douglas's interests and ambitions were completely constrained by those of the king. The lands he received, which included the Forest, won for the king by Douglas's own efforts, no more than reflected these ambitions. James was certainly prepared to assert his rights as lord and landowner and was involved in litigation to this end. He may even have clashed with Melrose Abbey, a house favoured by the king, over the respective rights of the monks and Douglas in Westerkirk in 1321.[66] Although Bruce gave James the responsibility for defending and reconstructing the house, Douglas did not do so at the expense of the rights he had obtained. The freedom from royal exactions and jurisdiction which Douglas received for his lands also point to a nobleman keen to protect his rights from the interference of the crown. The charter of 1324 which granted him wide powers of justice was an act of royal favour in return for the king's release of French knights captured by Douglas. While the readiness of Douglas to forego over 4000 marks owed for their ransom shows an obedience to a royal request, James received full recompense in control of justice and its profits across all his lands.[67] Like his descendants in the 1450s, it was natural for Douglas to think in terms of the rights and status of his dynasty. He was from a class which was, by nature, competitive and ambitious, and Barbour's suggestion of such a character to relations between Douglas and Randolph is not impossible.[68] The two magnates closest to the king, who had shared command in war and the exercise of power in the same region without apparent friction, could still compete for fame and reward.

In the 1320s James Douglas held the power and rank of a great magnate. His actions and expectations fitted with this status. However, Bruce clearly trusted Douglas and Randolph to work together with or without his presence and intended to pass the powers of regency to the two men. The reality of Robert I's relationship with Douglas was that the interests of king and subject coincided with the ideals of service and reward trumpeted by Barbour. Douglas fulfilled a vital role for Bruce in the war in the south but was rewarded on a scale otherwise confined to the king's close kinsmen. Their relationship was based on a strong bond of personal lordship, which the king himself wished to express. In the charter of 1324, itself a wide grant of powers to Douglas, King Robert accompanied the document with a gift: 'From our hand we personally invest the hand of James with a ring [set] with … an emerald in token of the sasine to James and the heirs of his name'.[69] The gift of rings as a token of lordship was an act with a centuries-old pedigree. In 1324 it symbolised something of Bruce's favour to James Douglas and those 'of his name'. The power of the Douglas name in the south was to prove the greatest aristocratic creation of the king's reign. It was little wonder that *The Howlat* looked back to the king as the patron of the dynasty.

To both Barbour and Holland the climax of the 'love' of Bruce and Douglas came at the king's deathbed in 1329. Bruce's request that Douglas go to war against 'Goddis fayis', the infidel, bearing his heart, was an act with special appeal to a medieval audience. The king's desire, after a life of war against fellow Christians, to strike a blow against Islam was a goal claimed by many contemporary rulers. The choice of Douglas to fulfill this goal was presented as the choice of the best man in the kingdom to stand in the place of the king. It was a mark of the highest honour and James was aware of it. Barbour describes the making of a silver casket in which the king's heart was hung round Douglas's neck, and other contemporaries describe James holding court in the style of the king on his journey. With him was a 'nobill cumpanny' which included followers of James and the king, the Sinclairs, Keiths and Logans. In his progress round the coasts of France, Flanders and Portugal, the fame of Douglas as well as Bruce was being spread.[70] The aim of his pilgrimage was Spain, where the Moorish kingdom of Granada represented the closest frontier with Islam for western Crusaders. The arrival of Douglas on the continental stage, where his successors were to follow, was to be short-lived. He joined the army of Alphonso XI of Castile and in early August 1330 was outside the walls of the Moorish town of Teba de Ardales. Three weeks later Douglas and the bulk of his men were dead. He was killed in the course of a Christian victory, cut off from the Christian army by error or rashness. James, who, to the amazement of Spanish knights who knew his reputation, had come through the English war with unscarred face, was killed in his first fight against the 'Sarazenis'. By the time of Barbour and Holland even this death was no defeat. Douglas had died in battle against the infidel, an act of spiritual merit in itself, but Barbour further describes Douglas dying to save a follower, William Sinclair, cut off by the Moors in battle. *The Howlat* goes further. Douglas, determined not to outlive the king, hurled the heart into the midst of the enemy, to die as he lived, following his lord. The 'bludy hert' of Bruce, borne with pride in the family arms after 1330, was in both the *Bruce* and *The Howlat* a symbol of unshakeable service to the royal hero of late medieval Scotland.[71]

During the next 120 years this symbol had its own political value. For the house of Douglas, the heart was a sign of their founder's qualities of loyalty, chivalry and patriotic bravery and the territorial rewards these earned. As early as 1342 Robert Bruce's son, David II, reminded James's heirs that the family's lands had been received on account of 'the faithful deeds and good services of James lord of Douglas in defence of our realm'. But Bruce's heart symbolised a relationship between king and magnate worth 'mair ... than ony lordschipe or land', as Holland put it.[72] The 'love' of Bruce was used as a warrant for the special place of James's successors in the kingdom. A tradition of loyalty and service to crown and community was to be balanced by reminders to the heirs of Bruce of the honour and status due to the heirs of Douglas. In his winning of the south James was to prove more than Bruce's adherent. He was also the founder of his family's dominance in the marches, a dominance which would be maintained, after 1330, not by royal patronage but by border warfare and dynastic ambition.

NOTES

1. *Rotuli Scotiae*, Records Comission, 2 vols (London, 1814–19), ii, 343; Hume of Godscroft, *Douglas and Angus*, 181; *Calendar of Documents Relating to Scotland, preserved in H.M. Public Records Office*, ed. J. Bain and others (Edinburgh 1881–88), 5 vols, iv, no. 1231; *E.R.*, v, lxxxv; Dunlop, *Bishop Kennedy*, 124 n. 3.
2. Another Scottish example of this is the 'Stewartis Originall' an account of the origins of the Stewart kings written by John Barbour. In England the Beauchamp earls of Warwick are the best examples of the practice (*The Original Chronicle of Andrew of Wyntoun*, ed. F. J. Amours, Scottish Text Society (Edinburgh, 1908), 6 vols, v, 256–57; E. Mason, 'Legends of the Beauchamps' ancestors: the use of baronial propaganda in medieval England', in *Journal of Medieval History*, x (1984), 25–40).
3. 'The Buke of the Howlat', in *Longer Scottish Poems*, vol. i (Edinburgh, 1987), ed. P. Bawcutt and F. Riddy, 43–84 (includes a list of secondary material on the poem).
4. 'Buke of the Howlat', 60–70, vv 30–49.
5. 'Buke of the Howlat', 60, v 30.
6. 'Buke of the Howlat', 67–68, vv 44–45.
7. 'Buke of the Howlat', 62–66, vv 34–42.
8. 'Buke of the Howlat', 61, v 31. A green tree was the background to the Douglas arms on the family's seals.
9. *Wyntoun*, v, 258–59; G. W. S. Barrow, *The Anglo-Norman Era in Scottish History* (Oxford, 1980), 56–57.
10. A. A. M. Duncan, *Scotland: The Making of the Kingdom* (Edinburgh, 1975), 137–38, 189, 279, 383–84; Barrow, *Anglo-Norman Era*, 43–46.
11. *Wyntoun*, v, 258–59; 'Buke of the Howlat', 67, v 43; *Registrum Episcopatus Moraviensis*, Bannatyne Club (Edinburgh, 1837), 61, 81. The founder of the Murrays, Freskin, was a Flemish lord and was referred to as uncle by Brice son of William of Douglas. Brice was made bishop of Moray and all William's sons gravitated to the north, suggesting strong ties with the house which dominated the region.
12. Douglas is referred to in the fourteenth century as the 'native land' and 'cuntre' of the heads of the house (J. Barbour, *The Bruce*, ed. W. M. Mackenzie (London, 1909), V, line 228; *Wyntoun*, vi, 192–93; *Scotichronicon*, ed. W. Goodall (Edinburgh, 1759) 2 vols, ii, 310–11).
13. *Cal. Docs. Scot.*, i, nos 2452, 2538; ii, nos 736, 950; R. Frame, *British Isles*, 53–60. James Douglas's mother was Elizabeth Stewart, whose family's interests spanned the Irish Sea. His stepmother was Eleanor Ferrers, an English widow.
14. William kidnapped and married Eleanor Ferrers, his second wife, without the permission of Edward I, the superior lord of her lands in Essex (*Cal. Docs. Scot.*, ii, 357, 358; J. Stevenson, *Documents Illustrative of the History of Scotland 1286–1306*, 2 vols (London, 1870), 83–85, 86, 155; *Rot. Scot.*, i, 2; *The Acts of the Parliaments of Scotland*, ed. T. Thomson and C. Innes, Record Commission, 12 vols (1814–75), i, 448).

15. A. Young, 'Noble Families and Political Factions in the Reign of Alexander III', in N. H. Reid (ed.), *Scotland in the Age of Alexander III 1249–1286* (Edinburgh, 1990), 1–30; Duncan, *Scotland: The Making of the Kingdom*, 142–64, 180, 420–23, 527–29.

16. G. W. S. Barrow, *Robert Bruce* (Edinburgh, 1976), 28–96; G. W. S. Barrow, *Kingship and Unity, 1000–1306* (Edinburgh, 1981), 155–69.

17. An example of this is provided by Gilbert Umfraville, earl of Angus, the overlord of the Douglases in their Northumbrian lands. In 1291 Gilbert had refused to hand over the castles of Dundee and Forfar entrusted to him by the community, the kingless kingdom. From 1296, though, he served Edward in war (*Foedera ...*, ed. T. Rymer, Records Commission, 20 vols (London, 1816–69), i, 756).

18. Stevenson, *Documents*, ii, 25, 204–205; *Chronicle of Henry Knighton*, Rolls Series, 2 vols (1889), i, 371, 376, 377; *Cal. Docs. Scots.*, ii, nos 960, 1054, 1055.

19. *The Bruce*, I, lines 353–406; II, lines 91–134; *Cal. Docs. Scots.*, v, no. 492. A very full study of James Douglas has recently been undertaken which analyses the sources for Douglas's life to considerable effect (S. Väthjunker, 'A Study of the career of Sir James Douglas – The Historical Record versus Barbour's *Bruce*' (unpublished Ph.D thesis, University of Aberdeen, 1992)).

20. *Cal. Docs. Scots.*, iii, no. 682.

21. *The Bruce*, I, lines 407–44; II, lines 112, 146–74.

22. *Cal. Docs. Scot.*, v, no. 492; Väthjunker, 'Douglas', 33–37.

23. *Cal. Docs. Scot.*, ii, no. 1979; iii, no. 682; Väthjunker, 'Douglas', 36.

24. *Cal. Docs. Scot.*, ii, no. 1979; Barrow, *Bruce*, 243–45.

25. 'Buke of the Howlat', 62–63, vv 34–35.

26. Barrow, *Bruce*, 261–65.

27. *The Bruce*, II, lines 240–41.

28. Barrow, *Bruce*, 243–45.

29. *Cal. Docs. Scot.*, iii, nos 3–7, 15, 69, 83–84; v, no. 655; *Chronicon de Lanercost*, ed. J. Stevenson (Bannatyne Club, 1839), 210, 212; *Bruce*, VIII, lines 1–122; Väthjunker, 'Douglas', 40–41.

30. Barrow, *Bruce*, 222.

31. *The Bruce*, V, lines 255–462; VI, lines 373–474; VIII, lines 437–498.

32. *The Bruce*, VIII, lines 425–27; XI, lines 672–76.

33. Duncan, *Scotland The Making of the Kingdom*, 364–66, 420–23. In 1299, the Scots maintained 100 horse and 1500 foot 'besides the foresters' in Selkirk Forest. A year before they had recruited men of the Forest, a region with a reputation for shielding robbers, as archers for the army defeated at Falkirk. The English sought to counter this by employing troops in the 'keeping of the Forest' from 1301 (*Cal. Docs. Scot.*, ii, nos 290, 1241, 1839, 1978; *Chronicle of Walter of Guisborough*, ed. H. Rothwell, Camden Third Series (London, 1957), i, 328).

34. *Cal. Docs. Scot.*, ii, nos 1317, 1839–40, 1782, 1978; iii, nos 11, 44; *Chron. Lanercost*, 204.

35. *Cal. Docs. Scot.*, iii, nos 28, 746; *Regesta Regum Scottorum*, v, ed. A. A. M. Duncan (Edinburgh, 1988), no. 108; *Liber Sancte Marie de Melros*, Bannatyne Club, 2 vols (Edinburgh, 1837), ii, no. 419; *Registrum Magni Sigilli Regum Scottorum*, ed. J. M. Thomson and others, 11 vols (Edinburgh, 1882–1914), i, appendix 1, no. 38.

36. *Chron. Lanercost*, 214; *The Bruce*, IX, lines 672–725; 'Buke of the Howlat', 68, v 45. Lanercost describes an English raid in 1311 which forced the inhabitants to submit. *The Bruce* tells a story of Douglas foiling an attempt by Thomas Randolph and two other Scots in English allegiance to capture him in the Forest which can be dated to about 1308 (*Cal. Docs. Scot.*, iii, no. 76). For a discussion of Douglas's motives and activities in a wider context, see A. A. M. Duncan, 'The War of the Scots, 1306–23', in *Transactions of the Royal Historical Society* (1992), 125–51.

37. *The Bruce*, X, lines 352–505; *Cal. Docs. Scot.*, iii, nos 347, 894, pp. 405–407; *Scalachronica of Thomas Gray of Heton Knight*, Maitland Club (Glasgow, 1837), 140; *Chron. Lanercost*, 223.

38. *Scalachronica*, 140; *Chron. Lanercost*, 223; *The Bruce*, X, lines 506–755.

39. *The Bruce*, I, lines 29, 381–406; XV, lines 537–38; Väthjunker, 'Douglas', 81–82.

40. *Cal. Docs. Scots.*, iii, no. 1636. For conflicting views of Douglas's role in the battle itself see Barrow, *Bruce*, 299–300, 316, 330; Väthjunker, 'Douglas', 51–55. Väthjunker places Douglas in a far more subordinate role than traditional accounts.

41. *Chron. Lanercost*, 228–32, 235, 239; *Scalachronica*, 143–44; *The Bruce*, XVII, lines 491–588. The best account of these raids, the English defences and the damage done by Scottish plundering is C. McNamee, *The Wars of the Bruces, Scotland, Ireland and England 1306–28* (East Linton, 1997), 72–165.

42. J. Scammell, 'Robert I and the North of England', in *English Historical Review*, 73 (1958), 385–403; McNamee, *Wars of the Bruces*, 75–76.

43. *The Bruce*, XVI, lines 30–34. Bower, writing in the next century, describes Douglas as warden of the marches, the title held by his descendants, but James's office was almost certainly a more *ad hoc* lieutenancy for the king. Structures of regular border defence only came later in the century. His partner in the role was his cousin Walter Stewart (*Scotichronicon*, ed. Watt, vi, 282–83).

44. See Väthjunker, 'Douglas', 83. The *Vita Edwardi* for example describes the 1319 attack being launched by 'James Douglas and his accomplices', while most accounts give Moray and Douglas shared command (*Chron. Lanercost*, 239; *Vita Edwardi*, 95).

45. *The Bruce*, XVI, lines 331–488; XVII, lines 505–88, XVIII, lines 291–332; *Chron. Lanercost*, 230, 239; *Scalachronica*, 143, 149.

46. *Vita Edwardi*, 95; *The Bruce*, XVIII, lines 333–568. The king actually led the raid of 1322.

47. *The Bruce*, XVII, lines 505–88; Stevenson, *Illustrations*, 5 (cited in Väthjunker, 'Douglas', 73); R. Nicholson, *Edward III and the Scots* (Oxford, 1965), 13–36; *Chronique de Jean le Bel*, ed. J. Viard and J. Déprez (Paris, 1904–5), 2 vols, i, 70.

48. G. Barrow, 'Lothian in the First War of Independence', in *Scottish Historical Review*, 55 (1976), 151–71, 171; *ibid.*, 'The Aftermath of War', in *Transactions of the Royal Historical Society*, 5th Series, 28 (1978), 103–25; *Chron. Knighton*, i, 460.

49. *The Bruce*, V, lines 271–302; *Registrum Honoris de Morton*, Bannatyne Club (Edinburgh, 1837), 2 vols, ii, no. 114; British Library, Harleian MSS, no. 6443, 19v–20r; Fraser, *Douglas*, iii, no. 374.

50. *Cal. Docs. Scot.*, iii, no. 15.

51. *The Bruce*, XIII, lines 553–60.

52. *Cal. Docs. Scot.*, iii, no. 470; *The Bruce*, XV, lines 320–68.

53. This status is clear both from Douglas's role in war and appointment as joint warden and the fact that he was made regent in the event of a minority, failing Randolph (*A.P.S.*, i, 465–66).

54. *Melrose Lib.*, ii, no. 421; Fraser, *Douglas*, iii, no. 289; *Registrum S. Marie de Newbattle*, Bannatyne Club (Edinburgh, 1837), no. 269. Douglas granted Murray, a local landowner in Peebles-shire, the lands of Fala and intervened on his behalf with the monks of Newbattle. The Douglas links with their distant cousins, the various Murray kindreds in the south, were to continue for the next 130 years.

55. *Newbattle Reg.*, no. 123; *Melrose Lib.*, ii, nos 421–24; *Morton Reg.*, ii, no. 50; Fraser, *Douglas*, iii, no. 289.

56. *R.R.S.*, v, nos 7, 384, 385, 388.

57. *R.R.S.*, v, *passim*; *Melrose Lib.*, ii, 419; *R.M.S.*, i, appendix 1, no. 38; *The Bruce*, XVI, 30–34. In 1319, when Randolph seems to have been made warden of the south (probably the same office given to James in 1317), Douglas was made keeper of Berwick Castle. From this point, Berwick seems to have been the regular residence of Douglas (*R.R.S.*, v, 146).

58. *R.R.S.*, v, nos 143, 166–67, 184, 267; *R.M.S.*, i, appendix 1, nos 38, 77; appendix 2, nos 117, 126, 671.

59. *R.R.S.*, v, nos 19, 20, 24, 389; Barrow, *Bruce*, 264–65, 278.

60. *A.P.S.*, i, 465–66; *R.R.S.*, v, no. 143.

61. *R.R.S.*, v, no. 267; Barrow, *Bruce*, 429–30; R. Nicholson, *Scotland: The Later Middle Ages*, 101–103; McNamee, *Wars of the Bruces*, 166–205.
62. *R.M.S.*, i, appendix 1, no. 38; *R.R.S.*, v, 269; *Registrum Monasterii de Passelet*, Maitland Club (Glasgow, 1832), 27–28.
63. *R.M.S.*, i, appendix 1, nos 34, 53, 88; appendix 2, no. 323; *R.R.S.*, v, no. 391.
64. For the development of the 'disinherited' as an element of northern English society see J. A. Tuck, 'The Emergence of a Northern Nobility', in *Northern History*, 23 (1986), 1–17.
65. *Cal. Docs. Scot.*, iii, no. 29.
66. *R.R.S.*, v, 151.
67. *R.M.S.*, i, appendix 1, no. 38. The knights had been captured two years earlier on a raid into Yorkshire. Bruce wished to release them to the king of France as a diplomatic gesture (Barrow, *Bruce*, 345–46).
68. *The Bruce*, XVII, lines 52–68.
69. *R.M.S.*, i, appendix 1, no. 38.
70. *The Bruce*, XX, lines 165–238, 303–307, 315–334, 470–74, 491; *Chron. Le Bel*, 85–87.
71. *The Bruce*, XX, lines 335–520; 'The Buke of the Howlat', 64–66, vv 38–42. For a full analysis of continental accounts of James's death, see Väthjunker, 'Douglas', 138–43.
72. *R.R.S.*, vi, ed. A. B. Webster (Edinburgh, 1982), no. 51; 'The Buke of the Howlat', 63, v 36.

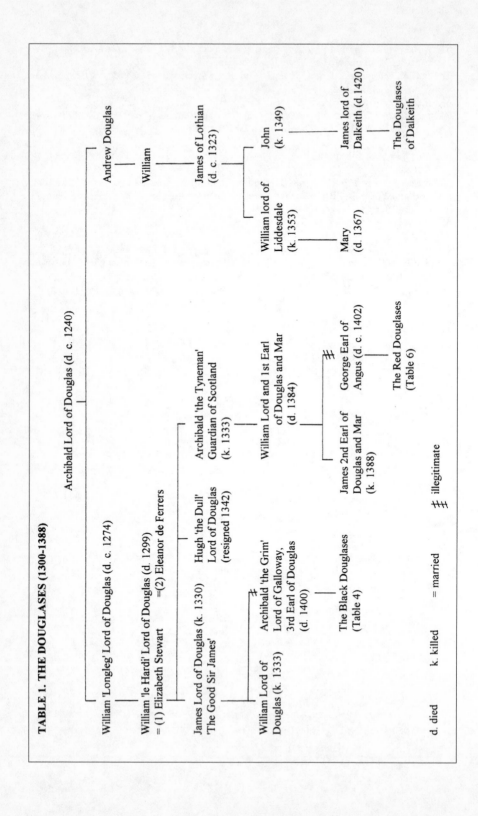

TABLE 1. THE DOUGLASES (1300-1388)

Archibald Lord of Douglas (d. c. 1240)

Andrew Douglas

William

James of Lothian (d. c. 1323)

William 'Longleg' Lord of Douglas (d. c. 1274)

William 'le Hardi' Lord of Douglas (d. 1299)
= (1) Elizabeth Stewart   = (2) Eleanor de Ferrers

Hugh 'the Dull' Lord of Douglas (resigned 1342)

Archibald 'the Tyneman' Guardian of Scotland (k. 1333)

William lord of Liddesdale (k. 1353)

John (k. 1349)

James lord of Dalkeith (d. 1420)

The Douglases of Dalkeith

Mary (d. 1367)

James Lord of Douglas (k. 1330) 'The Good Sir James'

Archibald 'the Grim' Lord of Galloway, 3rd Earl of Douglas (d. 1400)

William Lord and 1st Earl of Douglas and Mar (d. 1384)

William Lord of Douglas (k. 1333)

The Black Douglases (Table 4)

James 2nd Earl of Douglas and Mar (k. 1388)

George Earl of Angus (d. c. 1402)

The Red Douglases (Table 6)

d. died   k. killed   = married   ‡ illegitimate

CHAPTER TWO

The Knight of Liddesdale
and the Lord of Douglas

THE DISINHERITED

The first crisis of the house of Douglas as great lords in southern Scotland followed hard on the heels of James's death in battle with the Moors. The final achievement of Douglas's master in winning a peace with England which recognised his authority in an independent Scotland outlived King Robert by less than five years. By 1334 this independent kingdom, the Bruce dynasty and the house of Douglas all seemed to have fallen before the renewed assault of the English crown and its allies. The heirs of both Bruce and Douglas had been forced to flee into French exile and their lands had been overrun by their enemies. The eventual survival and revival of both the Bruce claim to Scotland and the house of Douglas's place in the marches was the product of warfare waged in the three decades after the death of the 'Good Sir James'. This second great Anglo-Scottish war brought no second triumph for royal leadership in Scotland. The principal fighters and victors in this war were the Scottish nobility. Across the whole kingdom, and especially in the front line of conflict in the marches, the years from 1330 to 1360 saw the forging of a new set of rules and relationships in Scotland which would last for a century. In the south, the Douglases were at the heart of these new relationships as magnates, and later earls, whose power dominated the central marches.

Despite James's death, the house of Douglas quickly resumed its place among the leaders of the Bruce regime when war returned to Scotland in 1332. Edward Balliol, claimant to his father John's lost Scottish throne, invaded the kingdom by sea with the backing of the 'disinherited', lords denied their lands in Scotland by the victory of King Robert.[1] These irreconcilable enemies of the Bruce party won a crushing victory over a huge Scots army at Dupplin Moor near Perth in August 1332. The defeat threw the Scots back on the resources of individual leaders. Among them was Archibald Douglas, half-brother of James and the man in control of the Douglas lands for his nephew, James's young son, William. The reputation and following of his brother made Archibald one of the leading Bruce partisans after Dupplin, but it was immediate military success which brought him wider powers. With the new guardian, his distant cousin, Andrew Murray, Archibald led a devastating attack on Galloway which had risen in support of 'its special lord' Edward Balliol. Four months later Douglas followed this by driving 'Edward king of Scots' in flight out of Annandale and into England.[2]

It was as 'the principal adviser in ... the confounding of the king' and as a leader and recruiter of 'a great band of men', as much as the heir to his brother's influence, that Archibald was made guardian of the kingdom after Murray's capture. The choice was probably made by those under arms in the south of Scotland rather than in any formal assembly, but Archibald's success in local raids did not prepare him for full-scale conflict. In early 1333 Edward III king of England entered the war on Balliol's behalf and laid siege to Berwick. Its loss was inconceivable to Archibald. He raised a huge army and, having failed to repeat his brother's feat of lifting a siege by raiding into northern England, he resorted to pitched battle to save the town. The result was a second huge defeat for the Bruce party on Halidon Hill. The guardian himself was killed in the fight. To later historians he was 'the Tyneman', or the loser, but Archibald's career was not simply about the carnage at Halidon. His few months of leadership had shown the link between military success and political and territorial power which would characterise aristocratic ambitions for the rest of the century. Defeat of Balliol at Annan brought more than just the guardianship. In early 1333 Archibald, lacking any legal right, had seized the vacant lordship of Liddesdale during his march through the borders. Leadership in war was allied to the search for personal lordship which no longer needed royal sanction. There were men with Archibald who remembered Liddesdale and did not forget the 'Tyneman's' brief dominance.[3]

After Halidon the house of Douglas was far from dominant. Young William of Douglas had been cut down alongside his uncle. In the face of English enemies and Scottish rivals, the new lord of Douglas was Hugh, James's surviving half-brother, a cleric without ability or energy. His nearest heirs were the young sons of Archibald, John and William, soon to depart for French exile alongside the child king, David Bruce. There was only one other male offshoot of the senior line of Douglas. The bastard son of the 'Good Sir James', a second Archibald, must have seemed a figure of limited importance even in the chaos after Halidon. Yet this Archibald, whose nickname of 'The Grim' was testament to both his character and his methods, was to rise to rule both the house of Douglas and much of southern Scotland, and, in a career which spanned seven decades, it was

Archibald the Grim who would be the real founder of the Black Douglas dynasty.[4]

There was an even greater threat to the Douglases than the deaths at Halidon. In the aftermath of the battle Edward Balliol fulfilled his agreement with his new overlord. In June 1334 Balliol ceded the six sheriffdoms of the south, Lothian, Berwick, Roxburgh, Selkirk, Peebles and Dumfries, to the control of the English king. For the landowners of the south, unlike the rest of Scotland, war was again about allegiance to English or Scottish kingdom. For men of the Douglas name the stakes were higher still. The greatest landed rewards of James Douglas, the forests of Selkirk, Ettrick and Jedburgh, were to become royal possessions of Edward III of England while the estates received by forfeiture of Bruce's enemies were reclaimed by their 'disinherited' heirs.[5] With their rights to virtually all the lands of the family challenged by the English king and his vassals, there was no place for the Douglases in any Balliol regime. While lords like the Steward and the earl of Fife could seek temporary agreements with Edward Balliol, Douglas survival depended on active and forcible resistance to both England and the 'disinherited'. Robert I's patronage had inextricably bound the Douglases to 'the Scottis fay', the phrase used by Scots to describe adherence to the Bruce regime and identify it as a patriotic or national cause. Between the 1330s and 1350s this allegiance, encouraged by self-interest, would see the Douglases emerge as natural leaders of the 'national' party in the new English 'pale' of southern Scotland.

THE RISE OF DOUGLAS OF LIDDESDALE

The source of this leadership was hardly obvious in early 1334. The losses of the family since 1329 were matched by the rest of the Bruce 'establishment'. Eight earls had been killed at Dupplin and Halidon and the houses of Bruce and Randolph had, like the Douglases, lost leading members. The deaths of William lord of Douglas, the younger Thomas Randolph and Robert Bruce of Liddesdale removed the second generation of Bruce leaders, the sons of the old king and his chief lieutenants.[6] In the south these losses created an acute crisis of lordship. The structure of power and military command set up by King Robert in the Borders was swept away. Robert's own place had fallen to his young son, David II. A child could not fulfill the role played by Robert I and in the 1330s leadership in war and politics passed of necessity to lesser figures. In the south, though, there was an equal vacuum in aristocratic leadership. While Robert Stewart, John Randolph earl of Moray, Andrew Murray and Patrick Dunbar earl of March all led Bruce forces in the region, both Stewart and Murray were based further north and Randolph was captured in 1335. Only Earl Patrick Dunbar remained as a magnate leader in the south. A man 'with no great liking for either side', Patrick's role in the war was, though far from insignificant, largely indirect.[7] Despite the losses of 1332 and 1333, there was no permanent collapse of support for the Bruce cause in the south. The English governors of southern Scotland recorded long lists of land-owners forfeited for their refusal to enter English allegiance in 1334 and 1335. The bulk of these 'rebels' were minor barons and knights, perhaps incited by hostility to English annexation. It was from this group, and especially from the knights of Lothian, a traditionally influential and inde-pendent community, that the leaders in the war against the English in the south were to be drawn. Men like John Haliburton, Laurence Preston and Alexander Ramsay, all forfeited Lothian barons, were identified by later fourteenth-century chroniclers as 'flowers of chivalry' whose 'labours for the liberty of the realm' took up the mantle of the Black Douglas. One knight sought the mantle of James Douglas more consciously. William Douglas of Lothian was a third cousin of the lords of Douglas. He had none of their lands and status, holding only scattered manors in Lothian.[8] He was, though, a man with the ambition and ability to see in the war a chance to emulate the rise of his kinsman. In a career characterised by war and murder and ending in his own violent death, William Douglas, known as 'the knight of Liddesdale', would contribute to both the later strengths and the weaknesses of the Douglas family.

Before 1334 William Douglas of Lothian had been a minor figure in Scottish politics. He had served the Randolphs in Annandale and main-tained links with his Douglas kinsmen. By 1341, though, he was among the most powerful magnates in the kingdom. His influence had all but sup-planted that of his cousins in the dales of the central Borders and he stood poised to appropriate the rank and rights of the lords of Douglas. Like the Good Sir James three decades earlier, the rise of William Douglas was achieved through his skill in a war of raid and skirmish against the English in southern Scotland. William's achievements were hailed as proudly as those of James by the chroniclers of the later fourteenth century and, in terms echoing those used about his cousin, William was 'the flail of the English and the wall of the Scots', a defender of 'the freedom of the realm'. This was far from empty praise. William Douglas was a mainstay of David II's cause from 1334. Captured in early 1333, William had escaped Halidon and on his release in 1334 immediately rejoined the fight. Initially he followed the lead of John Randolph earl of Moray, raiding Galloway and capturing the count of Namur alongside the earl in 1334 and 1335 and receiving Randolph's protection against his political enemies at parliament. The capture of Randolph near Jedburgh, which William almost shared, severed this link of service. Douglas may already have been seeking his own power-base in the south and from 1335 emerged as a leader in his own right. Later in that year, he led his own men, 'the floure of the south half the Scottis se', to the north-east, commanding them in the defeat of David earl of Atholl (Balliol's principal Scottish supporter) at Culblean. Four years later William again gave support to the Bruce party north of Forth by leading a Franco-Scottish force to the siege of Perth and negotiating the surrender of Cupar with its keeper William Bullock.[9] It was, however, in the south that Douglas was most active. In 1336 and 1338 he harassed the advance of major English forces through the region and in 1337 he ambushed and captured John Stirling, the head of the English administration in Lothian and constable of Edinburgh Castle. The crowning success of the war in the south was also won by William Douglas. In April 1341 William and his allies captured Edinburgh Castle in a surprise assault returning the kingdom's richest burgh to Bruce control. The base for these activities was, once again, the Forest. Following the example of James Douglas, by 1337 William had seized control of upper Teviotdale, the strategic hub of the south. Soon afterwards his capture of Hermitage Castle extended William's reach into the neighbouring lordship of Liddesdale. The central spine of the Borders provided a stronghold from which to challenge the English hold on the whole of their 'pale'. It also provided another Douglas with the means to establish personal power in the marches.[10]

In 1334 William had returned to the ravaged lands of Lothian and 'gaderit hym a companny'. Unlike James Douglas in 1307, William had no core of tenants from his estates to provide the backbone of his band. Instead men took service with leaders like William Douglas as captains 'namyt of prowes', known for their valour. The armed bands led by Douglas, his contemporary Alexander Ramsay and others lived 'in poverty' and 'like shadows', fighting a guerilla war against the English. Ramsay based his followers in a network of fortified caves at Hawthornden in Midlothian, while Douglas, operating from lairs in the Forest or the Pentland Hills, was wounded twice and risked frequent capture ambushing larger English forces. But these leaders engaging in small-scale warfare were the only active opponents of the English in the south. Later chroniclers saw Douglas and Ramsay as attracting 'schools of knighthood' made up of young Scots keen to follow famous captains. Beneath the chivalric gloss, such knights won support as the only leaders of the Bruce party in the south. Able to offer protection and enforce obedience from fellow Scots, they attracted bands of thirty to a hundred men, numbers sufficient to challenge English garrison troops and harry larger forces.[11] William Douglas found the core of his 'companny', not through territorial bonds, but in his extensive family. His uncle, Andrew, three brothers and five nephews all served William in his military exploits. These kinsmen provided Douglas with trusted deputies, keeping his castles and maintaining his influence in his absences. The risks of such support exacted a significant toll. Two of his brothers, James and John, were killed by William's English and Scots enemies, but John's sons, 'five handsome brothers of the name of Douglas', continued to be mainstays of their kinsman's cause.[12]

The power of William Douglas was based on the 'kindred, men and adherents' he had attracted to his service. Beyond his '*cognati*', military success was the key to Douglas's lordship in the south. In a land of war like southern Scotland local control was achieved through winning over or forcing the submission of the men of the region. In Lothian William's leadership won him the support of neighbouring landowners like James Sandilands and Andrew Ormiston. But more 'adherents' were won through direct success in war. The capture of Laurence Abernethy, a former follower of the Good Sir James who had entered English allegiance, led to the incorporation of Abernethy's own kinsmen into William's 'company'. Breakthrough in Lothian came with the fall of Edinburgh Castle in 1341. Control of the stronghold gave William access to 'the mycht of the bur-gesses', the significant military manpower of Edinburgh, which had aided in the capture. A dozen years later men from the burgh were still described as 'merchants of William Douglas', counting themselves among his fol-lowers. Beyond Edinburgh itself, the castle was a base for William's influence. Local men who had been members of the English garrison now turned to Douglas for lordship and protection.[13] This link between military success and political control also functioned further south. William's repeat of James Douglas's success in winning the Forest 'magre all his innemys' was achieved by enforcing the return of locals 'to the king's faith'. The position of the men of Teviotdale remained unenviable. Caught between Douglas in the Forest and the English in Roxburgh, allegiance was a matter of survival. Locals like the Kerrs and Eustace Lorraine might serve William Douglas as good Scots in the early 1340s, the Kerrs being described by the English as their 'greatest enemies ... in the Forest or the company of William Douglas'. In the dark days of the 1330s and after 1346, however, they adhered to the English cause. For such borderers, William Douglas, backed by his largely Lothian-based following and espousing the cause and claims of the Bruce party, appeared less clearly a freedom-fighter than another warlord, imposing personal control by raid and enforced lordship.[14]

War leadership had brought William Douglas wide political control in southern Scotland. The castles of Edinburgh and Hermitage in Liddesdale were 'stuffit with men' and provided bases for his military activities. Adherence to the 'Scottis fay' may have reflected personal loyalties, but more significantly it allowed William to fulfill his own ambitions. As leader of the Bruce cause in much of the south, William was building a position as the dominant magnate in the marches, whose actvities had won much of the region from the English in the late 1330s. In these terms, William's early career was a repeat of James Douglas's success in the marches between 1307 and 1314, but in reality there was a crucial difference between the aims and experience of the two men. William neither sought nor earned his kinsman's reputation as a trusted royal servant. While James was linked to Robert I by bonds of personal loyalty, there was no such tie between William and either the exiled David II or his lieutenant, Robert Stewart. In the 1330s, William received no military leadership or aid from the lieutenant. It was by his own efforts that William Douglas built his power in these years and the rest of his life was to be spent in attempts to secure and maintain this dominance by whatever means necessary.

Initially, however, a link to the young king seemed to be the best way to achieve this goal. In 1339 Douglas visited David II in Château Gaillard on the Seine, returning to Scotland with a force of French knights and crossbowmen.[15] The promise of royal reward was given in exchange for William's support in the preparations for the king's return. This alliance was strengthened the following year by the release of John Randolph, David's friend and Douglas's patron, who quickly assumed a major role in the marches. 'Care and custody of the west march' was given to Randolph, while William held the middle march and Alexander Ramsay the east. For the first time the marches had been recognised as a special region of the kingdom. Local military and political powers were officially handed to wardens, two of whom had entrenched ambitions and followings in the region. The grant emphasised the achievements of local warleaders. Douglas now had a formal commission to exercise the leadership in border war and politics which he had assumed in previous years.[16]

Such a commission was, however, only temporary. From 1339 William clearly undertook a search for a secure title to his position in the borders. In the next seven years he obtained over twenty grants of new estates in the south from a variety of landowners. Behind these grants lay the support of the king and the local power and ambitions of Douglas himself. Among William's 'benefactors' were a number of minor lay and ecclesiastical landowners, including two heiresses, who lacked the means to hold estates to advantage in a war-zone, especially against the will of a magnate like Douglas. Like James Douglas a generation earlier, William also received lands forfeited by Balliol supporters from a grateful king. Over half of these grants lay in the middle march, the key to William's regional power. Although mostly minor estates, Roberton in Clydesdale, Moffat and lands in Eskdale gave Douglas holdings which straddled the main routes between east and west borders and between Lothian and the middle marches.[17] Douglas clearly regarded grants of landed title as a means of cementing existing physical occupation in the region. In gaining title to the lordship of Liddesdale, William displayed his aims blatantly. Liddesdale had been won in war by William in 1337. Legal right was less certain. When David II gave it to his former lieutenant and nephew, Robert Stewart, in February 1342, Douglas immediately put in a counter-claim. His action was formally on behalf of his ward and godson, the son of the tyneman of Halidon, Archibald Douglas. Archibald had seized Liddesdale in 1333, but had no greater formal right than William. Although the Douglas claim was rejected, two days later a deal was reached in which William relinquished his title to the earldom of Atholl in exchange for receiving Liddesdale for himself. Atholl lay in Stewart's area of influence and the transaction reflected the formalisation of the gains made by both men in the warfare of the 1330s. From 1342 it was as lord of Liddesdale that William Douglas appeared in official records, clearly proud of the title which displayed his power in the marches.[18]

The ambition of the new lord of Liddesdale for formal power in the south was to have huge implications for the history of the Douglas line. Liddesdale aside, there were no great landed rewards available for distribution to reflect William's dominance in the borders. The principal lordships of the middle march were the legal possessions of William's kinsman, Hugh, lord of Douglas. Hugh, known as 'the Dull', was no man to defend his lordships against either the English or the ambitions of a man like Douglas of Liddesdale. He had probably been in France since 1337, leaving his lands and rights to his kinsman. If he could not obtain them wholesale, William set out to secure control of the Douglas lands piecemeal. He exercised the family's church patronage, securing the appointment of his associate, William Bullock, as rector of Douglas, but in late May 1342 he encroached directly on his kinsman's estates. A series of charters from Hugh gave William control of Westerkirk, Stablegordon and other lands in the marches. Exploiting royal support, William went further. Hugh resigned the bulk of the Douglas estates, including Douglasdale, the Forest, Lauderdale and Buittle, to his nephew, William. This second William was the son of the guardian, Archibald Douglas, and in 1342 was a youth in exile in France. He was also the ward of Douglas of Liddesdale, whose rights the latter had already used to cover his acquisition of Liddesdale. The resignation of Hugh had given the lord of Liddesdale legal rights in the bulk of the family lands, but there was more to the resignation than short-term administrative powers. In the terms of the settlement, should the new lord of Douglas die without male heirs, the succession would miss out his sister and other kinswomen and pass to Liddesdale. The elder William had an interest in preventing his charge obtaining his inheritance.[19]

The entail of 1342 was to become a crucial document in the history of the Black Douglases. Douglas of Liddesdale's influence was probably behind the inclusion of a second beneficiary in its terms. Should Liddesdale also die without a son, the heir to the lands of Douglas would be Archibald, the Good Sir James's bastard. For the first time and with major repercussions, the rights of Archibald were acknowledged by his family. That this was forced on them by Douglas of Liddesdale may reveal an early coolness between the young Archibald and his closest kin. His rights as ultimate heir would, though, bear fruit against the descendants of these kinsmen nearly fifty years later. Archibald's succession to the lordship of Douglas in 1388 would mark the culmination of a rivalry begun over four decades before.

The entail was the attempt of Douglas of Liddesdale to hijack the legal rights of the main Douglas line. The takeover by an active military magnate of estates held by inactive 'absentees' was not confined to the marches. In the increasingly military atmosphere of lordship throughout northern Scotland and in much of Ireland, legal rights were secondary to physical possession.[20] For both David II and Liddesdale this provided some justification for their actions. Robert I had given James Douglas his lordships in the south as the man best able to protect them. As Douglas of Liddesdale had inherited James's role, he could also claim rights to hold those lands he and his adherents had recovered from English rule. The 1342 entail which restricted possession of the principal Douglas lordships to the male line was not simply self-interested. It also provided for the succession of a line of lords who could best provide the military leadership essential for the Bruce cause in the region. The history of the Douglas line is a history of such warlords whose place in the marches would continue to depend on their command in war as much as their legal claims.

David II confirmed the Douglas entail as he had confirmed much of William's position between 1340 and 1342. He may have hoped that such patronage would tie William into the royal following and help the king's resumption of his father's role in the war, signalled by a series of royal raids into England in the early 1340s. If so, he underestimated the nature of Douglas of Liddesdale's aims. Unlike the 'Good Sir James', William was not content to fit into a royal-led establishment in the south, his powers in the region confirmed by David 'until peace is restored'. Royal campaigning in the borders was a threat to Douglas's own lordship and William may have refused to serve with the king when he crossed the border under Randolph's banner in February 1342. The next month, the king received an opportunity to counter William's dominance in the middle march. In the early hours of Easter Sunday a band of Scots led by Alexander Ramsay scaled the walls of Roxburgh Castle, repeating James Douglas's feat of eighteen years before. Ramsay, whose activities had been centred in the east marches, had captured the great fortress of the middle borders from under William's nose. The king seized upon Ramsay's success. He confirmed Ramsay as both keeper of the castle and sheriff of the surrounding region of Teviotdale. With official powers over royal lands and justice in the sheriffdom, and with his own armed retinue, Ramsay was a counterweight to Douglas.[21]

William could not ignore such a challenge. A powerful sheriff within 'his' march would weaken his hold on the men of Teviotdale. Ramsay, although a former ally, was now a rival source of lordship in a region where power depended on purely personal ties. Disputes between Ramsay and Douglas erupted almost at once. They were settled at first, perhaps in late May when David confirmed the Douglas entail, but any 'renewed friendship' ignored William's refusal to accept a rival. Within a month, Ramsay was testing his new powers by holding his sheriff court at Hawick on the borders of Liddesdale. When he heard that William was approaching, Ramsay 'suspected no evil' and greeted Douglas in the church before dismissing his own followers into the town. Immediately the sheriff was assaulted by Douglas and his men who wounded Ramsay and led him, bound on a mare, over the fifteen miles of moor to the grim keep of Hermitage Castle, deep in Liddesdale. Once there, Ramsay was doomed. According to the chronicle of Walter Bower, he was starved to death in seventeen days. Ramsay had been William's servant in a famous joust against the English at Berwick, but chivalric reputations had their limits. Douglas killed Ramsay to end a threat to his personal dominance, which counted more than chivalry or even allegiance to king and kingdom.[22]

The death of Ramsay revealed David II's weakness. The king sought to replace the sheriff with John Barclay, a man from his own household, and 'sent an armed band to capture William'. Douglas's ally, the chamberlain William Bullock, was arrested and put to death by the king's men. Douglas's response was to rely on his local power and wider connections. His 'kindred, men and adherents' were engaged in a private war with Ramsay's followers and, to counter royal hostility, Douglas turned to Robert the Steward, the ex-lieutenant and no friend of the king, his uncle. An alliance between Stewart and Douglas of Liddesdale threatened the king with 'unceasing discord in the realm'. David backed down, restoring William to his peace and giving him both Roxburgh and the office of sheriff, while the feud with the Ramsay affinity was settled in parliament. The king had found out an unpalatable truth. Despite his attempts to provide effective leadership in war, he lacked the authority and resources to challenge an entrenched local magnate on his own. Ramsay's local following had made him a threat to Douglas, but without Ramsay, David was even more dependent on William to provide leadership in the marches. Not only was Douglas warden, sheriff and keeper of Roxburgh, he 'exceeded all in the skill of arms' and 'of the weris mast wesy wes'. When the king planned his invasion of England in 1346, it was to Douglas's experience that he turned. Characteristically, William turned the offensive to his private goal of clearing the English from Liddesdale. Having 'fillit fullely his baggis' with English loot, Douglas wished to return home. Overruled, he was caught up in the crushing defeat of the Scots army at Neville's Cross near Durham and captured along with his king. An English prisoner for the second time, Douglas of Liddesdale's release would now come at a high price.[23]

THE EARLDOM OF DOUGLAS

Neville's Cross brought the threat of renewed occupation to the south. The English and Edward Balliol swept into Scotland in late 1346 and even the Scots chroniclers admit the loss of the key areas of the marches, Annandale, Galloway, the Forest and the dales of the Tweed and Teviot. Without William Douglas of Liddesdale the defence of the borders collapsed. His deputy in Roxburgh, a local man, Eustace Lorraine, surrendered the castle to secure the release of his son who had been captured with Douglas. Hermitage in Liddesdale followed suit and many of Douglas's border partisans clearly made peace with the invaders to guarantee their survival. The power of a border magnate like Douglas lasted only while he could provide protection and leadership to his followers.[24]

It was into this land of war that the young William lord of Douglas returned to seek his inheritance. The nephew of the Good Sir James and godson and ward of William Douglas of Liddesdale, he had spent almost his whole childhood and youth in France. Liddesdale would hardly have wished such a rival, and it was only in 1347, after the capture of his guardian, that the lord of Douglas made his appearance in Scotland. If Neville's Cross had been a disaster for Douglas of Liddesdale, for the younger William it was an opportunity which he was determined to seize. French upbringing or not, the lord of Douglas was aware of the means by which his kinsmen had built their power in the borders. Like James Douglas, his first target was Douglasdale itself, 'his native land'. He drove the English from the lordship in an act which not only won him a landed base but, more potently, drew a powerful link with the past and announced the return of the heir of the Black Douglas to the war. To recover the great estates he claimed, William could not rely on his name alone. Power in the border lordships depended on armed force and the new lord of Douglas needed to find his own followers. Like Douglas of Liddesdale, the key to this support came from his kin. In 1347 the lord obtained men and a secure base from his uncle, David Lindsay of Crawford. Lindsay was a powerful Lanarkshire neighbour of Douglas and keeper of Edinburgh Castle. He rendered up the castle to his nephew and Douglas used the position to raise 'a gret companny' from the burgesses and funds from the burgh customs. With this force he turned to the traditional road of his family to power in the south, control of the Forest. While in the Forest the Douglas fame worked well and 'the folk coyme to the fay, for luffande folk til hym war thai', the men of neighbouring Teviotdale did not see the lord of Douglas as their saviour. Instead they 'war agaynnys hym al haille' supported by the English garrison of Roxburgh. Douglas routed both forces and, faced with devastation, 'the mast part of that cuntre to sauf thar lywis, coym hym til'. The men of Teviotdale took oaths to the lord of their return to 'the Scottis fay'. Once more military power and Scottish allegiance were the means to the dominance of a Douglas lord. Fear of such a magnate and support for his name and cause were both weapons in the winning of men to his 'companny'.[25]

By winning the Forest, the lord of Douglas had made himself the new leader of war in the marches. For the next decade he matched his guardian's energy in taking the war to the English and their allies. The attack on Teviotdale was followed by a series of campaigns aimed at widening his lordship as well as disrupting the English hold on the south. However, while William lord of Douglas established his power in the marches, there remained a check on his influence. Douglas of Liddesdale was neither dead nor had he abandoned his ambitions in the marches. From English captivity he retained influence and contacts in Scotland. He received regular visits from his close adherents led by James Sandilands and Andrew Ormiston. His interests were represented in the lord of Douglas's councils by his brother, John, and uncle, Andrew, who may have deputised for William in a more general sense. Most importantly, Douglas of Liddesdale himself came north on parole with increasing frequency and witnessed his godson's charters in 1348 and 1350. Such contacts did not signify harmonious relations between the two. Like Liddesdale and Ramsay, the two Douglases were rivals for predominance. The lord of Douglas would not relinquish his power, the lord of Liddesdale would not accept its loss. The former was building a base of support which would squeeze out his guardian's influence. Liddesdale's kinsmen were balanced by the lord of Douglas's own connections, Lindsay of Crawford, Thomas earl of Angus and, more ominously, former adherents of Alexander Ramsay. The marriage of Liddesdale's ally, James Sandilands, to the lord of Douglas's sister was, in this context, an attempt to win over the remaining adherents of the prisoner. Douglas of Liddesdale was forced to witness the erosion of his own support by his young kinsman who had taken on the military leadership of the marches.[26]

The result was increasingly violent rivalry. In 1349 John Douglas, Liddesdale's brother, was killed in what may have been a direct attack on his family's position in the south by David Barclay of Brechin, a leader in the royal assault on Douglas after Ramsay's death. Liddesdale's response was equally forceful. Barclay was cut down in Aberdeen in a planned assassination on orders issued by William from England. Despite this impressive display of power, a challenge to the lord of Douglas was a different prospect. To win back lost ground, William looked for help to a man in a similar position, David II. David had been offered his freedom if he agreed to a possible English succession to his kingdom. He wanted Douglas of Liddesdale to put his case and win support from the Scots but the rejection of the proposal in a general council at Scone in early 1352 ended David's immediate hopes. For Liddesdale the release of the king was not the issue. His efforts on David's behalf had allowed him to return to Scotland in 1348, 1350–51 and 1352 when he had used the opportunity to build his own influence as much as his master's. He used his adherence to the Bruce king to justify growing links with England. In February 1352 he received a licence to recruit English aid against 'a certain party who wish to rise up or rebel against David'. The cause of Bruce kingship was being used by both William Douglases to support their rival ambitions. By the summer the issue was clear. In July Douglas of Liddesdale was using English-held Teviotdale as a base, sending his men south for support. The final step was taken on 17 July when Liddedale became Edward III's man, promising to perform military service for the English king against anyone except the Scots. In return he received his freedom and title to Liddesdale, Moffatdale and lands in Eskdale, vital areas of Douglas's control. Liddesdale's act was not the betrayal of a 'patriotic' career but the maintenance of his long-standing goals in the marches. The indenture released him to challenge his godson for regional dominance. It was recognised in these terms by both his friends and enemies. Despite his 'treason' Liddesdale's men still adhered to him in 1352–53. In response, the lord of Douglas made his own approaches to the English.[27]

The presence of two ambitious leaders in the middle marches was a recipe for conflict. As with Ramsay and Douglas of Liddesdale in 1342, there was little room for comfortable co-operation between the two William Douglases in 1352–53. The lord of Douglas clearly feared Liddesdale's proven ability to win local support and his designs on the lands of the main Douglas line to which he was heir. Liddesdale must have hoped for a way to recover his lost leadership in the march. Despite English support, Liddesdale's goal was difficult. He sought to challenge a magnate who had won a string of successes since 1347 and had built up an impressive following. In the summer of 1353 the lord of Douglas continued this by forcing the men of Galloway to submit to him. His success in the west may have prompted a last attempt to pacify Douglas of Liddesdale. The lord of Douglas granted his rival lands near Buittle in Galloway, an act witnessed by men from across the south with connections to both Douglas magnates.[28] The grant may have been intended to draw Liddesdale into his kinsman's service, but it only prepared the way for a final act of violence. Within weeks, in August 1353, the final clash between the two men took place in the heartlands of their power, the Forest. Hunting in the hills between the Ettrick and the Tweed, Douglas of Liddesdale fell into an ambush laid by his godson and was 'cruelly and wretchedly killed'. William Douglas of Liddesdale died as he had lived. According to Bower, he was slain 'in revenge for the deaths of Ramsay and Barclay or for other other emnities which arose between those ambitious for lordship'. Though adherents of Ramsay may have sought Liddesdale's death, it was ambition for unchallenged lordship in the middle marches which spurred the lord of Douglas to eliminate his rival.[29]

The death of Liddesdale removed the last Scottish rival to Douglas's position. Within six months the lord had overturned his godfather's gains of the 1340s. In early 1354, he gained recognition of his title to all the lands of his uncle, Sir James, and father, Archibald, including the lordship of Liddesdale itself, which the lord took from the English in 1358.[30] If this reflected the lord of Douglas's clear leadership of the Douglas connection, the old lord of Liddesdale did not pass without trace. Douglas of Liddesdale had made James, son of his brother John, heir to many of his estates. James Douglas, lord of Dalkeith, and his brothers were to retain something of their uncle's ambitions and would not work easily with his killer. However, the lord of Douglas sought to win over other partisans of the man he had killed. In 1356 he led a company of forty knights, including kin and adherents of both Alexander Ramsay and Douglas of Liddesdale, on pilgrimage. His aim may have been to repeat his uncle's crusading feats in Spain, but he got no further than Poitiers where he fought in the King of

France's defeat at English hands. Douglas escaped only through the efforts of his own men who showed their loyalty in battle. In 1360 the lord of Douglas endowed a mass for his dead rival at Melrose Abbey where Liddesdale was buried. Through pilgrimage and bequest, the lord of Douglas was seeking to heal the conflicts within his family and following.[31]

The mid-1350s brought an end to the conditions which had been in force for twenty years and had seen the creation of the Scottish marches. In early 1356 Edward Balliol sold his claim to the Scottish throne to his English overlord and within weeks Edward III crossed the border in a hugely destructive attempt to enforce them. The invasion of 1356 was to be the last major English attack on Scotland for nearly thirty years. Prospects of French glory once more seized Edward's attention and the following year he released David II in return for a ransom and a series of truces which would last until 1384. However, the end of large-scale conflict and the return of the king did not spell the end of Douglas power in the south. The 1350s also saw the consolidation of William lord of Douglas's lordship in Lothian and the middle march. The invasion of 1356 and its aftermath had seen the last English attempt to break his hold in the marches. William led the Scottish attacks on Edward III's army and had followed its retreat by once more compelling the '*Scotos Anglicanos*' to submit to him.[32]

This stabilising of relationships between Douglas and lesser lords in the south must have been helped by the beginning of a period of Anglo-Scottish truce. William's military achievements since 1347 had drawn men to his banner. Peace allowed him to turn short-term connections into more secure lordship. The same men who had backed Douglas on his return to Scotland provided him with councillors, agents and local adherents for the next thirty years. The core of his following remained a close family group. His cousins, the sons of David Lindsay who had welcomed William into Edinburgh Castle, were key allies, barons of local power and influence in their own right. A similar place was held by the connections of Thomas earl of Angus. His daughter, Margaret, was to become William's mistress in the 1370s and her kinsmen, the Sinclairs, would be among Douglas's closest supporters. This pattern was completed by the successive marriages of Douglas's sister, Eleanor, with three Lothian barons, James Sandilands, William Towers and Patrick Hepburn. Lothian remained central to Douglas's power. His construction of the massive castle of Tantallon near his town of North Berwick displayed William's power in the region. Like all Douglas lords, William saw the rich knights of Lothian as an essential part of his following. The kin and descendants of the captains of the 1330s, Ramsay, Abernethy, Haliburton and Preston were all drawn into the lord of Douglas's following in the 1350s. Like his successors, he also found his hold on this group to be hard to maintain, especially in the face of effective kingship. The return of David II would undermine William Douglas's power in Lothian.[33]

William's real legacy to his successors in the house of Douglas was the creation of a following drawn from the marcher sheriffdoms of Selkirk, Peebles and Roxburgh where the family's landed power was concentrated. The ending of major warfare gave Douglas a chance to turn alliances and submissions based on the tide of war into more lasting bonds of lordship and landholding. English influence in Teviotdale was largely confined to their last strongholds at Roxburgh and Jedburgh. The local families which dominated Roxburghshire and the Forest and had switched allegiances between 1332 and 1356 became an increasingly important element in the Douglas following. The Kerrs, the Rutherfords and the Turnbulls of Minto, the Gledstanes, the Pringles and the Glendinnings, and by the 1380s the Colvilles of Oxnam, all came from the war-zone of Teviotdale and Eskdale. Most were tenants of the great lordships of the Forest, Jedburgh, and Eskdale which Douglas held, and in time of truce William may have been able to regularise relations with these vassals for the first time. Members of all these families served William or his son, James, in warfare, as local officials or as councillors between the 1350s and 1380s, and over the century from 1356 the links between these border kindreds and their Douglas lords would continue to provide the latter with their principal strength in armed followers. The establishment of secure lordship over these families, the culmination of the previous fifty years of local warfare, gave a firm basis for the Douglas leadership in the marches and confirmed their position of dominance in the central Borders.[34]

The 1350s also confirmed the limits of Douglas's regional interests. His power in Lothian and the marches functioned alongside that of other magnates. In 1356 John Stewart of Kyle, eldest son of the Steward, recovered parts of Annandale while the local landowner, Roger Kirkpatrick, brought Nithsdale to 'Scottis fay'. Douglas was not without interests in these areas which separated his principal lands from Galloway and in the 1350s Douglas influence made its first inroads into the west march. With his eastern neighbour, Patrick earl of March, Douglas seems to have had a more clear-cut relationship. March's influence in Lothian seems to have functioned alongside that of Douglas. His followers, the Ramsays, the Haliburtons, the Hepburns and the Maitlands, also appeared in William's affinity and the two magnates co-operated in warfare and politics from as early as 1350. In general, though, the activities of Douglas and March reflected different zones in the marches. While the Douglases concentrated on Teviotdale, Patrick's interests remained centred on Berwickshire and Dunbar.[35]

As throughout Scotland in the 1350s, regional power in the marches reflected the ambitions of individual magnates. King David found himself once more forced to confirm existing arrangements in the Borders. The killing of his erstwhile ally, Liddesdale, in 1353 had left him with no alternative but to confirm the killer in power. In 1354 the king gave William lord of Douglas a charter which gave him title to all the lands of his father and uncle and reversed David's judgement of 1342 on Liddesdale. Once again, legal right was a product of physical power. The 1354 charter went beyond confirmation of landed title. David also granted Douglas 'the leading of the men of the sheriffdoms of Roxburgh, Selkirk, Peebles and the upper ward of the Clyde'. The military leadership of the Douglases in the central Borders had been recognised and turned into a hereditary power which cut across the rights of landowners and royal officials between Lanark and Berwickshire.[36] When David returned to his kingdom, he went further than simple confirmation in an attempt to win Douglas's support. On 4 February 1358 William was created earl of Douglas. Only the third earldom to be created since the twelfth century and the second in the south since the 1070s, the creation of an earldom for William altered the composition of the southern Scottish nobility. Unlike earlier creations, the earldom of Douglas had no provincial basis or legal unity and was in one sense simply a mark of status. However, the title of earl in fourteenth-century Scotland was no empty honour but gave the holder regional powers as war leader and justiciar. The 1354 charter had already sanctioned Douglas's leadership of the middle march; his promotion to earl emphasised his rights over the communities of these sheriffdoms. As earls of Douglas, the importance and achievements of the dynasty in the years of warfare since 1300 were given formal recognition. William now stood in the top rank of the Scottish nobility, his power as local lord matched by his status in the kingdom. His eight successors in the title would inherit both the fame of the Douglas name and the regional authority which William had built around it.[37]

NOTES

1. For the details of Balliol's followers and a full account of the war in 1327 and from 1332 to 1335, see R. Nicholson, *Edward III and the Scots*. For a briefer view over a longer period, see B. Webster, 'Scotland without a King, 1329–41', in A. Grant and K. Stringer (eds), *Medieval Scotland, Crown, Lordship and Community* (Edinburgh, 1993), 228–38. For the European context, see J. Sumption, *The Hundred Years War, Trial by Battle* (London, 1990).
2. Nicholson, *Edward III and the Scots*, 91–104; *Chron. Lanercost*, 269–71.
3. Nicholson, *Edward III and the Scots*, 105–38; *Chron. Lanercost*, 274; *Scotichronicon*, ed. W. Goodall, 2 vols (Edinburgh, 1759), ii, 310–11; *Scotichronicon*, ed. D. E. R. Watt, 8 vols (Aberdeen, 1987–96), vii, 82–85, 88–91;*Wyntoun*, vi, 12; *R.R.S.*, vi, nos 44–45; *Morton Reg.*, ii, nos 61–63.

4. Fraser, *Douglas*, iii, nos 290, 315; *R.R.S.*, vi, no. 51; *Wyntoun*, vi, 192–93.
5. *Cal. Docs. Scot.*, iii, no. 1127. According to the *Lanercost Chronicle*, Robert Clifford was restored to Douglasdale by Edward Balliol (*Chron. Lanercost*, 271).
6. The elder Thomas Randolph had died in 1332.
7. *Chron. Lanercost*, 275, 278, 282–83; Nicholson, *Edward III and the Scots*, 143–44. Dunbar's kin and adherents were the mainstay of Alexander Ramsay's following (*Wyntoun*, vi, 148–49; *Scotichronicon*, ed. Watt, vii, 146–49).
8. *Cal. Docs. Scot.*, iii, 326–47; *Wyntoun*, vi, 49–52, 100–107, 114–23; *Scotichronicon*, ed. Goodall, ii, 317, 329–30, 333–34; *Scotichronicon*, ed. Watt, vii, 138–39, 146–49; *Morton Reg.*, ii, nos 13, 15, 16, 19.
9. *Wyntoun*, vi, 49–57, 63–66, 114–125; *Scotichronicon*, ed. Goodall, ii, 317, 319, 321–331; *Scotichronicon*, ed. Watt, vii, 88–89, 106–109, 114–17, 138–41, 274–75.
10. *Chron. Lanercost*, 285–87, 292–93; *Wyntoun*, vi, 114–23, 138–43; *Scotichronicon*, ed. Goodall, ii, 329–30, 332–33; *Scotichronicon*, ed. Watt, vii, 136–40, 144–47.
11. *Wyntoun*, vi, 147, 150–51; *Scotichronicon*, ed. Goodall, ii, 316, 329–30; *Scotichronicon*, ed. Watt, vii, 104–105, 146–49. Douglas's capture of John Stirling, head of Edward III's administration of Lothian, was carried out with forty men. Stirling was accompanied by fifty men from the Edinburgh garrison which totalled 140 men in 1340 (*Wyntoun*, vi, 116–17; *Chron. Lanercost*, 293, 295; *Cal. Docs. Scot.*, iii, no. 1323; v, no. 809).
12. *Morton Reg.*, ii, nos 66, 70, 71; Fraser, *Douglas*, iii, nos 315–19; *Wyntoun*, vi, 56–57; *R.R.S.*, vi, no. 17; *Chron. Lanercost*, 351; *Scotichronicon*, ed. Goodall, ii, 320, 348; *Scotichronicon*, ed. Watt, vii, 114–115, 146–47, 274–75; *Rot. Scot.*, i, 678; Jean Froissart, *Chronicles of England, France and Spain*, ed. and trans. T. Johnes, 5 vols (London, 1803–10), iv, 30.
13. *Historical Manuscripts Commission*, 11th report, sixth appendix, Hamilton Muniments, no. 21; Fraser, *Douglas*, iii, 316–18; *Rot. Scot.*, i, 706, 715, 729, 730, 734, 752; *Wyntoun*, vi, 109, 138–45, 192–93; *Scotichronicon*, ed. Goodall, ii, 322–23, 330; *Scotichronicon*, ed. Watt, vii, 144–47; *E.R.*, i, 507; *Cal. Docs. Scot.*, iii, nos 1291, 1323, 1369; *Morton Reg.*, ii, no. 66, 70, 71; *Newbattle Cart.*, no. 275; *R.R.S.*, vi, no. 411; British Library, Additional Manuscripts, no. 6435.
14. *Wyntoun*, vi, 100–101, 120–22, 147; *Cal. Docs. Scot.*, v, no. 809.
15. J. Sumption, *The Hundred Years War*, *Wyntoun*, vi, 123–25.
16. *Wyntoun*, vi, 152–55; *Scotichronicon*, ed. Goodall, ii, 333–34; *Scotichronicon*, ed. Watt, vii, 148–49; *Calendar of Close Rolls* (London, 1901), 1339–41, 540.
17. *Morton Reg.*, ii, nos 31, 57, 61–63, 67–68, 114–117; *H.M.C.*, 11th report, app. 6, no. 21; *Melrose Liber*, i, 418–19; Fraser, *Douglas*, iii, no. 17. For the strategic importance of Douglas's lands, see H. R. G. Inglis, 'Ancient Border Highways', in *Proceedings of the Society of Scottish Antiquaries*, lviii (1924), 203–23. Douglas also received the rich Lothian lordship of Dalkeith.
18. *R.R.S.*, vi, nos 44–45; *Morton Reg.*, ii, nos 61–63. William's right to the earldom of Atholl and interests in Perthshire possibly stretched back to 1335 when he was involved in the death of David of Strathbogie, the previous earl, at Culblean. For the regional power of the Steward, see S. Boardman, *The Early Stewart Kings, Robert II and Robert III* (East Linton, 1996).
19. *Morton Reg.*, i, nos 41, 114–17; *H.M.C.*, 11th report, app. 6, no. 21; *R.R.S.*, vi, no. 51; *Melrose Liber*, i, 418–19.
20. S. Boardman, *Early Stewart Kings*; S. Boardman, 'Lordship in the North-East: The Badenoch Stewarts, I. Alexander Stewart, earl of Buchan', in *Northern Scotland*, 16 (1996), 1–30; M. Brown, 'Regional Lordship in North-East Scotland: The Badenoch Stewarts II. Alexander Stewart, earl of Mar', in *ibid.*, 31–54; R. Frame, 'Power and Society in the Lordship of Ireland 1272–1377', in *Past and Present*, no. 76 (1977), 3–33; G. O. Sayles, 'The rebellious first Earl of Desmond', in *Medieval Studies presented to Aubrey Gwynn*, edd. J. A. Watt, J. B. Morrall and F. X. Martin (Dublin, 1961), 203–29.
21. *Wyntoun*, vi, 160–65; *Scotichronicon*, ed. Goodall, ii, 334–35; *Scotichronicon*, ed. Watt, vii, 150–51; *E.R.*, i, 508.

22. *Wyntoun*, vi, 102–109, 164–67; *Scotichronicon*, ed. Goodall, ii, 335–36; *Scotichronicon*, ed. Watt, vii, 152–57.

23. *Wyntoun*, vi, 166–69, 172–75; *Scotichronicon*, ed. Goodall, ii, 336–46; *Scotichronicon*, ed. Watt, vii, 154–57, 252–65; J. M. Thompson, 'A roll of the Scottish Parliament, 1344', in *S.H.R.*, no. 35 (1912), 235–40, 240; *Chron. Lanercost*, 345–48; *Scalachronica*, 112, 115; *Chron. Knighton*, ii, 23, 32–33, 41–43. Douglas's alliance with Stewart linked him to a growing body of opposition to David in the kingdom. David clearly removed Edinburgh from William at this point but Wyntoun and not the *Scotichronicon* is correct in saying John Barclay was only briefly keeper of Roxburgh in 1342. William held it from then until it was lost to the English in 1346.

24. *Rot. Scot.*, i, 675–78, 685, 688, 761; *Wyntoun*, vi, 182–83; *Scotichronicon*, ed. Watt, vii, 268–69; *Scalachronica*, 115. De Lorraine and his son were in English allegiance into the mid-1350s (*R.R.S.*, vi, no. 157; *R.M.S.*, i, no. 463).

25. *Wyntoun*, vi, 192–93; *Scotichronicon*, ed. Goodall, ii, 346–47; *Scotichronicon*, ed. Watt, vii, 270–71. William lord of Douglas's elder brother had died in France during the early 1340s (*E.R.*, i, 540; *Wyntoun*, vi, 192–93). David Lindsay's sister, Beatrix, was the lord of Douglas's mother. The lord maintained his close connections with Edinburgh over the next decade (Fraser, *Douglas*, iii, nos 20, 315). While *Scotichronicon* describes the men of Teviotdale as rising in support of Douglas, Wyntoun's distinction between the Forest and Teviotdale suggests a more detailed reading of their common source which reflects the loyalties of local men.

26. *Rot. Scot.*, i, 706, 715, 727, 729, 730, 734, 737, 740, 746, 752, 758; Fraser, *Douglas*, iii, nos 18, 20, 291, 316, 317, 318. Ramsay's nephew, William, witnessed for the lord of Douglas as sheriff of Lothian, suggesting close links with the magnate.

27. *Rot. Scot.*, i, 748, 752–53, 758; *Chron. Knighton*, ii, 356. A full analysis of the Anglo-Scottish negotiations and Douglas of Liddesdale's role in them is given in A. A. M. Duncan, 'Honi soit qui mal y pense: David II and Edward III, 1346–52', in *S.H.R.*, 67 (1988), 113–41.

28. *Morton Reg.*, no. 13; British Library, Add. Ms., no. 6433; *Wyntoun*, vi, 206–207, 222–23; *Scotichronicon*, ed. Goodall, ii, 350, 356; *Scotichronicon*, ed. Watt, vii, 296–97. The dating of the Galloway attack is suggested by the forfeiture of Dougal MacDowell by the English in August 1353. MacDowell had submitted to Douglas during the attack (*Rot. Scot.*, ii, 761). The charter to William Douglas (called Douglas of Lothian), though undated, was witnessed by many key men in contemporary politics, Richard Small, Andrew Douglas and Fergus MacDowell of Galloway among them.

29. *Scotichronicon*, ed. Goodall, ii, 348–50; *Scotichronicon*, ed. Watt, vii, 274–75; *Rot. Scot.*, i, 752.

30. *R.M.S.*, i, no. 123; *Rot. Scot.*, i, 826; *Scotichronicon*, ed. Watt, vii, 296–97. Hermitage castle was taken by Douglas in breach of the newly-made Anglo-Scottish truce.

31. *Scalachronica*, 125; *Wyntoun*, vi, 231–32; *Scotichronicon*, ed. Goodall, ii, 357–58; *Scotichronicon*, ed. Watt, vii, 296–301; Fraser, *Douglas*, iii, no. 23.

32. *Wyntoun*, vi, 192–95, 206–207, 222–23; *Scotichronicon*, ed. Goodall, ii, 354, 356; *Scotichronicon*, ed. Watt, vii, 296–97; *Knighton Chron.*, ii, 85; *R.R.S.*, vi, no. 137.

33. Fraser, *Douglas*, iii, nos 20, 22, 23, 317, 318, 323; *Registrum Cartarum Ecclesie Sancti Egidii de Edinburgh*, Bannatyne Club (Edinburgh, 1859), 7–8; *Melrose Liber*, ii, nos 461–62, 490; *Royal Commission on the Ancient and Historical Monuments of Scotland: East Lothian* (Edinburgh, 1924), 61–67.

34. Fraser, *Douglas*, iii, nos 22, 23, 25, 293, 323, 330, 332, 333, 334, 335; *H.M.C.*, Drumlanrig, 2 vols, i, no. 2; *Carte Monialium de Northberwic*, Bannatyne Club (Edinburgh, 1847), no. xxxvii; *Melrose Liber*., nos 461, 462, 490; *Newbattle Reg.*, no. 276; *Scotichronicon*, ed. Goodall, ii, 350; *Scotichronicon*, ed. Watt, vii, 280–81; Froissart, *Chroniques*, iv, 15–30.

35. *Wyntoun*, vi, 222–25; *Scotichronicon*, ed. Goodall, ii, 350–51, 356, 361; *R.R.S.*, vi, no. 379; Fraser, *The Book of Carlaverock*, 2 vols (Edinburgh, 1873), ii, no. 15.

36. *R.M.S.*, i, appendix 1, no. 123.

37. *R.M.S.*, i, appendix 2, no. 1222; A. Grant, 'Earls and earldoms in late Medieval Scotland

1310–1460', in *Essays presented to Michael Roberts*, ed. J. Bossy and P. Jupp (Belfast, 1976), 24–41; A. Grant, *Independence and Nationhood* (London, 1984), 123. Dr Grant's view that the earldom of Douglas had 'no provincial connotations' ignores the region of military leadership stated in the 1354 charter and almost certainly included in the creation charter of 1358.

CHAPTER THREE

Archibald the Grim (1358–1388)

THE STARS OF THE MURRAYS

While William earl of Douglas was laying the foundations of lasting Douglas dominance in the middle march, the man who would eventually succeed to his titles and power was a servant in his following. Archibald Douglas, the son of the Good Sir James, had spent his youth and manhood in the service of greater kinsmen. In his late thirties in 1358, Archibald seemed destined to remain a landless and illegitimate knight, an important but dependent relative of the new Douglas earl. However, by his death forty-two years later, Archibald had risen from these origins to hold unparalleled lands and power in southern Scotland and to set the descendants of the Black Douglas at the head of their kindred and of the Scottish nobility.

Archibald's origins were unpromising. Born in the 1320s, as a bastard he had no rights to succeed to the Douglas estates after the death of his father and brother in the early 1330s. Instead he may have been sent with his cousin, the future earl of Douglas, to join King David II in French exile.[1] He probably spent his youth with king and kinsman but returned to Scotland before either of them. In 1339 William Douglas of Liddesdale visited King David's court at Château Gaillard in Normandy, recruiting both French troops and some of the royal household to his service. Archibald may have been in this latter group as, in the years that followed, his appearances were made as a 'servant' and 'kinsman' of the powerful and ambitious lord of Liddesdale. It was the ambition of Liddesdale which gave Archibald formal rights within the Douglas kindred for the first time when he was acknowledged in the family's entail of 1342, and Archibald repaid Liddesdale with service which survived the decline in the latter's fortunes after Neville's Cross and which did not cease with William's entry into English allegiance in 1352. For the rest of his long life, Archibald's connection with Liddesdale influenced his loyalties and won him allies in the creation of his own power and following.[2]

For great lords, lesser kinsmen like Archibald had a significant role as agents in diplomacy, castle constables and subordinates in war. Liddesdale had relied on Archibald and his other kin in these roles, and the future

Earl of Douglas was quick to seek the support of his cousin after Liddesdale's murder. Whatever his personal doubts, Archibald entered the service of the earl as the only source of protection and reward in the marches and in 1356 he was one of the knights who covered Douglas's flight from the battle of Poitiers, only escaping capture himself by posing as a common soldier. Although denied lands and independent power, Archibald was always more than an ordinary knight in the earl's entourage and in the mid-1350s could claim to be his cousin's heir.[3] More importantly, as the son of Sir James Douglas, Archibald succeeded to his father's fame and reputation, if not his landed status, and appearances emphasised the connection. Archibald had his father's dark looks and was remembered in the fifteenth century as 'Blak Archibald'. To contemporaries, though, he was Archibald the Grim or the Terrible, inheritor of his father's terrifying prowess on the battlefield. The great chronicler of fourteenth-century Europe, Jean Froissart, described Archibald as 'much feared by his enemies', a 'gallant knight' who 'wielded before him an immense sword ... which scarcely another could have lifted', with which 'he gave such strokes that all on whom they fell were struck to the ground'.[4] By the late 1350s the war which had dominated the marches since 1332 was over but lordship in the region was to remain bound up with warfare and military leadership, and Archibald's skill with his great sword was not detached from the lands and status he would achieve within Scotland. While his skills in fighting and in leading his followers had been learned by Archibald in the war against the English, from 1357 they were to be employed in the internal politics of the kingdom, and the man who would employ them was not a Douglas magnate but King David Bruce. In the 1360s Black Douglas and Bruce would consciously re-create the link between their two fathers, but their relationship would rest on more than just sentiment. Like his father, Archibald's service to his king would bring rewards in land and power, but, unlike James Douglas, the career of Archibald the Grim would flourish after his master's death. The change from servant to lord was slow for 'Blak Archibald' but the passing of the Bruce dynasty would see his emergence, not as a royal lieutenant, but as the dominant magnate in the western marches.

When he returned from a decade of English captivity in 1357, David II needed the support of men like Archibald Douglas. Though rival claims on his throne were dormant, effective power in the kingdom remained in doubt. The concessions of 1354 and 1358 to William Douglas were mirrored by similar grants to Robert Stewart and others, leaving royal authority severely limited in scope and scale. Like Liddesdale, the new earl of Douglas saw kingship in this context, as a source of legitimation and patronage, not of active rule or interference with local rights and relationships. Initially David II fulfilled this role. In the south Earl William's provincial leadership was given royal sanction. The creation of the earldom of Douglas was followed by William's confirmation as justiciar, warden of the east march and appointment as sheriff of Lanark. Memories of the outcome of his earlier efforts to interfere in the marches led the king to avoid direct confrontation.[5]

Instead, the road to royal dominance began at a lower level in political society. Despite the years of exile and captivity, David was still the King of Scots and heir of Robert Bruce and his office was still the potent source of lordship exploited with such skill by his father. From 1357 onwards David Bruce used the prestige and powers of the crown to win the direct support of knights and barons who had previously looked to great lords for leadership and protection. This policy had most to offer among the rich and politically independent nobility of Lothian where David already had the basis for a following. Though ultimately disastrous, the king's alliance with Alexander Ramsay in 1342 had left its mark on loyalties in Lothian. Ramsay's kin and adherents turned to David for lordship in the 1340s and four of the family had shared the king's capture at Neville's Cross. In 1358 this group, led by Alexander's nephew, William, and two other Ramsay adherents, Walter Haliburton and John Preston, were quickly installed as royal councillors, and their support must have increased David's influence in Lothian. This increase took place at the Earl of Douglas's expense. Before 1358 many of these Lothian knights had accepted the earl's lordship. William Ramsay, the sheriff of Lothian, was in Douglas's retinue at Poitiers and had served, along with Haliburton and Preston, as a councillor and armed follower of the earl in Southern Scotland.[6] The attendance of these men on the king spelled a decline of the earl's influence in an area which had previously been a centre of his interests, and the consolidation of Douglas power in the marches may have been prompted by the weakening of the earl's position further north. The same process was probably at work in Edinburgh itself. The regular presence of king and court in the burgh must have loosened Douglas's influence with the burgesses, though he was still regarded as the principal local magnate in 1360 when a townsman, John Allincrum, endowed masses for both the king and the earl at St Giles'. Ominously for Earl William, the final person to be remembered in prayer was Archibald Douglas. Archibald's significance in Edinburgh was not as an adherent of the earl. By 1360 he was the king's man in the burgh, both keeper of the castle and sheriff of Lothian. Though he witnessed charters of the earl in 1360, Archibald had followed the example of his friend, William Ramsay, in abandoning Douglas for royal service. From early 1362 until the end of David II's life, Archibald was to stand among the key councillors and lieutenants of the king.[7]

David looked for support from men like Archibald Douglas who could be drawn from ties to great magnates. The king valued the military experience and skill of knights like Douglas, Ramsay and others who had won fame in continental warfare, and his household had a military character. The leading adherents of the king were men trained for war. David was drawing on the structures of military lordship forged in the early years of his reign to provide physical backing for his own authority. In the process the king aimed to weaken the power of great lords, like Douglas and the Steward, from whose followings such knights had been detached.[8] To Archibald the Grim and others, the rewards of royal service were greater than anything to be won from their old lords. While, at most, Earl William may have offered fees and gifts designed to retain Archibald's adherence without creating a rival, David Bruce held out a richer prospect. Like his father, David saw the promotion of a Douglas lieutenant, not as a source of rivalry, but as a means to increase his own hold on his kingdom. The king opened the way for his servant to obtain the lands, status and power of a great lord.

In 1362 Archibald married Joanna Murray, heiress to one of the kingdom's senior noble dynasties. As later writers proclaimed, the match re-united two parts of the same kindred, both branches of which bore three silver stars on blue as their arms. The marriage also brought Archibald 'tresour untald, towris and towns ... with rent and with riches'. The union was the king's work. Since the death of Joanna's first husband in 1361, both she and her mother had been under royal supervision.[9] The king wanted to ensure that the Murray inheritance would pass into safe hands and manipulated land law to extend the estates involved. Joanna was declared to be the heiress of, not just her father, Maurice Murray, lord of Drumsergard, but also of her first husband and kinsman, Thomas Murray, lord of Bothwell. The accumulated estates of both lines included lordships across the kingdom from Lanarkshire and Roxburghshire in the south to Aberdeenshire, Moray and Ross in the north. In the seventy years before 1362, the Murray lords had been leaders of resistance to English domination, providing two guardians and numerous local leaders.[10] The insertion of Archibald Douglas, as the king's man, into the whole Murray inheritance showed that the king was ready to bend the rules and structures of the Scottish nobility to his advantage. At stake was not just royal patronage, but also David's own exercise of power as king.

Archibald Douglas's marriage made him heir to the Murrays' influence in central and southern Scotland and, in both regions, this worked to the king's benefit. In Perthshire David was putting pressure on Robert Stewart's position in Strathearn and Menteith. Archibald's new wife was the daughter and granddaughter of previous earls of Strathearn and, although Douglas did not inherit his father-in-law's lands in the region, he could be counted on to use the connections of his new kindred in central Scotland to the advantage of the king. More significantly Archibald had been given a base from which to dominate northern Lanarkshire. In the 1330s and 1340s Joanna's father, Maurice Murray, was referred to as sheriff, even lord, of Clydesdale and led the men of that district in warfare. Through his marriage, Archibald had obtained not just Maurice's lordships of Drumsergard, Carmunnock, Stonehouse and Strathaven in the area, but also the lordship of Bothwell.[11] This concentration of estates and the political influence of the Murrays transformed Archibald's prospects at a stroke and made him a figure of significance in Scottish politics. It also identified him unequivocally with the king and his policies and made him a target for the king's enemies, a group led by the Steward and the Earl of Douglas. Both magnates had grievances concerning Archibald's promotion. The Steward's son, Robert, lord of Menteith, had his own claim to a share of the Murray lands through his wife, while the Earl of Douglas may have feared the return of a powerful rival within his kindred less than a decade after Liddesdale's death. The threat posed by such men was hardly lost on Archibald. At some point, perhaps within months of his marriage, Archibald refortified the castle at Bothwell, which had been in ruins since 1337. The result was a crude, walled enclosure, designed not as an aristocratic residence but as a military base. To a man seeking to cement his local authority in the face of powerful enemies, a simple but defensible stronghold made obvious sense.[12]

Any such preparations in late 1362 were well-timed. Within months the king's enemies were in rebellion. The Steward and his sons, and the earls of Douglas and March, all put their seals to a petition which demanded payment of the ransom and 'wiser government'. These complaints concealed more diverse and personal goals. Robert Stewart's concerns were with his own place as the king's heir, while Douglas was pushed into revolt by the king's attack on his brother-in-law, the Earl of Mar, in January 1363. The threat this posed to Douglas's hopes to succeed to Mar's estates was the last in a series of grievances which also included the death of the earl's associate, Thomas earl of Angus, in royal custody. Against the background of royal interference with the followings of both Douglas and Patrick earl of March, such clashes were unwelcome marks of increased royal dominance in Scottish politics.[13]

The Earl of Douglas dominated the military revolt. While Stewart quickly came to terms, Earl William, backed by his '*adherentes*', probably men from his marcher following, waged a private war against the king. Avoiding a direct clash, 'the earl rode ... against those who were of the king's party, imprisoning the king's people wheresoever he could take them'. Attacks against the Haliburtons' castle of Dirleton, and the capture of Robert Ramsay at Inverkeithing, were moves against men who had abandoned their ties to the earl. Earl William's next move was against Lanarkshire. His obvious targets there were the lands and men of his cousin, the new lord of Bothwell, but Archibald was not present to defend his estates. Since late 1362, Archibald had been with the king, first in the force which had been led against Mar and which later received the Steward's surrender, then as one of the leaders of the paid royal army which assembled at Edinburgh in late May. Archibald was almost certainly with his royal master in the rout of Earl William at Lanark which closed the rebellion, forcing both the Earls of Douglas and March to submit to the king.[14]

In the aftermath of his armed defiance of the king, Earl William of Douglas suffered no major royal retribution. David II's principal aim in late 1363 was to detach the earl from his ally, Robert Stewart. The king won Douglas's support for his plans for the succession which threatened the Stewart's hopes of the throne with an English rival. Though, ultimately, these English negotiations were rejected in Parliament, there could be no mistaking the dominance of the king and his party after 1363. The rebellion had emphasised the success of the king in undermining the lordship of his opponents. Douglas found himself opposed by many of his former servants including, not just Archibald Douglas, Haliburton and the Ramsays, but probably and more damagingly his cousins, Alexander and James Lindsay. After the revolt, the Lindsays were central participants in the king's knightly entourage, and, though Earl William also acted as a regular councillor of the king from 1363 onwards, he could hardly count on the support of his cousins and former adherents in any future quarrel with David II.[15]

The support of Archibald Douglas for the king in 1363 secured his place in the 'king's party' and, indirectly, established him as a rival to Earl William within the Douglas kindred. By August 1364 he was warden of the west march, a natural counterbalance to the earl's dominance in the eastern Borders, and five years later he was given powers within William's heart-lands in Roxburghshire. In September 1369 David II reviewed a grant of his father to Melrose Abbey, which allowed the monks £2,000 from the profits of the sheriffdom of Roxburgh to repair war-damage to their buildings. Robert I had given the powers to raise the money to Sir James Douglas. Renewing the obligation to Melrose, David gave these same powers not to James's legal heir, Earl William, but to Archibald. Archibald

could seize goods within Roxburghshire and interfere with the earl's adherents in the sheriffdom. He had also been made the agent of the Melrose
monks, a role previously played by William, with both physical and symbolic
significance. In the late 1360s the earl was being reminded of Archibald's
closer ties of blood with the dead hero of the Douglas line. It may have
been at this time that Archibald built an impressive tomb 'of alabast' for
Sir James in Douglas church, displaying both filial piety and pride for the
dead hero. Though in lands and power, Archibald could hardly challenge
his cousin, with royal support the new Black Douglas could seek to use his
father's reputation to his own ends.[16]

Archibald was not the earl's only rival within the Douglas kindred in the
1360s. Earl William's fears about his cousin must have been increased by
Archibald's association with a more distant kinsman, James Douglas,
nephew and co-heir of the Lord of Liddesdale. The link between Archibald
and James went back to Liddesdale's following and, after his release from
English custody in late 1354, James may have gravitated naturally to his
kinsman's company. This alliance between Archibald and James proved
crucial to the rise of both men and, in the 1360s, Archibald knighted his
kinsman and backed his claims to lands and influence. Unlike Archibald,
though, James had made no peace with Earl William. The earl had been
involved in the deaths of James's father and uncle, and had obtained title
to, or possession of, many of the latter's estates, including Liddesdale and
the rich lands of Dalkeith in Midlothian. In the 1360s James Douglas sought
the recovery of this inheritance and, through the friendship of Archibald
and the connection between his uncle and the king, James turned naturally
to David II to achieve this end.[17]

The link between the two Douglas lords and the king was no isolated
alliance. Instead it was part of a concerted royal-inspired attempt to alter
the balance of power in southern Scotland, extending David's reach beyond
Lothian and into the marches. To achieve this the king sought allies within
the second great border house, the Dunbars. In 1363, just as Douglas
adherents had backed the king, so too did many of Earl Patrick Dunbar's
tenants and kinsmen. Chief among them was his great-nephew and heir,
George Dunbar. George's background was similar to that of Archibald and
James Douglas. His father had been a leader of the Lothian men in the
1330s and 1340s and had fought alongside Archibald at Poitiers. George
was also drawn to David II's affinity in the 1360s and was a natural ally
of Archibald and James Douglas. After 1363 the king exploited the connection to his and their advantage. In 1368 George Dunbar succeeded to
the childless veteran of border politics, Earl Patrick Dunbar, and the
following year his influence was further increased by the liaison between

the king and his sister, Agnes. From 1369 Dunbar and his allies represented
a major force in royal councils.[18]

Archibald and James Douglas gained from Dunbar's influence. In 1369–
70 James's rights as heir to many of his uncle's lands were confirmed, and
he also received estates from both the king and Dunbar. Much of this
patronage was at the expense of the Earl of Douglas, who had been forced
to relinquish Dalkeith and other lands. In 1370 the new lord of Dalkeith
received letters giving him 'leadership of all the men of his lands' in terms
which echoed and challenged Earl William's rights to lead the march-men.
The letters specifically referred to 'riding in the marches', a clear licence
for James to renew his uncle's lordship in the south.[19] Alongside this direct
challenge, the king's support of Dunbar and Douglas lords signalled a wider
challenge to the earl's regional dominance. These men were no court-based
faction. They were magnates with lands and ambitions in the marches and,
during the 1360s, it was in the marches that they sought to establish
themselves as rival sources of lordship to Earl William. In particular, from
1363 Archibald Douglas laid the foundations of his own regional power in
the borders. Though directed and supported by royal and aristocratic allies,
once again Douglas power would be built in the local warfare and politics
of frontier society.

THE LION OF GALLOWAY

When, in September 1369, David II granted 'Archibald Douglas, for his
labours and services to us ... all our lands of Galloway between the waters
of Cree and Nith', he was rewarding his 'dear and special knight'.[20] The
gift of lordship over half of the great sub-kingdom of the south-west was
not simply a further display of royal patronage. It also marked the end of
the special problem Galloway had posed for the king. Once again Douglas
lordship was to be the means of enforcing adherence to the Bruce dynasty.
Galloway had created trouble for Scottish kings long before 1306. An
independent Norse-Celtic realm, Galloway was never fully absorbed into
Scottish society, retaining its distinct laws, identity and political structure.
Although its native rulers were replaced by Anglo-Normans after 1234, the
province was still dominated by powerful local kindreds whose role in law
and warfare was accepted by the crown.[21] For the Bruces, however, Galloway posed greater difficulties than local particularism. As 'special lords'
of Galloway, John and Edward Balliol could rely on the support of the
great native houses, the McCullochs, the McLellans, the Adairs and the
MacDowells, and between 1306 and 1356 the Galwegians repeatedly took
up arms against the Bruces. The only means of controlling this independent

and disaffected frontier lordship were the savage raids which Robert and his brother Edward, the Bruce lord of Galloway, launched to force local submission.[22]

Not surprisingly, this failed to win Galwegian sympathy for the Bruce cause. In 1332 the province again rose in support of its old lord, Edward Balliol, and when David II returned to Scotland in 1341 most of the 'great men of Galloway' were in English allegiance. David gradually won support in Galloway. Like his father he handed power in the province to close supporters. To the west of the river Cree he made Malcolm Fleming earl of Wigtown. To control the east he relied on his friend John Randolph earl of Moray and local lords from Annandale, the old Bruce lordship. With these agents, David won allies among the war-weary natives and, in particular, within the MacDowell kindred which dominated the Dee valley and may have established a role as 'captains of the army of Galloway'. David won over junior members of the family and put pressure on its head, Dungal, who was still in English service. When Dungal was arrested by his suspicious English masters in 1345, resistance to the Bruce party seems to have collapsed. David's ally Fergus MacDowell was made steward of Kirk-cudbright and the king confirmed both the rights of four other 'captains' of native kindreds and the laws and liberties of Galloway.[23]

However, as in the marches, Neville's Cross saw the reversal of these gains. In Galloway, Edward Balliol returned to renew his leadership of the principal local families and establish himself in castles at Burned Isle in Loch Ken, Hestan Island off the coast and at the traditional centre of the lordship at Buittle which he held into the 1350s.[24] The final blows to Balliol power in Galloway would be dealt by the house of Douglas, who would ultimately succeed to their enemies' role in the whole lordship. Douglas lords had long been involved in royal attacks on the region, and Robert I's grant of Buittle to James Douglas signalled a role as the opponents of Balliol influence in Galloway. In the early 1350s, William lord of Douglas renewed this purpose, leading one, and probably several, campaigns into the lordship. He won the support of David II's adherent, Fergus MacDowell, who had 'the leding of the land ... to do skaith (to) Inglismen'. The submission of Fergus's senior kinsman, Dungal, to William in 1353 probably involved family politics as well as wider issues, but it also marked the surrender of 'the men of Galloway'. Balliol was reduced to Hestan Island off the coast, cut off from England by the submission of Nithsdale; and in 1356 'Edward, king of Scots' finally abandoned his kingdom and his lordship of Galloway.[25]

Despite the end of Balliol influence and of major war with England, David II still had reasons for anxiety about his authority in Galloway and the neighbouring dales of the west march. In the dales too, the king's supporters of the 1340s had been replaced by kin and allies of the Earl of Douglas. Neville's Cross had been followed by the English occupation of Annandale and Nithsdale and, although the sheriff of Dumfries, Roger Kirkpatrick, had recovered Nithsdale in 1355–6, he was treacherously killed by Sir James Lindsay of Crawford. Lindsay's ambitions in the south-west had the backing of his cousin, the earl of Douglas, and his murder of Kirkpatrick in Caerlaverock Castle removed one of the king's men from their path.[26] David's response indicated the importance of the region to him. He ordered Lindsay's execution, re-appointed another local adherent, John Stewart of Dalswinton, as march warden and sought his own allies among the kindreds of Galloway. However, his visit to Annandale in 1362 may have convinced him of the need for a powerful lieutenant in a region where royal influence was challenged by Scottish magnates, Galwegian lords and English enemies.[27]

In these circumstances, the king turned to Archibald Douglas. As king's man, veteran of border politics and, most recently, great southern lord, Archibald was an obvious royal lieutenant in the south-west. By August 1364 he was warden of the west march, responsible for maintaining the king's peace and authority in a region from Eskdale to the Cree. This office would be held by Archibald until his death thirty-six years later and by his descendants until 1455.[28] The roots of Black Douglas regional dominance stemmed from Archibald's early years as warden and there is reason to think that his commission on the marches always extended beyond border discussions. The king also gave his warden licence to enforce his will on eastern Galloway, a course which led directly to the grant of 1369. The character of his rule is only given in fifteenth-century accounts. In his epitaph for Archibald, Walter Bower stated that 'when Galloway rebelled, he subdued it for the king, whereupon the king conferred it on him ... [and] his successors for ever'. *The Buke of the Howlat* again elaborates in heraldic terms. 'The lyon lansand', the crowned rampant lion on the Douglas arms, 'is of Galloway':

> Quhen thai rebellit the croune and couth the king deir
> He gaif it to the Douglas heretable ay,
> On this wys, gif he couth wyn it on weir,
> Quilk for his soverane saike he set till assay,
> Kelit doune thar capitanis and couth it conquer,
> Maid it ferme, as we fynd, till our Scottis fay.

According to *Howlat*, like the Forest and Teviotdale, Galloway was won in war by a Douglas lord. The family's propaganda claimed regional power as just reward for military conquest. David II may have shared this view. He stated in the charter which granted eastern Galloway that, 'for the pacification and bringing to justice of which, the same Archibald made no modest payment and supported his efforts forcefully in person'. With his own resources Archibald had forcefully pacified a province which was still regarded as needing 'bringing to justice' by the crown.[29] Some kindreds, such as the McCullochs, remained in English service but, more importantly, sixty years of local warfare had allowed the great captains of Galloway to acquire new lands and power in the province. They would have been unwilling to accept a dependent role under the lordship of the Bruce king and his lieutenants. In western Galloway the earl of Wigtown faced similar 'feuds' with 'the great men and inhabitants of the earldom', and it was by cutting down these captains, as *Howlat* put it, that Archibald the Grim made his new lordship secure.[30]

Archibald used the methods of marcher lordship. He was remembered as a lord who 'had in his following a large company of knights and men of courage' and, like his father and cousin, Archibald won land and power by maintaining and leading an armed retinue. In Galloway this 'company' may have included traditional Bruce loyalists from Annandale and Nithsdale, but at its core were the knights from Lothian and the marches who were Archibald's kinsmen and friends. Once again it was James Douglas of Dalkeith who gave his kinsman greatest support, 'riding in the marches' with his own retainers. As Archibald cemented his hold on Galloway, James's power in the region also increased. In 1367 James was named heir to Douglas of Liddesdale's lands in Dumfriesshire and in Galloway and in 1370 he received further lands near Buittle, which the king had taken from the Earl of Douglas. In 1369 the earl of March added the rich barony of Morton in Nithsdale to James's regional interests. Most directly, before 1371 Archibald himself rewarded James for his service with the lands of Borgue near Kirkcudbright.[31] By the end of 1370, the lands between Nith and Cree had been transformed into an area dominated by the two Douglas adherents of the king. But to local leaders, the establishment of these king's men in the region must have felt like military subjection by forces raised in Lothian and the marches. A community largely in Scottish allegiance had been 'pacified' by Douglas warlords. Though the MacDowells and McLellans were apparently at peace by 1366–67, both kindreds lost lands to their new lord. James Douglas's lands of Borgue, taken from the MacDowells, were just the first territorial redistribution in the lordship by Archibald the Grim.[32]

When three years later Archibald's lordship was extended to Wigtown, the whole of the ancient province had passed under firm Douglas control. His victory over the native kindreds was symbolised by his adoption of two 'wild men' as the supporters of his arms. For the next two decades, Galloway was the basis of Archibald's power. Despite the wealth and security of the Murray lands and Archibald's frequent presence at his castle of Bothwell on the Clyde, the character and goals of Black Douglas lordship derived from Galloway and the west march. In the 1370s Archibald built his own base in the province at Threave on an island in the Dee. A great tower seventy feet tall surrounded by outer works was both a stronghold in the heart of his new estates and a symbol of Douglas domination in Galloway. By the late 1370s the Galwegians recognised this dominance and Archibald was able to act as not just an occupying warlord but as the leader of a well-defined provincial lordship. The advantages of this role were clear. Archibald faced no further internal challenges to his rule in Galloway and, if in financial terms the province remained devastated after nearly a century of warfare, Archibald presented himself as the leader and protector of the 'Galwegian men' in border warfare. Control of the kindreds of Galloway confirmed Archibald as a military force on the marches to rival even the Earl of Douglas and secured his place among the great men of the kingdom.[33]

ARCHIBALD THE HONOURABLE

When David II suddenly died in February 1371 Archibald lost his lord. It had been as the king's agent and through royal patronage that Archibald had risen in power and rank. Without King David, Archibald's place was less certain. The new king, Robert Stewart, was no friend of David or his adherents and many royal officers, including Archibald's comrade William Ramsay, entered exile rather than face the Steward as king. Powerful in lands and friends, Archibald was better able to defend his existing rights in 1371. In return for acting as an ambassador to France and 'for the many labours in which [he] ... sweated for our lord uncle and predecessor', Archibald received letters from Robert II which ensured his rights to the Murray inheritance, even if his marriage to Joanna should remain childless. As the new king's son, Robert, now earl of Fife and Menteith, had challenged Archibald's tenure of the whole inheritance before David's death, such a protection was essential under a Stewart ruler who, as the letters stated, identified Archibald with his predecessor's regime.[34]

Robert II's readiness to accept Archibald's possession of the Murray lands reflected the king's attitude to the south of the kingdom. As before 1358, when he was guardian, Stewart's principal interests lay north of Forth. He was prepared to leave the marches and Lothian to the direct control of others. As the earl of Douglas had shown in the 1350s, and would show

again in the 1370s, such kingship created opportunities for ambitious magnates to extend and entrench their regional predominance. Archibald the
Grim, well-versed in such predatory lordship as a follower of power-hungry
kinsmen, was quick to exploit the change in political atmosphere. His first
target was obvious. The earldom of Wigtown was both weakly held and
strongly connected to eastern Galloway. The lands and power of Galwegian
kindreds like the McCullochs, Adairs and MacDowells did not stop at the
Cree but extended into Wigtown.[35] As the captains of these families accepted
Archibald's lordship, the latter's ambitions were drawn westwards into the
neighbouring earldom. There had been no 'pacification' of Galloway beyond the Cree. The Fleming family lacked the resources to control their
tenants and act as effective earls. The first earl, Malcolm, remained based
in Dumbarton Castle far to the north, while his grandson and successor,
Thomas, only returned from England to inherit Wigtown in 1366 after a
four-year vacancy. Short of funds and followers and without support from
his king, Thomas found himself unable to dominate 'the great men ... of
the earldom', whose resistance may even have been supported by an
ambitious neighbour like Archibald the Grim. By 1370 Thomas was ready
to abandon the unequal struggle and resigned the earldom to the king.
David II may well have been reluctant to see Archibald's grip tighten on
the whole of Galloway but his death in early 1371 changed everything. In
the south-west, Archibald quickly moved in. Ignoring the earlier resignation,
Archibald bought the earldom of Wigtown from Thomas for £500 sterling
in February 1372. He was probably already the man in possession, and
experience of recent politics would reassure him that this was at least
nine-tenths of the law. Although Fleming was paid off, Archibald, like his
first lord, Douglas of Liddesdale, was seeking legal justification for a military
and political takeover of the lands of an 'absentee' lord. This was based
on Archibald's mastery of the Galwegian kindreds, and at least one family,
the Agnews of Lochnaw, maintained a tradition that they had been driven
from the earldom by the hostility of their new master, who replaced them
with a Douglas lord.[36]

The 1370s also witnessed a more gradual growth in Archibald's influence
in the lands to the east of Galloway. Through the early fourteenth century
the men of Nithsdale and Annandale had found their fate in war bound
up with that of Galloway. However, in contrast to the Galwegians, Annandale in particular was a local community with strong ties to its lords of
Bruce blood. From the 1330s to the 1360s, the Carruthers of Mousewald,
Kirkpatricks of Closeburn, Herries of Terregles, Johnstones, Jardines,
Corries and Crichtons, all minor Dumfriesshire landowners, showed personal adherence to David Bruce, despite continued English occupation of

Lochmaben Castle, the centre of the lordship of Annandale. David responded by reclaiming his ancestral homeland in the 1340s and acting as its direct lord. Robert II had no similar interest in being lord of Annandale. In 1372 he granted it to George Dunbar, earl of March, in return for concessions north of Forth.[37] Dunbar did not seek to dismantle the influence which Archibald had built up in the 1360s as David's lieutenant. The connection between the two David II partisans was strengthened in 1372 by the marriage of Dunbar's sister, Agnes, the old king's mistress, to Archibald's close ally, James Douglas of Dalkeith. Moreover, March's interest in the west was limited. Between 1369 and 1372 he had resigned his baronies of Morton and Tibbers in Dumfriesshire, the former to James Douglas, and when March met his Annandale tenants in July 1372, it was at Moffat, held from the earl by his new brother-in-law. Dunbar's interests remained concentrated on his earldom of March in the east.[38] He was content to exercise lordship in the west alongside Archibald Douglas, and the two men co-operated to great effect in local warfare to undermine the English hold on Annandale. In dealings with England, though, the lordship of Annandale was firmly in the 'bounds' of Lord Archibald and it was Douglas who handled local truces with the English lieutenant of Lochmaben. This role in local war and peace cemented Archibald's existing links with local men like Thomas Kirkpatrick, John Crichton and John Murray of Ae, and with the burgesses of Dumfries. From low-key beginnings, these links with local men would become an essential part of the Black Douglas affinity. The maintenance and, ultimately, the weakening of the family's hold on the region would do much to determine Douglas strength and decline up to the 1480s. A century earlier it was natural for Annandale and Nithsdale to look to Archibald for leadership. Their Bruce lords had died with David II and, while March was clearly accepted as the king's formal successor, Archibald's role as David's march warden, the association of his family with the Bruce cause and his growing resources as lord of Galloway made him influential as the maintainer of peace and defence for the men of Dumfriesshire.[39]

The contrast between Archibald's takeover of Galloway and the growth of his influence in Annandale and Nithsdale has some parallels with William earl of Douglas's contrasting receptions in the Forest and in Teviotdale in 1347. The different attitudes of neighbouring communities towards allegiance and lordship illustrate the complexity of local political society in southern Scotland and the recognition of the need for differing styles of aristocratic leadership between communities, like Annandale and Galloway, which had vastly different traditions of government. In both areas, though, while David's death allowed Archibald to extend powers first exercised as the king's lieutenant, it also brought the end to royal support and protection for Archibald. Robert II's attitude to the south was conditioned by his early settlement with William, earl of Douglas, the greatest magnate in the region, which shaped relations between the new royal dynasty and the Douglas earls for the next fifty years. Robert symbolised his recognition of William's predominance by restoring him to the justiciarship south of Forth and Douglas was linked to the extended Stewart kindred by the marriage of his son and heir, James, to Isabella, Robert's daughter, in 1371. Until 1388 William and then James would dominate the south as members of the royal house which now formed the top rank of Scottish political society.[40]

The renewed predominance of his cousin spelled danger for Archibald. Neither king nor earl had reason to back the extension of Archibald's lordship and his actions in Wigtown and Dumfriesshire did not go unchallenged. In April 1372, two months after selling his rights to Archibald, Thomas Fleming repeated his resignation of Wigtown to the crown. Robert bestowed the earldom on his nephew, James Lindsay of Crawford. This was no isolated patronage; the next year Lindsay was to receive the New Forest in Galloway and was clearly marked out for regional power in the south-west. James Lindsay was from a family with the connections and ambitions in the region to achieve this goal. His family had risen alongside the Douglases. His great-grandfather was a companion of James Douglas, his grandfather was Earl William's first ally in 1347 and his father had been the earl's cousin and councillor and had sought to impose his leadership on Nithsdale in the 1350s. In the 1370s James Lindsay, sheriff of Lanarkshire and a major lord in the south, could count on both Douglas earl and king to support his own ambitions. Lindsay also had local friends. His brother-in-law was John Maxwell of Caerlaverock, the major lord in Nithsdale. His alliance with James and his backers in 1371–72 may also have renewed an alliance from the 1350s, and with such allies, Lindsay represented a potent threat to Archibald's hard-won influence in the south-west.[41]

However, despite the apparent strength of opposition, Archibald's position proved unassailable. Lindsay's challenge may even have tightened Archibald's links with local men in Galloway and Dumfriesshire. In particular, Thomas Kirkpatrick, whose father, Roger, had been killed by Lindsay's father in Maxwell's castle of Caerlaverock in 1357, would oppose the ambitions of his father's enemies. Archibald probably won further support from local men engaged in a dispute which had lasted at least one generation and probably went back to conflicting loyalties in the years of major Anglo-Scottish conflict. The Maxwells had repeatedly entered English allegiance during the early fourteenth century and were conspicuously absent from the local following of David Bruce, which included many of their close neighbours. The support of these men allowed Archibald to neutralise the challenge to his lordship without major violence. In October 1372, backed by former adherents of David II, Archibald upheld his rights to Wigtown in general council. Lindsay's resignation of the New Forest to Maxwell in 1376 marked the end of his ambitions in the south-west, though not of his influence or his hostility to Archibald. King Robert accepted Archibald's hold on Wigtown as the only alternative to conflict in a region vulnerable to English attack. The arch political realist, Robert recognised existing regional interests rather than trying to compete with them.[42]

So, in this case, did Earl William Douglas, though relations between the earl and his cousin remained chilly. Archibald experienced difficulties in enforcing his rights over his lands in the middle march, finding himself, for once, an 'absentee' in terms of local clout and, as late as 1381, the earl provocatively led an entourage, including James Lindsay and two of his kinsmen, to Wigtown. Archibald pointedly confirmed the earl's grant to nearby Whithorn priory, reminding all present of his place as Earl William's superior in the previously disputed earldom of Wigtown.[43] The lord of Galloway did not return to the service of the earl. Instead William surrounded himself with relations who did not bear the Douglas name. Chief among these were the Lindsays, nephews and great-nephews of the earl's mother and magnates in their own right. The group also included Malcolm Drummond, husband of the earl's daughter, Isabella, and Patrick Hepburn, latest husband of William's much-married sister, Eleanor. Finally, Douglas also drew support from the Lothian connections of his mistress, Margaret Stewart, countess of Angus, led by her half-brothers, the Sinclairs of Herdmanston. The roots of this affinity lay back in the 1340s and '50s when the fathers of many of these associates had supported William's dominance in the south, but in the 1370s the network of support reflected wider power, mastery in the middle marches, influence in Lothian and landed interests which extended to the north-east earldoms of Angus and Mar. In the council chamber and on the battlefield these kinsmen would give constant support to William and his son, James, in the maintenance of the power and prestige of the Douglas earls.[44]

There was no place for Archibald in this group of Douglas adherents after 1371, but neither was there any return to the violent and potentially disastrous struggle for power within the Douglas kindred which had lasted from 1347 to 1353. Both Archibald and William had been participants in that struggle and may have wished to avoid a repetition. However, rather than any change of heart from men whose local powers had so clearly stemmed from a readiness to use violent means against opponents, regardless of allegiance, there were more concrete reasons for their acceptance of the *status quo* in the 1370s. Firstly, unlike the 1350s when Earl William and Douglas of Liddesdale competed for leadership of the same men and region of the marches, Archibald dominated areas which were never more than peripheral to the earl. As Lindsay discovered, Archibald's position in Galloway was secure from an essentially external challenge. In addition, the two magnates exercised lordship over fairly well-defined regions, roughly corresponding to the division between west and middle marches. When the English talked, in border negotiations, about the 'bounds' of the earl and those of Archibald Douglas, they showed a clear recognition of the areas under the leadership of the two men, not simply as wardens, but as lords. Across Scotland, regional lordship functioned best when it corresponded to clear geographical or political units. Where, as in the marches in the 1340s and '50s, and in Moray after 1371, there were rival lords in the same region, the result was conflict.[45]

The growing influence in the south of the king's son and heir, John, earl of Carrick, also eased regional tensions. John had long been the only Stewart magnate active in the region, campaigning in Annandale in the 1350s and inheriting the Stewart family lands in Ayrshire. After 1371 this role was increased. Carrick was made sheriff of Edinburgh and keeper of its castle, and by 1381 was 'lieutenant for the marches', representing his father in war and diplomacy. This apparent royal interference with the established place of border magnates did not arouse hostility. By the 1380s Carrick had clearly made himself acceptable to the major lords of the region. His influence centred on his complete association with the earls of Douglas, brokered by mutual kinsmen, Lindsay of Crawford and Malcolm Drummond. Carrick diverted royal patronage to Douglas adherents and supported their interests in war and diplomacy with his family's resources. In return, the huge connection of Earl William and, after 1384, Earl James was used to promote Carrick's rise in the kingdom, culminating in his appointment as lieutenant-general in 1384, removing his father from political power. For the first, but hardly the last, time the power of the Douglas family had been used in central political conflict.[46] Despite the lieutenant's ties with the earls of Douglas, Archibald Douglas saw Carrick as an ally. Alone in his family, Carrick had been a protégé of David II, receiving his earldom and his wife, the king's niece, from David. His following included men, like William Cunningham and Thomas Erskine, who were Bruce adherents, and to other partisans of the old king, Archibald and James Douglas among them, Carrick was the acceptable face of the Stewart dynasty. The lieutenant deliberately sought to build connections with these men. In 1378, Douglas of Dalkeith married Carrick's middle-aged aunt and arranged a match for his elder son with the earl's daughter, Elizabeth.

Finally, in about 1387, Archibald's family was drawn into the Stewart nexus. His son by Joanna Murray, and his designated successor, also named Archibald, was married to Carrick's daughter, Margaret. By binding all the main branches of the Douglas kindred to the royal dynasty, Carrick was copying his father's favoured means of cementing alliances and influence. The earl also ensured that royal politics and Douglas ambitions would be inextricably linked for the next sixty years, ultimately heightening tensions rather than controlling them.[47]

In the mid-1380s, however, the marriage of Margaret to the heir to Bothwell and Galloway must have seemed a successful move. Archibald, never fully part of Carrick's affinity, was tied more closely to the earl, and the appearence of Archibald and Douglas of Dalkeith among the councillors of James, 2nd earl of Douglas, suggests better relations with the head of their kindred.[48] The value of both Dalkeith and Archibald to Carrick and Douglas was primarily military. During the 1370s and 1380s there was slowly escalating warfare with England. While the nature of this conflict will be examined fully later, it is clear that the prosecution of the war was vital to Carrick's influence in the kingdom and his relations with southern magnates. As lieutenant, Carrick delivered exceptional rewards to war leaders like Douglas of Dalkeith's brother, Henry, and Archibald's bastard son, William. William received pensions and cash grants for his service and was named lord of Nithsdale, a title which had limited value in lands but gave substance to his father's dominance in the district. Moreover, he was given, 'on account of his skill' in war, Carrick's 'beautiful' sister, Egidia, as his bride. Such rewards, added to the possibility of winning land, influence and booty, made military and political co-operation hugely attractive to magnates like Douglas, March and Archibald the Grim and produced a largely concerted war effort. During the early 1380s the attractions of warfare kept other tensions in the region under control.[49]

However, Carrick's wooing of the whole Douglas kindred only papered over old divisions. Despite becoming a paid retainer of Earl William, Douglas of Dalkeith had not abandoned his claims to the lost lands of his uncle. That the earl's son was known as lord of Liddesdale simply emphasised Dalkeith's grievances. Similarly the presence of James Lindsay as the Douglas earls' constant councillor may have perpetuated their distrust of Archibald as a rival in the marches and in the Douglas kindred. Friction between Lindsay as sheriff of Lanark and Archibald as an aggressive local magnate continued up to 1388 when old and simmering quarrels would be resumed with new intensity. For a variety of reasons, though, much of the 1370s and 1380s was a period of internal stability for southern Scotland. Great magnates exercised effective control over clearly defined areas of the kingdom and apparently accepted the leadership of Carrick, a lieutenant who rewarded and did not seek to undermine his principal allies. The death, in early 1384, of Earl William, the architect of Douglas dominance in the middle march, during a period of intense war with England, had no effect on his family's authority. Within a month his son, the second Earl of Douglas, had assumed the leadership of his father's affinity in war and politics to great effect.[50] The dynasty of Douglas earls seemed secure in its regional powers. Similarly, barons like Douglas of Dalkeith found it easy to serve several magnates without conflict, protecting their lands and winning patronage. This southern political society, the product of war before 1358 and aggressive kingship in in the 1360s, seemed to have established a balanced regional structure capable of defending and extending allegiance to the 'Scottis fay'.

By early 1388 Archibald the Grim had carved out a place for himself in this society. Although, like his cousin, Earl William, he was not exclusively a border magnate and had lands in the north-east, it was in Clydesdale and Galloway that his power rested. His castles of Bothwell and Threave gave him the reputation of being 'honourable in habitaciounis', and while war-torn Galloway was primarily valuable for its manpower, the Murray lands provided 'riches'.[51] From being a landless, if not friendless, knight in 1358, Archibald was a figure of wealth and power, second only to the Douglas earls on the marches. His rise, though less spectacular than that of his father or cousin, combined Sir James's service to his king with Earl William's domination of marcher communities in war. The result was the creation of a separate Douglas lordship in the south-west, dominated by Black Archibald. Now in his sixties, he was still active and still ambitious for new rights and powers and, through the fortunes of war, which had made the Douglas dynasty, the servant would finally become the master.

NOTES

1. Although, as a minor figure, Archibald's presence in France is nowhere specified, it would not be surprising if he accompanied his cousins, William and John, and the young King David to France in 1334. His later relations with both the king and with the French suggest significant early contacts (*Wyntoun*, vi, 192–93; *Scotichronicon*, ed. Goodall ii, 346–47; *Scotichronicon*, ed. Watt, vii, 270–71).

2. *Scotichronicon*, ed. Goodall, ii, 330–31; *Scotichronicon*, ed. Watt, vii, 140–41; *Chron. Lanercost*, 318; *Morton Reg.*, ii, no. 70; *Rot. Scot.*, i, no. 752; *R.R.S.*, vi, no. 51.

3. *Wyntoun*, vi, 231–32; *Scotichronicon*, ed. Goodall, ii, 357; *Scotichronicon*, ed. Watt, vii, 300–301; Fraser, *Douglas*, iii, nos 23, 323; *Melrose Lib.*, ii, nos 461, 462, 463, 490. It is doubtful whether Earl William accepted the 1342 entail at any point, and his nephews by his sister Eleanor clearly saw themselves as his heirs later on. William's own son and successor James (the future second earl) was born in about 1359.

4. *A.P.S.*, i, 715; *Scotichronicon*, ed. Watt, viii, 34–35; Froissart, *Chroniques*, ii, 225.

5. S. Boardman, *Early Stewart Kings*, 13–14; *R.M.S.*, i, app. 1, no. 123; app. 2, no. 1222; *E.R.*, ii, 77, 82; *R.R.S.*, vi, nos 137, 221.

6. *R.M.S.*, ii, app. 2, nos 962–64, 998, 1070, 1072; *Rot. Scot.*, i, 678; Fraser, *Douglas*, iii, no. 20, 291, 318; *Scotichronicon*, ed. Goodall, ii, 350, 357; *Scotichronicon*, ed. Watt, vii, 278–281, 301; *R.R.S.*, vi, nos 163, 168, 169, 171. Haliburton's father, John, was with Alexander Ramsay and was killed with Douglas in 1350. William Ramsay was made earl of Fife in 1358 through David II's influence. Though there were two William Ramsays active in the 1360s, the knight with Douglas in 1350 and 1356 and the sheriff of Lothian in 1357 are all probably Ramsay of Colluthie, who was the uncle of the later William of Dalhousie. Bower's designation of William in 1350 as 'of Dalhousie' is a much later addition (S. Boardman, *Early Stewart Kings*, 13–15; *Newbattle Chart.*, 308–309; *Scotichronicon*, ed. Watt, vii, 474).

7. *St Giles Chart.*, 7–8, 12–14; *E.R.*, ii, 92, 131, 166, 176; *R.R.S.*, vi, nos 210, 240, 267, 269, 270, 272, 274. Ramsay had saved Archibald from captivity at Poitiers in 1356 and thereafter Archibald 'luffit richt weil this Ramsay' (*Wyntoun*, vi, 231).

8. S. Boardman, *Early Stewart Kings*, 25, 45–48.

9. *Cal. Papal Letters*, iv, 76; 'Buke of the Howlat', 67, v. 43. In a document issued by Joanna in 1361 and confirmed by her mother, both women are in company with members of David II's household (*Laing Chrs.*, no. 379 (1–2)).

10. *Scots Peerage*, ii, 120–31; viii, 255–58; *R.M.S.*, i, no. 401; app. 1, no. 130; app. 2, nos 775–76, 822, 850, 871, 886, 896, 904, 1015, 1097–99, 1244, 1306, 1333, 1406; *R.R.S.*, vi, nos 216, 217. Maurice Murray had been a close associate of David II who had been killed at Neville's Cross. Thomas Murray of Bothwell had died in London in 1361 while acting as a hostage for the king's ransom.

11. *R.M.S.*, i, nos 775–76, 822, 850, 853, 904, 1862; W. Fraser (ed.), *The Red Book of Menteith*, 2 vols (Edinburgh, 1880), ii, 281; A. Grant, 'The Higher Nobility in Scotland and their estates, 1371–1424' (unpublished D.Phil thesis, Oxford University, 1975), 154–62, 383–86; S. Boardman, *Early Stewart Kings*, 15–17, 24; *Wyntoun*, vi, 47–48; *Scotichronicon*, ed. Goodall, ii, 316–17; *Scotichronicon*, ed. Watt, vii, 140–41.

12. *A.P.S.*, i, 505. Robert Stewart's claims were for the terce owed to his wife, Margaret of Menteith, as widow of a previous lord of Bothwell.

13. *Scotichronicon*, ed. Goodall, ii, 365–66; *Scotichronicon*, ed. Watt, vii, 322–23; *Scalachronica*, 172. For analyses of the grievances of Douglas and Stewart, see S. Boardman, *Early Stewart Kings*, 14–17; A. A. M. Duncan, 'The Laws of Malcolm MacKenneth', in Stringer and Grant, *Medieval Scotland*, 239–273, 262–65.

14. *Scalachronica*, 172–74; *E.R.*, ii, 154, 164; *Scotichronicon*, ed. Goodall, ii, 369; *Scotichronicon*, ed. Watt, vii, 330–33; *R.R.S.*, vi, nos 277, 278, 281, 287, 291, 292, 294, 296, 297.

15. S. Boardman, *Early Stewart Kings*, 45; *Cal. Docs. Scot.*, iv, no. 93; *R.R.S.*, vi, 402, 418, 420, 423, 426, 430, 436, 438, 451, 459. The Lindsay brothers were absent on crusade in the mid-1360s and on Alexander's return he immediately received the favour of the king who valued such activities. The Lindsays' half-brother, Walter Leslie, had been both a crusader and a royal retainer through the 1360s (S. Boardman, *Early Stewart Kings*, 45–47).

16. *Cal. Docs. Scot.*, iv, no. 100; *R.R.S.*, vi, no. 450; Barrow, *Robert Bruce*, 446; Väthjunker, 'Douglas', 143; *The Bruce*, XX, lines 597–600.

17. *Morton Reg.*, ii, no. 70; *Rot. Scot.*, i, 752, 772; *R.R.S.*, vi, no. 435; *Chron. Fordun*, i, 370, n. 12. The chief stronghold of Liddesdale, Hermitage castle, had been surrendered to the English by Douglas of Liddesdale's widow in 1353 on her marriage to the English borderer Hugh Dacre, but the lordship was claimed and eventually recovered by the earl of Douglas (*Rot. Scot.*, i, 761, 771, 826, 896; *R.M.S.*, i, app. 1, no. 123).

18. S. Boardman, *Early Stewart Kings*, 23–25. George Dunbar and a number of other vassals of Earl Patrick were given royal rewards in the immediate aftermath of the 1363 rebellion. George received his great-uncle's lands in July 1368. Archibald and James Douglas witnessed several of March's charters between 1369 and 1371 (*R.M.S.*, i, nos

149, 150, 152, 159, 160, 291, 292; *R.R.S.*, vi, nos 379, 506, 508–509; *Morton Reg.*, ii, no. 100; *Yester Writs*, no. 28; *H.M.C.*, Milne Hume, 79, no. 173).

19. *R.R.S.*, vi, nos 417, 419, 455, 457, 459. James's prominence from 1368 also related to his successful claim to be the heir of Mary, daughter of William Douglas of Liddesdale, who died without offspring in 1367. This allowed him, not without difficulty, to reclaim Dalkeith from the earl of Douglas in 1369–70. More open royal support was involved in the grant to James of lands taken by David II from the earl (*R.R.S.*, vi, nos 417, 424, 465, 469).

20. *R.R.S.*, vi, no. 451. The Cree was the traditional division between east and west Galloway.

21. For a history and studies of medieval Galloway, see R. Oram and G. Stell, *Galloway, Land and Lordship* (Edinburgh, 1991); D. Brooke, *Wild Men and Holy Places* (Edinburgh, 1994). For the role of kindreds in south-western lordship, see also H. L. MacQueen, 'The Kin of Kennedy, 'Kenkynnol' and the Common Law', in Grant and Stringer, *Medieval Scotland*, 274–96.

22. *Chron. Lanercost*, 205–206, 210, 212, 269; *Cal. Docs. Scot.*, iii, nos 15, 218, 278.

23. *Chron. Lanercost*, 205–206, 269, 278; *R.M.S.*, i, app. 2, nos 743, 755, 834, 835, 910, 912–915, 1006, 1007, 1012; *R.R.S.*, vi, nos 30, 39, 52, 78, 81; R. C. Reid, 'Edward de Balliol', in *Transactions of the Dumfriesshire and Galloway Antiquarian and Natural History Society*, 35 (1956–7), 38–63; *Cal. Docs. Scot.*, iii, nos 1462, 1469.

24. Reid, 'Edward de Balliol', 45–47; *Cal. Docs. Scot.*, iii, no. 1578; D. Brooke, *Wild Men and Holy Places*, 162–64.

25. *R.R.S.*, v, no. 269; *Scotichronicon*, ed. Goodall, ii, 356; *Scotichronicon*, ed. Watt, vii, 296–97; *Wyntoun*, vi, 186–87, 222–223; *Morton Reg.*, ii, no. 13; *Legends of Saint Ninian and Saint Machar*, ed. W. M. Metcalfe (Paisley, 1904), 65–75 (I am grateful to Dr Richard Oram for this reference). Buittle surrendered sometime in 1354 (*Cal. Docs. Scot.*, iii, no. 1578).

26. *Rot. Scot.*, i, 678; *Scotichronicon*, ed. Goodall, ii, 346, 361; *Scotichronicon*, ed. Watt, vii, 308–309, 368–69; *Wyntoun*, vi, 182–83, 222–23, 241; *R.R.S.*, vi, no. 81. The Kirkpatricks were long-standing Bruce adherents in Annandale and Roger had appeared among David's local councillors in the 1340s (*H.M.C.*, Drumlanrig, i, nos 69, 71, 75, 76, 77).

27. *Cal. Docs. Scot.*, iv, no. 47; *R.M.S.*, i, app. ii, nos 1147, 1176; *R.R.S.*, vi, no. 282. Stewart of Dalswinton was warden and one of the king's local adherents in 1344 before he was captured at Neville's Cross (*R.R.S.*, vi, no. 81). In January 1364 a marriage was negotiated between Fergus MacDowell and the daughter of David's adherent, William Cunningham, designed to prevent 'bloodshed' (*Wigtownshire Charters*, no. 131).

28. *Cal. Docs. Scot.*, iv, no. 100; *Rot. Scot.*, i, 913–914, 957; ii, 73.

29. *Scotichronicon*, ed. Watt, viii, 34–35; 'Buke of the Howlat', 67, v. 44, lines 560–567; *R.R.S.*, vi, no. 451.

30. *Rot. Scot.*, i, 821, 824, 881, 931; *Charter Chest of the Earl of Wigtown*, Scottish Records Society (Edinburgh, 1910), no. 7.

31. *Morton Reg.*, ii, nos 83, 99, 100, 124; *R.R.S.*, vi, nos 251, 445, 469; British Library, Harleian Ms, no. 6439; *Scotichronicon*, ed. Watt, viii, 34–35; A. Fraser (ed.), *The Frasers of Philorth* (Edinburgh, 1879), no. 16.

32. Members of the MacDowell kindred had held Borgue in the 1340s and 1350s (*R.M.S.*, i, app. 2, nos 835, 1147, 1193). For a full discussion of Douglas lordship in Galloway, see Chapter 8.

33. *R.C.A.H.M.S.*, Galloway, 2 vols (Edinburgh, 1914), ii, 28–34; *E.R.*, vi, cx, 193, 643–44; *Scotichronicon*, ed. Goodall, ii, 397; *Scotichronicon*, ed. Watt, vii, 394–95. For the state of Galloway in the 1360s and 1370s, see *Calendar of Papal Letters to Pope Clement VII of Avignon 1378–94*, ed. C. Burns, Scottish History Society (Edinburgh, 1976), 67; *Morton Reg.*, ii, no. 83; H. Laing, *Impressions from ancient Scottish seals*, Bannatyne Club (Edinburgh, 1850), 44.

34. *R.M.S.*, i, no. 401; *A.P.S.*, i, 505; *Rot. Scot*, i, 903. For Archibald's French embassy, see

Chapter 10. An excellent analysis of Scottish politics between 1371 and 1406 has been provided by S. Boardman, *The Early Stewart Kings*.

35. The McCullochs were coroners of Galloway west of Cree in 1358. Dungal MacDowell led the Gallovidians from 'beyond the Cree' in 1334 and had lands at Elrig near Wigtown (*R.M.S.*, i, no. 1303; *Chron. Lanercost*, 278; P. H. M'Kerlie, *History of the Lands and their Owners in Galloway*, 2 vols (London, 1906), i, 280–93).

36. *Newbattle Chart.*, 230–31; *R.R.S.*, vi, nos 335, 368, 399; *Wigtown Charter Chest*, nos 4, 7; *R.M.S.*, i, nos 108, 129, 250, 371, 414, 466; A. Agnew, *The Hereditary Sheriffs of Galloway*, 2 vols (Edinburgh, 1893), i, 233–34. Though there are serious doubts about the Agnews' view of their origins they may derive from Archibald's forceful handling of Wigtown-shire kindreds.

37. *H.M.C.*, Drumlanrig, nos 69, 71, 72, 75, 76, 77; *Cal. Docs. Scot.*, iv, no. 100; *Rot. Scot.*, i, 678, 957; *Scotichronicon*, ed. Goodall, ii, 346, 361; *Scotichronicon*, ed. Watt, vii, 296–97, 308–309; *R.R.S.*, vi, nos 78, 81, 282; *The Miscellany of the Scottish History Society* (Edinburgh, 1893-), v, no. 17; S. Boardman, *Early Stewart Kings*, 51–52, 67. David II had, however, granted Annandale to John Logie, his stepson, in the late 1360s (W. Fraser (ed.), *The Red Book of Grandtully*, 2 vols (Edinburgh, 1868), ii, 131–32). Dunbar was heir of John Randolph, the lord of Annandale who had resigned the lordship to his friend, King David, before 1343.

38. *Morton Reg.*, ii, nos 98–101, 108, 130–34; *S.H.S. Miscellany*, v, no. 17. James Douglas obtained other lands in Annandale with the earl's assistance in 1371- 72. Though George Dunbar issued a number of charters to his new tenants in Annandale, the locations and witness lists of these indicate that he was firmly based in his earldom of March (*Morton Reg.*, ii, nos 122, 136; *H.M.C.*, vii, 710; Drumlanrig, i, 74; W. Fraser (ed.), *The Book of Carlaverock*, 2 vols (Edinburgh, 1873), ii, no. 24; S.R.O. GD 158/1; GD 246/1).

39. *Rot. Scot.*, ii, 3, 38–39; *R.M.S.*, i, no. 507; *Melrose Liber*, no. 455; Fraser, *Frasers of Philorth*, no. 16; *Scotichronicon*, ed. Goodall, ii, 397; *Scotichronicon*, ed. Watt, vii, 394–95; A. Mac-Donald, 'Crossing the Border, a study of the Scottish Military Offensives against England c. 1369–c. 1403', unpublished Ph.D thesis University of Aberdeen, 1995), 18–37.

40. *Wyntoun*, vi, 192–95; S. Boardman, *Early Stewart Kings*, 39–45; *E.R.*, ii, 394.

41. *R.M.S.*, i, nos 414, 446, 450, 451, 469, 471, 507, 590, 608, 696, 763; *Wigtown Charter Chest*, no. 7; Fraser, *Douglas*, iii, nos 330, 333; S.R.O. GD12/1; *E.R.*, ii, 418.

42. *Scotichronicon*, ed. Goodall, ii, 361; *Scotichronicon*, ed. Watt, vii, 308–309; *Chron Lanercost*, 269, 290–91; A. B. Webster, 'Scotland without a King', in *Medieval Scotland*, 229–30; A. B. Webster, 'English Occupations of Dumfriesshire', in *Dumfriesshire Transactions*, 35 (1956–57), 64–80; *Cal. Docs. Scot.*, iii, 317; *A.P.S.*, i, 560; *R.M.S.*, i, nos 507, 576. A fortnight after the council broke up, Archibald was accompanied by a following which included Douglas of Dalkeith, Patrick Hepburn and Alexander Fraser, all established allies of Archibald and ex-adherents of David II. Lindsay's attempts to establish control of the New Forest may have antagonised Roger Gordon of Stichill, a significant landowner in Galloway who had a claim to the estate. Gordon and his descendents were to act as close local adherents of the Black Douglases (*Morton Reg.*, ii, no. 124; *Melrose Liber*, ii, no. 455; J. B. Paul (ed.), *The Scots Peerage*, 9 vols (Edinburgh, 1904–14), v, 99).

43. *Calendar of Papal Letters to Pope Clement VII of Avignon, 1378–94*, 69; *R.M.S.*, ii, 450 n.

44. Fraser, *Douglas*, nos 330, 332–35; *H.M.C.*, Drumlanrig, i, no. 2; *H.M.C.*, Milne Hume, no. 582; *Melrose Liber*, ii, no. 491; S.R.O. GD12/1; S.R.O. B30/21/3; S.R.O. RH6/191; S. Boardman, *Early Stewart Kings*, 57, 81–82, 120–21, 142.

45. *Rot. Scot.*, ii, 38–39. For examinations of Moray after 1371, see S. Boardman, *Early Stewart Kings*, 51–52, 71–107, 130–35, 168–86, 206–15, 256–67; A. Grant, *Independence and Nationhood*, 198, 200–20.

46. *John of Gaunt's Register, 1379–83*, ed. E. C. Lodge and R. Somerville, Camden Society (London, 1937), no. 564; *E.R.*, ii, 364, 393, 435; *Rot. Scot.*, ii, 3; Boardman, *Early Stewart Kings*, 55, 108–25.

47. S. Boardman, *Early Scottish Kings*, 22, 121; *Morton Reg.*, ii, nos 162, 166–67, 174; Fraser, *Menteith*, i, 157–58. Douglas of Dalkeith's first wife, Agnes Dunbar, died before 1378. William Cunningham, Thomas Rait and other old servants of David II appeared with Archibald in 1372 (*Morton Reg.*, ii, no. 124).

48. *H.M.C.*, Drumlanrig, i, no. 2. Archibald and James witnessed a charter of the 2nd earl to his bastard son, William, accompanied by James Lindsay and other key Douglas adherents like Alan Lauder, Adam Pringle and William Borthwick.

49. *Scotichronicon*, ed. Goodall, ii, 403; *Scotichronicon*, ed. Watt, vii, 410–15; *R.M.S.*, i, nos 752, 753, 770; *Morton Reg.*, ii, 158–59; Fraser, *Douglas*, iii, no. 338. See also A. Grant, 'The Otterburn War from the Scottish point of view', in A. Tuck and A. Goodman (eds), War and Border Societies in the Middle Ages (London, 1992), 30–64. The value of Archibald to the Scottish war effort and his significance in border society is clear from the £5,500 given to him by the French in 1385. This was the same sum received by Carrick and greater than all other magnates except Earl James Douglas (Archives Nationales J677, no. 15).

50. S. Boardman, *Early Stewart Kings*, 119–22, 136–39; *E.R.*, iii, 162–64.

51. 'Buke of the Howlat', 67, v. 43.

CHAPTER FOUR

The Douglas Inheritance

THE HEIRS OF DOUGLAS

In early August 1388 Archibald Douglas was outside Carlisle, leading a major raid on the English west march when news reached his force of a battle sixty miles away at Otterburn in Redesdale. The continental chronicler, Jean Froissart, described how the Scottish army in the west, commanded jointly by Archibald and the king's second son, Robert earl of Fife, responded to the news with mixed emotions. Joy at the Scottish victory was tempered with disappointment at a missed opportunity for glory. If Archibald shared the chivalric reaction described by Froissart, other emotions must have quickly taken over. Tidings of the battle were followed by the information that the Scottish leader, James 2nd earl of Douglas, had been killed in a confused night battle on the moors. In the words of the Otterburn ballad, a dead man had won the fight but, in Scotland, the death of Douglas would overshadow the earl's last victory.[1] Earl James's death meant more than the loss of a 'ferocious knight' who was 'everywhere a danger to the English'. The young earl had died without children from his marriage to the king's daughter and without formal plans for the succession to his vast estates. From the great border lordships, James's inheritance stretched through Douglasdale and North Berwick to the lands of his mother, the earldom of Mar in Aberdeenshire. This huge transregional collection of lands was now left without a lord. The next decade would witness a struggle for shares in the Douglas inheritance by the kinsmen, allies and rivals of the fallen earl which would revive old conflicts within the family. From this struggle, Archibald the Grim and the Black Douglases would emerge as dominant in the dynasty and in the south of Scotland.

More was at stake than just lands and rents. Since the 1340s the rise of the Douglas earls had been built on the creation and extension of a network of adherents, kinsmen, tenants and friends, whose support and service gave force to their lords' policies and brought men to follow the Douglas banner of the bloody heart. The army of several thousand men which Earl James led into England in 1388 was dominated by these adherents, members of a massive political connection drawn from the marches, Lothian, Clydesdale and north of the Tay. With the earl's death in Redesdale, these men were lordless and leaderless. Bonds of service established during four decades were broken and, despite the victory, the death of the earl brought changes to the balance of power in the marches as drastic as those which followed Halidon and Neville's Cross. Though 1388 would cause no return to English allegiance, Douglas adherents would turn to new patrons and protectors, rivals for the fragmenting power of the Douglas earls. The competition to succeed to the Douglas inheritance was primarily a struggle for the loyalties of the men who had followed the earl to Otterburn.[2]

The death of the Earl of Douglas had immediate impact. The Scottish war effort came to a grinding, if temporary, halt and, within a fortnight of Otterburn, on 18 August, a general council assembled at Linlithgow, primarily to consider the Douglas succession. Though no heir was formally named, in the month which followed the council the choice of the lieutenant, John earl of Carrick, became apparent. The best claim to succeed to the Douglas estates under normal inheritance law belonged to Malcolm Drummond of Concraig. He was the husband of Isabella Douglas, only sister of Earl James, and in her right Malcolm was given legal possession of Selkirk Forest in September 1388. Recognition of his rights to the crucial Douglas lordship in the middle marches indicated acceptance of Drummond's claims to the bulk of his brother-in-law's estates and to the title of earl of Douglas. The formal succession of Malcolm Drummond in the autumn of 1388 represented an outcome which would have satisfied most of the leading men in the kingdom and, in particular, the Earl of Carrick. Drummond's sister, Annabella, was married to Carrick, and the lieutenant saw in Drummond's succession a means of maintaining his alliance with the Douglas earls which was crucial to Carrick's position in the kingdom.[3] Malcolm could also be counted on to uphold other Stewart interests in the Douglas estates, which included the rights of Carrick's widowed sister, the countess of Douglas, to a 'reasonable terce', a widow's portion, from the Forest, and the claims of Robert, earl of Fife, to take custody of the barony of North Berwick as its superior lord. Fife, Carrick's ambitious and opportunistic younger brother, wanted possession of Tantallon castle, the huge, sea-girt, red stone fortress built by Earl William in the 1350s. In late 1388, in return for the backing of the great Stewart princes, Drummond accepted these demands. However, Malcolm Drummond was no Stewart-backed outsider to the Douglas affinity. He had been a councillor of Earl James, closely involved in the preparations for the Otterburn campaign and present at the fight itself. He must have believed he could win the support of many of the key adherents of the Douglas earls with whom he had close contacts.

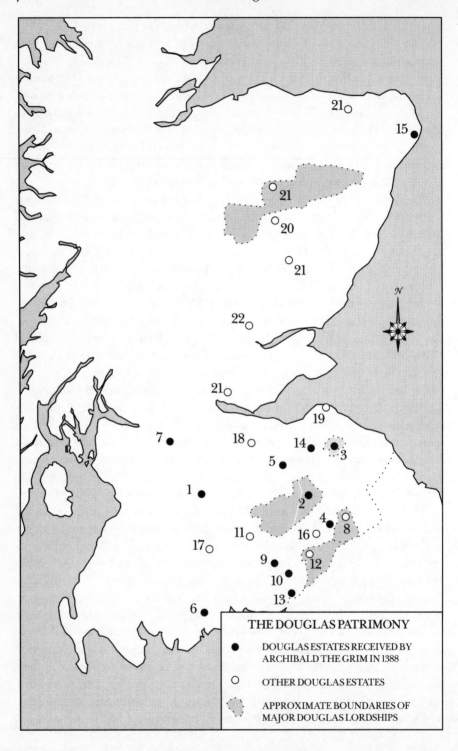

THE DOUGLAS PATRIMONY

● DOUGLAS ESTATES RECEIVED BY
   ARCHIBALD THE GRIM IN 1388

○ OTHER DOUGLAS ESTATES

⬭ APPROXIMATE BOUNDARIES OF
   MAJOR DOUGLAS LORDSHIPS

KEY TO MAP 2: THE DOUGLAS PATRIMONY

The lands of the First and Second Earls   *Post 1388 Owners*

A) *The lands of James Lord of Douglas*

i) **Estates included in the Douglas Entail of 1342**

1	Douglas	Archibald 3rd earl of Douglas
2	Selkirk Forest	Archibald 3rd earl of Douglas
3	Lauderdale	Archibald 3rd earl of Douglas
4	Bedrule	Archibald 3rd earl of Douglas
5	Romanno	Archibald 3rd earl of Douglas
6	Buittle	Archibald 3rd earl of Douglas
7	Fermes of Rutherglen	Archibald 3rd earl of Douglas

ii) **Estates omitted from the Douglas Entail of 1342**

8	Jedworth Forest	Isabella Countess of Mar (to 1408) George 1st earl of Angus as heir
9	Westerkirk (Eskdale)	Held by Douglas of Dalkeith from Archibald 3rd earl of Douglas
10	Staplegordon (Eskdale)	Held by Douglas of Dalkeith from Archibald 3rd earl of Douglas
11	Polmoodie	Archibald 3rd earl of Douglas ?

B) *The Lands of Archibald Douglas 'the Tyneman'*

12	Liddesdale	Disputed by Douglas of Dalkeith and Isabella, Countess of Mar (to 1400) George 1st earl of Angus (from 1400)
13	Kirkandrews (Eskdale)	Archibald 3rd earl of Douglas
14	Heriot	Archibald 3rd earl of Douglas
15	Rattray	Archibald 3rd earl of Douglas (with Isabella Countess of Mar as superior?)
16	Cavers	Isabella Countess of Mar (to 1405) Archibald Douglas of Cavers (from 1405)
17	Drumlanrig	William Douglas of Drumlanrig
18	West Calder	James Sandilands
19	Tantallon	George 1st earl of Angus
20	Coull and Oneil	Resigned to superior ?

C) Lands acquired by William Earl of Douglas

21	Mar and appurtenances	Isabella Countess of Mar
22	Strathord	Isabella Countess of Mar

TABLE 2. THE DOUGLAS SUCCESSION

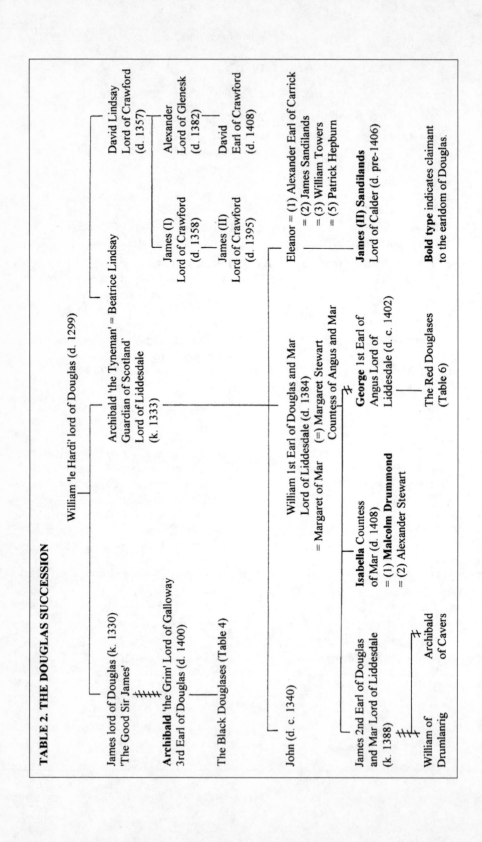

William 'le Hardi' lord of Douglas (d. 1299)

James lord of Douglas (k. 1330)
'The Good Sir James'

Archibald 'the Grim' Lord of Galloway
3rd Earl of Douglas (d. 1400)

The Black Douglases (Table 4)

Archibald 'the Tyneman' = Beatrice Lindsay
Guardian of Scotland
Lord of Liddesdale
(k. 1333)

David Lindsay
Lord of Crawford
(d. 1357)

James (I)
Lord of Crawford
(d. 1358)

Alexander
Lord of Glenesk
(d. 1382)

James (II)
Lord of Crawford
(d. 1395)

David
Earl of Crawford
(d. 1408)

John (d. c. 1340)

William 1st Earl of Douglas and Mar
Lord of Liddesdale (d. 1384)
= Margaret of Mar
(=) Margaret Stewart
Countess of Angus and Mar

Eleanor = (1) Alexander Earl of Carrick
= (2) James Sandilands
= (3) William Towers
= (5) Patrick Hepburn

George 1st Earl of
Angus Lord of
Liddesdale (d. c. 1402)

James (II) Sandilands
Lord of Calder (d. pre-1406)

Isabella Countess
of Mar (d. 1408)
= (1) **Malcolm Drummond**
= (2) Alexander Stewart

The Red Douglases
(Table 6)

James 2nd Earl of Douglas
and Mar Lord of Liddesdale
(k. 1388)

William of
Drumlanrig

Archibald
of Cavers

Bold type indicates claimant
to the earldom of Douglas.

TABLE 3. DOUGLAS AND STEWART

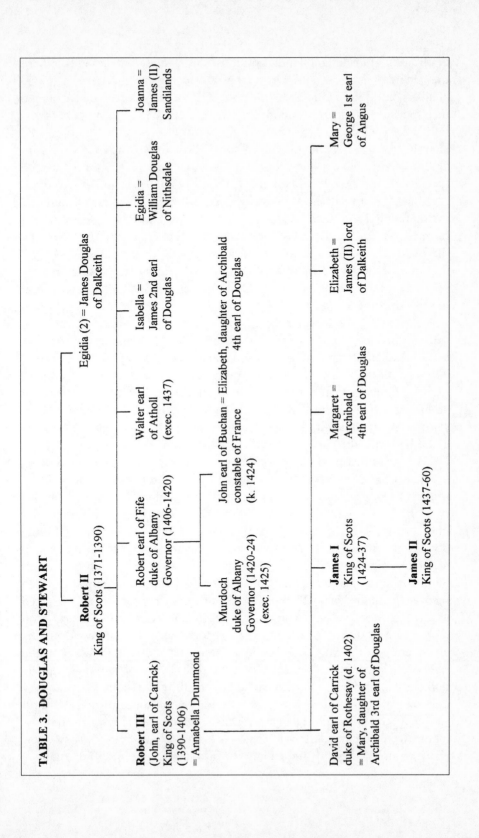

Several, like James Sandilands, cousin of Earl James, who stood to inherit the Douglas lands should Drummond's twenty-year marriage continue childless, and John Haliburton, the powerful Lothian baron, quickly offered their backing, and Drummond must have hoped for the support of the influential Lindsay family. The loss of James Lindsay of Crawford in 1388 after his capture by the English at Otterburn was to prove a significant blow to Drummond's hopes but, even without Lindsay, Drummond and his wife seemed set to obtain the earldom of Douglas and its principal lordships. To strengthen his authority in the marches further, Drummond was appointed sheriff of Roxburgh, an office probably held before Otterburn by the Earl of Douglas. However, the lands and powers granted by the lieutenant had to be won at local level and, as the autumn of 1388 drew on, it became increasingly clear that Drummond's claims, readily recognised by his brother-in-law, Carrick, met a more hostile response in the lordships he sought to inherit.[4]

As Drummond and Carrick must have feared, the principal challenge came from the old connection of the knight of Liddesdale in the shape of his political and territorial heirs, Archibald the Grim and James of Dalkeith. Throughout their already long and successful careers the two old comrades in war, diplomacy and politics were bound together by the legacy of ambitions left by their first lord, William Douglas of Liddesdale. These centred on the lands and power of the senior line of Douglas. The goals of both Archibald and James in 1388 arose directly from the knight of Liddesdale's attempted appropriation of the Douglas estates forty-six years previously. The entail of 1342, designed to give Liddesdale a claim to the principal Douglas estates, included Archibald as ultimate heir to Douglasdale, the Forest, Lauderdale, Buittle and Eskdale. With no surviving 'heirs male of the body' of either Earl William or Liddesdale in late 1388, the document justified Archibald's own ambitions to succeed to the leadership of the Douglas kindred. James Douglas of Dalkeith also sought to gain his full inheritance. As the nephew of Douglas of Liddesdale, James hoped to recover the lands his uncle had won in war from the English and extracted in law from the lords of Douglas. Claims to Westerkirk and Staplegordon in Eskdale and, most importantly, to Liddesdale itself had rested on the ascendancy of the knight. With his fall and the rise of Earl William, the rights of his heirs were denied out of hand in favour of the often equally dubious claims of the Douglas earls. The death of Earl James brought the dominance of the earls to a sudden end. In the autumn of 1388 both Archibald the Grim and Douglas of Dalkeith produced long-rejected documents from their charter-chests, aware that they would justify their challenge to the legality and extent of Malcolm Drummond's inheritance.[5]

This challenge did not rest simply on dubious inherited claims. Just as in the 1340s and 1350s, the leadership of the house of Douglas was to be decided, not by a legal judgement in the king's court, but by the establishment of local bases of power from which to defy rival claimants. Across the southern Scottish lands of Earl James this process was rapidly underway in the autumn of 1388. The clearest indication of such a local challenge to Drummond came from a third Douglas claimant, George, the young bastard son of Earl William and the dead Earl James's half-brother. George's rights were championed by his mother, Margaret countess of Angus, from a secure base in Tantallon castle. Within weeks of Otterburn and for the rest of 1388, the countess defied repeated attempts to remove her from the castle. Her aim was to secure Tantallon and a share of the Douglas lands for her son. Margaret's bargaining power was not just provided by the massive wall of Tantallon. With her in the castle were many of the principal Lothian adherents of the Douglas earls, including her half-brothers, William and John Sinclair from nearby Herdmanston, Alan Lauder, the constable of Tantallon, William Borthwick and William Lindsay of the Byres, a cousin of the Lord of Crawford. All these men had attended Earls William and James, witnessing their charters, and the Sinclairs, Lindsay and Lauder had all been with James at Otterburn. Malcolm Drummond must have expected to win the support of Douglas loyalists in Lothian, where Carrick's influence was strongest. The support of John Haliburton of Dirleton, a near neighbour of Tantallon, does suggest Drummond won some local followers but, by December, the core of the old Douglas affinity in East Lothian was backing the claims of Countess Margaret and her son.[6] The support of the Sinclairs, Lauder, Lindsay of the Byres and others secured the position of George Douglas as one of Earl James's heirs.

In the marches Drummond's attempts to establish himself as the heir to the Douglas earls met similar opposition. Between August and October 1388 Archibald the Grim and James of Dalkeith began to turn their claims on the Douglas lands into reality by winning or enforcing support from the principal adherents of Earl James in the middle marches. By the time the vital political deals were struck over the Douglas inheritance in April 1389, and probably by the time of the lieutenant's first concessions in November 1388, Archibald and James, not Malcolm Drummond, were the effective masters of the Forest, Liddesdale and Eskdale, and perhaps even of Douglasdale itself. The extension of lordship which allowed this mastery is suggested in evidence from April 1389 and from the 1390s. Firstly, by the April negotiations, Archibald and James Douglas were closely associated with Matthew Glendinning, a link which suggests more than just ecclesiastical support for the two men. Matthew's career was based on service to the earls of Douglas. He was named as Earl William's clerk in 1374, and his church benefices and activities as a diplomat were products of this link, which stemmed from the close bonds between the earls and the family of Glendinning, their principal vassals in Eskdale. Two of Matthew's brothers, Adam and Simon, fought at Otterburn, where the latter was killed alongside his lord. Bishop Matthew retained strong connections with his family both before and after 1388, and his early adherence to Archibald and James suggests the support of the locally powerful Glendinning kindred for them as their new lords. Such support would have been crucial for the acquisition of Eskdale, including Stablegordon and Westerkirk, before any legal grant was made.[7]

Archibald the Grim also secured support from Teviotdale kindreds. In 1389, Robert Colville of Oxnam and Adam Turnbull of Foulton were acting in place of a sheriff of Roxburgh after Drummond had been stripped of the office. The two men were specifically empowered to hand formal possession of Liddesdale to James Douglas of Dalkeith. Their appointment in place of Drummond and their actions in office suggest that they had already transferred their loyalties to James and their own new lord, Archibald Douglas. The winning of local loyalties by the two Douglases extended into Liddesdale where James Douglas sealed indentures with two local men, Richard Broun and 'Alix Armstrang', the latter the head of an already powerful local kindred in the bleak, upland districts of the marches. Certainly Armstrong and two of his kinsmen were among those who acted for Archibald in border negotiations in the 1390s, a group which also included Rutherfords, Scotts, Turnbulls and Nixons. The military kindreds of the middle marches, whose allegiance and service determined the frontier between English and Scottish lordship and the power of the Douglases in the borders, probably first accepted Archibald's leadership of frontier society in late 1388.[8]

The flow of these kindreds to Archibald Douglas began before the formal recognition of his rights and stemmed, not from the support of lieutenant or estates, but from his possession of the same strengths which had been the foundation of the knight of Liddesdale and Earl William Douglas's dominance in the marches. Archibald was a proven war-leader with a military retinue drawn from similar border kindreds in the west and from the 'savages' of Galloway, which had been consolidated in the warfare of the 1380s. This retinue and the men of James Douglas of Dalkeith provided a forceful reason for local landowners to accept their lordship, which Drummond and his allies could not match. In late 1388 clashes clearly took place. James Douglas received a remission for an attack on James

Sandilands, perhaps raids on his lands in Lothian and Douglasdale. Violence and the threat of violence clearly underlay James and Archibald's success. When Malcolm Drummond refused to attend the council which met at Edinburgh in April 1389 unless given secure lodgings in Edinburgh castle, he feared that his enemies would copy Douglas of Liddesdale's assassination of David Barclay in Aberdeen three decades earlier. Borderers too feared Archibald Douglas. Like their fathers, they saw submission to a Douglas lord as a means to escape destruction at his hands.[9] Like their fathers too, they regarded a Douglas lord as best able to protect the marches from English attack. In the face of English preparations to avenge Otterburn, the men of Teviotdale, whose old lord was dead and buried, saw in Archibald a magnate capable of defending them. To emphasise this, it was almost certainly Archibald who organised a raid on the English west march in February 1389, winning booty and captives and, more importantly, showing his credentials to lead the war in the marches.[10] Archibald was the natural successor to the military lordship of his kinsmen. He was a Douglas lord whose surname itself strengthened his claims to the family lands. The 1342 entail, the use of the name as a war-cry at Otterburn, as under the Good Sir James sixty years earlier, and the specification by Douglas of Dalkeith that his heirs hold the Douglas name, all demonstrate its importance. Even Froissart, hostile to Archibald's succession, thought in terms of heirs to the earldom from men of 'the name of Douglas'. As kinship based on the surname became increasingly important in fourteenth-century Scotland, only Archibald the Grim possessed both the name and reputation of his great predeccessors. Name and reputation were vital in securing his control of Douglas retainers in the marches.[11]

The success of Archibald and his allies in the marches can also be measured by the disintegration of Drummond's position. The first signs of this came in early November 1388 when James Douglas of Dalkeith was given wardship of Westerkirk and Staplegordon in Eskdale until 'the true and legitimate heirs' claimed them. This temporary recognition of James's control of the two baronies may have been designed to buy him off, but it failed to satisfy either James or Archibald, who was also present. Carrick's inability to settle the dispute was fatal to his lieutenancy. In December, his southern power-base in shreds, Carrick was supplanted as lieutenant by his brother, Robert earl of Fife. Fife's predominance in the Stewart dynasty would last for much of the next thirty years. Though he was never a close ally of Archibald and his family, Fife's ascendancy would be crucial to Black Douglas success in these decades. Unlike Carrick, he had limited interests south of Forth and was repeatedly ready to accept the status quo in border politics. In 1389, Fife's interests in the Douglas inheritance were fixed on the barony of North Berwick. In January he accepted Countess Margaret's occupation of Tantallon castle in return for gaining temporary access himself. Drummond accompanied Fife to Tantallon and may have hoped that the agreement would guarantee support for his claims from the Lothian adherents of the countess. However, Fife was increasingly keen to settle the whole dispute and was reluctant to take on Archibald, whose February raid marked him as the active defender of the marches. By April Drummond found himself deserted by his backers and physically menaced by his Douglas enemies. There would be no border inheritance for Malcolm Drummond in 1389.[12]

Instead the settlement which was formalised in parliament at Holyrood in early April divided the southern Douglas lands between three lords of the Douglas name, two of them bastards. James of Dalkeith received custody of his uncle's lands of Staplegordon, Westerkirk and Liddesdale. George Douglas, Earl William's bastard, received only custody of Tantallon from his father's estates, but was acknowledged as earl of Angus in succession to his mother. The real victor was Archibald the Grim. Drummond was stripped of his formal rights to Selkirk Forest in his absence and on 7 April Archibald produced the 1342 entail which was taken as proof of his claim to most of the Douglas inheritance in the south. To these lands was added the title of earl of Douglas, which Archibald was using by 10 April. The heir of the Black Douglas had finally claimed the lands of his father.[13]

THE RED AND THE BLACK

The legality of Archibald's succession was superficial. The weight given in parliament to 'entail and hereditary right' was balanced by the lieutenant's statement that the settlement was 'to avoid discord ... to the destruction of people and homeland'. Whatever his doubts about the dominance of Archibald in the marches, Fife had more direct concerns. To challenge Archibald risked chaos in the face of English invasion. The verdict of lieutenant and parliament was hardly impartial, nor, as events since Otterburn had shown, was it necessarily final. In June 1389 Malcolm Drummond, still claiming lordship of the Douglas estates and accompanied by his allies, sought English military aid in overturning the April settlement. Within a week, Fife and the new Earl of Douglas faced and successfully turned back an English invasion in the east march. The repulse of the English must have confirmed Fife in the wisdom of his decision. If Archibald and not Drummond had been pushed into seeking English support in 1389, the prospects for Scotland would have been far worse.[14]

Despite the English retreat, there was to be no end to the 'discord' over the Douglas inheritance. During the 1390s Archibald would face sustained political and physical challenges to his new lordship. At the heart of these challenges was a new rival dynasty, born from old tensions. The Red Douglases, the name given to the line of George, bastard son of Earl William, had designs on the regional lordship of the Douglas earls and would shadow and covet the power of their Black Douglas cousins until the latter's extinction in the 1490s. A century earlier, however, the border lands and influence of Earl William seemed securely in the grasp of the new earl of Douglas and his ally James Douglas of Dalkeith. Drummond's attempts to obtain English help ended with the conclusion of an Anglo-Scottish truce in July, part of a cessation of hostilities between England and France. The truce was extended through the 1390s and seemed to offer Archibald the opportunity to stabilise his authority in the middle march. From 1389 Archibald and Douglas of Dalkeith played a leading role in border diplomacy. The nine Scots appointed to oversee the truce in 1390 included Archibald, his bastard son William of Nithsdale, Douglas of Dalkeith and Bishop Glendinning, and in 1393 Richard II of England conducted direct negotiations with Archibald about the relations between his 'subjects' and 'the lands, lordships and subjects' of the Douglas earl. Archibald's influence was also recognised on the Scottish king's council, where the earl, James Douglas of Dalkeith and Bishop Glendinning formed a bloc, guaranteed to defend the 1389 settlement in the south. In the kingdom as in march diplomacy, Archibald appeared successfully to have retained the influence of the Douglas earls.[15]

In reality, this ascendancy overlay anxiety for Archibald and his allies which still stemmed from the unfinished business of the Douglas succession. While the earl had a copy of the 1342 Douglas entail drawn up, suggesting doubts were still being raised about its validity, James Douglas of Dalkeith produced a will which named Archibald as tutor of his heir. This may have been designed to secure a strong protector for James's lands should he meet the same violent end as his uncle and father. Against a background of growing pressure on James and Archibald, their presence on the council was not a mark of influence but an attempt to defend interests which were under threat from enemies led, once again, by Malcolm Drummond and James Lindsay of Crawford. In the early 1390s, these enemies could count on royal support, as James Lindsay dominated the household of the young heir to the throne, David earl of Carrick. Despite his youth, David was the focus of efforts, backed by his father, Robert III (the royal title adopted by John earl of Carrick), to reverse the losses of 1388–89 in the south. With this support Drummond obtained formal recognition of his rights to the unentailed Douglas lands, including the earldom of Mar and the border estates of Jedworth Forest and Cavers, but any full reversal of the defeat of 1388 in the south required a direct challenge to the new regional leadership of Archibald Douglas in the marches.[16]

By 1394 this challenge was underway. When Archibald's bastard son, William Douglas, met a violent death on crusade, his lordship of Nithsdale was bestowed on David earl of Carrick. Archibald had good reason to fear that James Lindsay would renew his old ambitions in the region, using David's rights to build his own influence.[17] The truce with England was exploited to re-assert Stewart leadership in border diplomacy. In 1394, Carrick, accompanied by James Lindsay and his cousin, David Lindsay of Glenesk, attended negotiations in the marches and, in March 1398, it was the 'yong Prynce', David, who was presented as the defender of Scottish interests in talks with the English wardens. Though Archibald was present at these talks, neither he nor his close adherents were named as commissioners in the discussions. Like Richard II across the border, the Scottish Crown was seeking greater control of the marches during the truce. The intention went beyond diplomacy. Prince David and his associates were also looking for adherents in the region, especially from the old Douglas affinity. Men like William Douglas, lord of Drumlanrig in Nithsdale and bastard son of James earl of Douglas, and the Berwickshire landowners, John Swinton and Patrick Hepburn, both connected to Earl James by marriage, were drawn into Carrick's following by grants of land and money. Even more valuable was the support of the warlike William Stewart of Jedworth, a Roxburghshire lord with close ties to the key kindreds of Teviotdale, the Turnbulls, Rutherfords, Colvilles and Kerrs. From 1391 he was a paid retainer of the king and Carrick, marking the intrusion of royal lordship into the old heartlands of Douglas power.[18]

Stewart of Jedworth's links with Carrick point to wider stresses within Archibald's newly-established dominance in the middle march in the 1390s. Local men had accepted Archibald in 1388 as the magnate best able to defend their lands and goods in war. They also expected their new earl to follow his predeccessor in leading them in attacks on the English. Since 1369 these raids had won lands in Scotland, which had been granted to successful captains like William Stewart and John Swinton, and brought captives and booty back from England. Earl James's death had not ended this run of victories and Archibald was expected to continue his familiy's traditional role in war. Not surprisingly, the truce of 1389 was unpopular with the border captains who had benefited from the profits of war for twenty years and the English complained that the 'principal cause of distroublance' on the marches was Scots born in English fealty. This implicated the men of Teviotdale, Stewart, the Colvilles and Kerrs, a group specifically antagonised by English demands for the surrender of Scottish gains since 1369.[19] Earl Archibald was conscious of the need to placate such men. A poem, written in the 1390s by one of Archibald's servants, glorified his predeccessor's death in terms designed to appeal to the men who had followed the Douglas banner to Otterburn. But poetic propaganda was hardly a substitute for the war lordship which had made Douglas power in the marches, and by 1397 there were clear cracks in the new earl's hold on the border adherents of his family.[20]

The problems faced by Archibald, with the march men, with the Crown and with his old rivals for the Douglas inheritance, crystallised in 1397 around the figure of George Douglas, earl of Angus. The son of Earl William Douglas, bearing the bloody heart on his arms, George had already won the loyalty of Douglas adherents in Lothian. He would seek to copy his father and use this Lothian following to establish his ascendancy in the middle march. George was backed in this ambition by other enemies of the Black Douglases. In April 1397 James Sandilands resigned his rights to Douglas estates in George's favour in a charter witnessed by the Lindsays. Sandilands was not simply giving up his place as heir to the unentailed lands held by Malcolm Drummond and named in the charter. As well as Jedworth, Cavers, Selkirk burgh and Liddesdale, Sandilands resigned to George, 'son of Earl William', his claim 'to all lands to which the true heir could succeed by law'. The ground was prepared for George to act as the 'true heir' of his father, a process continued two years later when Drummond similarly exchanged his right to Liddesdale in return for the lands of George's mother in Mar. The claimants of 1388 were selling their rights in the marches to a magnate with the support to win them from Earl Archibald and his allies.[21] George also had royal backing. During the summer of 1397 he married the king's daughter, Mary, in a match which Robert III and Prince David hoped would re-create the alliance with the Douglases, broken in 1388. In return, the king promised to ratify any deals struck by his new son-in-law concerning the Douglas lands. In early November he fulfilled this promise when the council recognised Sandiland's resignation, but the process clearly alerted Archibald and James Douglas, who were present. The king's charter did not call George the son of Earl William and omitted the clause about any wider rights as 'true heir'. For Archibald this was a minor success in the face of the coalition of king, prince and Lothian barons which backed Earl George. Fears of George Douglas and the men around him were behind Douglas of Dalkeith's preparations to enter England in the aftermath of the council and, in the new year, these fears would be realised.[22]

During the spring and early summer of 1398 George Douglas led a series of destructive attacks on the estates of James Douglas of Dalkeith. The targets of these raids were not the disputed lands in the march but James's rich Lothian estates, most vulnerable to George and his allies. James's lands in West Lothian, Mid-Calder, Kincavil and Bondington, his estates at West Linton, Esshiels and Bordlands in Tweeddale, and the barony of Dalkeith itself, were attacked and plundered. As in border war, the aim of this campaign of destruction was not the deaths of opponents but the destruction of their wealth and will to fight and the submission of tenants and neighbours whose lands would also suffer. By wasting James's richest properties, George hoped to force his surrender of the disputed lands. Given his family's history, George must have recognised that the only means to enforce his claims was warfare. The preparations of the previous year, which included his acquisition of Calder castle, the probable base for the West Lothian raids, were designed with these physical assaults in mind. The attacks were carried out by a following which indicated the strength of George's challenge to both James and Archibald Douglas. Alongside the men who had supported his claims since 1388, kinsmen like the Sinclairs of Herdmanston and William Stewart of Angus, and neighbours like the Borthwicks and Lindsay of the Byres, was a powerful coalition of Lothian lords. Henry Sinclair, earl of Orkney, James Sandilands, William Seton, the Ramsays, Prestons and Hepburns all participated in George's campaign against his enemies. These men, and raiders like David Fleming and John and James Hamilton, were old adherents of the Douglas earls and their ally, Carrick. 1398 represented a full-scale, militarised counter-attack by this royal-backed faction against the men who had taken leadership in the south from its grasp ten years earlier. Most worrying for Earl Archibald was the support for George Douglas from the marches. Rutherfords, Scotts and Nisbets were all involved in the attack on James Douglas, and the attack on Tweeddale was led by two Roxburghshire men, Andrew Roule and John Turnbull. Early in 1399 George stated his intention 'to recover from James Douglas all mails and rents of Liddesdale which he wrongfully occupies'. The lordship of Liddesdale, won from James's uncle by George's father, was again within reach. By early 1399, though, the moment for a wider challenge to the Black Douglases had passed.[23]

Although directed against his ally and not himself, the shock of George's attack was felt strongly by Earl Archibald. His own position in the marches was shaken by the support given to his enemy by his own tenants and neighbours. The march men were possibly attracted by the opportunity for plunder as much as by George's cause, and Archibald sought to provide the same opportunities in Black Douglas service and against the English. In his seventies and faced by his young Red Douglas rival, Archibald the

Grim passed the role of war leader to his elder son by Joanna Murray, also called Archibald. It was probably in 1398 that Earl Archibald resigned Douglasdale, the Forest, Lauderdale and the other entailed Douglas lands to his heir, retaining Galloway and the Murray lands. This division would ease his son's succession to the Douglas inheritance and provide aggressive leadership for the middle march. The younger Archibald, who was a 'giant warrior' in the image of his father, quickly put his new powers into practice. In early autumn, 'the Erle son of Douglas' and his 'cumpany' launched a major raid on English-held Roxburgh. The town was despoiled and torched, the wells filled, the stored hay from the harvest burned and the bridge across the Teviot broken. £2,000-worth of damages were done by a force which included William Stewart of Jedworth and probably came from Teviotdale. Stewart clearly appreciated a lord who could both cause mayhem and argue that it did not breach the truce because Roxburgh 'was and ys Scoct mennys heritage', words gleefully repeated by Stewart to the English later in the month. Stewart took a major role in these talks as a substitute for Earl Archibald, and the appearance of Richard Rutherford, Walter Scott, Thomas Turnbull and Robert Lauder in the same capacity suggests the recovery of Black Douglas authority over kindreds involved earlier in George Douglas's attacks. Stewart's relationship with the younger Archibald was crucial to this, and during the negotiations in late October the lands and castle of Abercorn in West Lothian were granted to William by his Black Douglas lords.[24]

Events beyond the march confirmed Black Douglas success. The earls of Carrick and Fife, now promoted to the dukedoms of Rothesay and Albany, planned to remove power from Robert III's shaky grasp. As in 1384, a Stewart coup needed the backing of the greatest lord south of Forth, and in November Earl Archibald met the two dukes at Albany's castle of Falkland. Though he agreed to Rothesay's appointment as lieutenant of the kingdom, the meeting can hardly have been comfortable. Since 1394, David duke of Rothesay had supported a concerted assault on Archibald and his allies, and the end of this backing must have been the key to Archibald's goodwill.[25] Significantly, Archibald was left to resolve the conflict between the lord of Dalkeith and George Douglas, the latter now deprived of the backing of Rothesay and his father. In 1399 or early 1400, a settlement was negotiated by Archibald earl of Douglas which probably gave George control of Liddesdale. In return for the loss of Liddesdale, Douglas of Dalkeith sought extensive compensation from the numerous lords who had ravaged his estates and retained possession of Westerkirk and Staplegordon, lands held from Earl Archibald as lord of Eskdale.[26] After over a decade of tension and sporadic violence involving Lothian and the middle march, the struggle for the Douglas inheritance was settled by the negotiation of the new head of the kindred.

In 1400, the last year of his life, Archibald Douglas had reason to feel satisfied. The marriage of his daughter, Mary, to David duke of Rothesay signalled Archibald's place in the new regime and raised the prospect of a Black Douglas queen. The settlement of the conflict with George Douglas, which would be confirmed during 1400, secured the gains of 1389. Though George's rights to Liddesdale and Jedworth forest would give the Red Douglases ambitions in the marches, in the short term this was balanced by his recognition of Black Douglas leadership in the kindred. When Archibald 3rd earl of Douglas died at Christmas 1400, it was at Bothwell, in the collegiate church he had founded, that he chose to be buried. His burial at Bothwell, rather than in the borders at Melrose abbey, the resting place of the first two earls, or in Douglas church, alongside the tomb he had built for his father, may suggest the earl's desire to be remembered as a pious benefactor and good lord in the rich valley of the lower Clyde. Bothwell had been his first possession and remained his favourite residence. But although his foundations at Bothwell and Lincluden, near Dumfries, were recalled in his epitaph, he remained Archibald the Grim, the tamer of Galloway and a leader in half a century of border war, and his greatest foundation was the Black Douglas dynasty.[27] Archibald's already exceptional rise from landless bastard to lord of Bothwell and Galloway was crowned by his recognition as earl of Douglas. The settlement of 1399 secured Black Douglas control of the chief lands and titles of the family after a bitter political and military struggle which lasted over a decade. Though the Red Douglases remained to trouble Archibald's successors as he had troubled Earl William, there would be no further challenge to the Black Douglases as heirs to the name, reputation and much of the following of the house of Douglas.

However, the end of competition for the Douglas inheritance did not mean the end of conflict in southern Scotland. Black Douglas success in 1399 rested on their continued control of a militarised following and, even before the old earl's death, this following was being employed by his son in a new campaign of family aggrandisement. During his last months, as for much of his life, Archibald Douglas was surrounded by the noise of war.

NOTES

1. Froissart, *Chroniques*, iv, 30; J. Reed, 'The Ballad and the Source: Some literary reflections on *The Battle of Otterburn*', in Tuck and Goodman, *Border Societies*, 94–123, 111. For excellent general accounts of the battle and its effect, see Goodman, 'Introduction', in

ibid., 1–29, Grant, 'The Otterburn War', in *ibid.*, 47–50, Boardman, *Early Stewart Kings*, 142–49.

2. Froissart, *Chroniques*, iv, 15; Wyntoun, *Chronicle*, ed. Laing, iii, 38; *Scotichronicon*, ed. Watt, vii, 414–43.

3. *A.P.S.*, i, 555, 557; *Liber S. Marie de Calchou*, Bannatyne Club, 2 vols (Edinburgh, 1846), ii, 408. For a full and coherent analysis of the Douglas inheritance from the perspective of Stewart family interests, see Boardman, *Early Stewart Kings*, 149–68.

4. *Kelso Liber*, ii, 408; *A.P.S.*, i, 555, 556, 557; N.R.A.S., Lauderdale Muniments, no. 832, 44/36; *Cal. Docs. Scot.*, iv, no. 391; *The Westminster Chronicle 1381–94*, eds L. C. Hector and B. Harvey (Oxford, 1982), 348–51.

5. *R.R.S.*, vi, no. 51; Fraser, *Douglas*, iii, no. 290; *R.M.S.*, i, appendix 2, no. 790; *Morton Reg.*, ii, nos 41, 61–63, 114–17.

6. Fraser, *Douglas*, iii, nos 36, 39–43, 294, 296, 329, 332, 340–41; S.R.O., RH 6/190; N.R.A.S., Lauderdale Muniments, no. 832, 44/36; Boardman, *Early Stewart Kings*, 150, 160.

7. S.R.O., RH6/195; *R.M.S.*, i, nos 792–800; *Melrose Liber*, ii, 478–80; *Rot. Scot.*, ii, 4, 81–82; *Foedera*, vii, 484–86; D. E. R. Watt, *A Biographical Dictionary of Scottish Graduates to A. D. 1410* (Oxford, 1977), 220–23; Froissart, *Chroniques*, iv, 15, 30; A. Teulet, *Inventaire Chronologique des Documents relatifs á l'histoire d'Ecosse* (Edinburgh, 1839), 29; Fraser, *Douglas*, iii, nos 330, 332, 333; S.R.O., GD 12/1. After 1389 the Glendinnings retained close ties with their new lords (*Morton Reg.*, i, lxxvi; W. Fraser (ed.), *The Scotts of Buccleuch*, 2 vols (Edinburgh, 1878), no. 22).

8. S.R.O., GD 150/78; *Morton Reg.*, i, lxx; *Foedera*, viii, 54–60; Boardman, *Early Stewart Kings*, 162–64; A. Grant, 'The Otterburn War', 51.

9. *Scotichronicon*, ed. Goodall, ii, 348, 397; *A.P.S.*, i, 557; S.R.O., GD 150/78.

10. *Westminster Chronicle*, 382–85.

11. *R.R.S.*, vi, no. 51; Froissart, *Chroniques*, iv, 12, 31; *Chronique de Jean le Bel*, i, 70; *Morton Reg.*, ii, no. 140.

12. *A.P.S.*, i, 556; *Morton Reg.*, ii, nos 183–84; Fraser, *Douglas*, iii, nos 40, 340; S.R.O., GD 212, Box 1, Book 25; Boardman, *Early Stewart Kings*, 151–66.

13. *A.P.S.*, i, 557; S.R.O., GD 150/78; *Morton Reg.*, i, lxviii–lxxvi; Fraser, *Douglas*, iii, nos 41, 296, 340–41.

14. *A.P.S.*, i, 557; *Cal. Docs. Scot.*, iv, no. 391; *Westminster Chronicle*, 396–97; Wyntoun, *Chronicle*, ed. Laing, iii, 40–41; *Scotichronicon*, ed. Watt, vii, 442–43.

15. A. Goodman, 'Introduction', 20; *Westminster Chronicle*, 402–405; *Rot. Scot.*, ii, 122, 126; *Cal. Docs. Scot.*, iv, 416; *R.M.S.*, i, nos 796–800, 801, 802, 805, 806, 828–46, 848–70, 872, 933. The origins of the name 'Red Douglas' for George's family are not clear. They may relate to physical charactersistics or to a desire to differentiate the two main lines of the kindred. Douglas of Dalkeith included extents of Liddesdale and his Eskdale lands in the rental of his estates (*Morton Reg.*, i, lxvii–lxxvi).

16. *R.R.S.*, vi, no. 51; A. Grant, 'The Otterburn War', 51; *Morton Reg.*, ii, 170–71; S. Boardman, *Early Stewart Kings*, 198–99; Fraser, *Douglas*, iii, no. 47; S.R.O., GD 124/1/118. Fife's lieutenancy had ended in 1393 and Robert III resumed formal control of government.

17. J. Stuart (ed.), *Spalding Club Miscellany*, v (Aberdeen, 1852), 250–52; *Westminster Chronicle*, 474–77. The Lindsays may have enjoyed some success in healing old disputes with local men like Thomas Kirkpatrick (S. Boardman, *Early Stewart Kings*, 222, n. 90).

18. *Rot. Scot.*, ii, 126, 502; Wyntoun, *Chronicle*, ed. Laing, iii, 65–67; A. Tuck, 'Richard II and the Border Magnates', in *Northern History*, 3 (1968), 27–52; S.R.O. GD 40/3/234; *E.R.*, iii, 287, 384. The claims of Drumlanrig and Swinton (the stepfather of Earl James Douglas) to shares of the Douglas inheritance were recognised by Carrick and Malcolm Drummond (S.R.O., GD 124/1/118; *H.M.C.* Drumlanrig, i, nos 2–3); Stewart was nephew of John Turnbull and was granted lands of Minto by the King in late 1391 (*R.M.S.*, i, nos 814, 850).

19. *R.M.S.*, i, no. 850; S.R.O., GD 12/1, GD 12/4; A. MacDonald, thesis, 40, 58; Thomas Walsingham, *Historia Anglicana*, 2 vols (London, 1863), ii, 182–83; *Scotichronicon*, ed. Watt, viii, 62–63; *Rot. Scot.*, ii, 123; *Foedera*, viii, 54–57.

20. *Scotichronicon*, ed. Goodall, ii, 406–13; *Scotichronicon*, ed. Watt, vii, 420–43. The author was Thomas Barry, provost of Archibald's foundation, Bothwell college (D. E. R. Watt, *Scottish Graduates*, 31–32).

21. Fraser, *Douglas*, iii, nos 43, 44; S.R.O., GD124/1/120; *H.M.C.*, Mar and Kellie, ii, 12. The death of James Lindsay in 1395 had not reduced his family's hostility towards Archibald.

22. *Spalding Misc.*, v, 252–53; Fraser, *Douglas*, iii, nos 45, 46, 48; *R.M.S.*, i, appendix 1, no. 154; *Rot. Scot.*, ii, 25. Sandilands was to be compensated with 200 marks of land from the Douglas estates should George recover them (Fraser, *Douglas*, iii, no. 44).

23. N.L.S., MS 72, 33r–39v; A. Grant, 'The Otterburn War', 63, n. 90; S. Boardman, *Early Stewart Kings*, 205; S.R.O., GD 124/1/120.

24. *Scotichronicon*, ed. Watt, viii, 59; *R.M.S.*, i, appendix 2, nos 1817, 1962; *Foedera*, viii, 54–57. The earl's substitutes, his 'borowis', in the west march were certainly men with established connections to Archibald Douglas (*Foedera*, viii, 58–60). On 12 October Earl Archibald issued a charter to the college of Bothwell to which Archibald, 'my first born son and heir', gave his consent, supporting the idea that the earl was increasingly transferring powers to his designated successor in 1398 (S.R.O., GD 120/76).

25. S. Boardman, *The Early Stewart Kings*, 214–15; *R.M.S.*, i, no. 886.

26. This mediation is referred to in the reference to a settlement reached 'be helpit of the Erle of Douglas confirmacon', in the Dalkeith cartulary (N.L.S., MS 72, 2v; A. Grant, 'The Otterburn War', 63, n. 90). George finalised his possession of Liddesdale in a deal with Malcolm Drummond in April 1400 while Douglas of Dalkeith was involved in Westerkirk in 1404. The long list of George's accomplices and the damages caused by their raids which were recorded by Dalkeith were clearly designed in a suit for compensation (Fraser, *Douglas*, iii, no. 51; *Calendar of Papal Letters to Pope Benedict XIII*, ed. F. McGuirk, Scottish History Society (Edinburgh, 1976), 122; N.L.S., MS 72, 32r–39v).

27. *Scotichronicon*, viii, ed. Bower, 30–31, 34–35; Fraser, *Douglas*, iii, no. 342.

CHAPTER FIVE

Ruler of the South:
Archibald Fourth Earl of Douglas

MAKER OF WAR

The breaking of Roxburgh bridge in 1398 marked the emergence of a new Black Douglas warlord. Archibald, master and soon to be fourth earl of Douglas, brought fresh aggression and ambition to the enlargement of his family's power. Over the next year the master and his border allies took their attacks onto English soil, provoking a major response. In late 1399 the new English king, Henry IV, told his parliament that he would exact personal revenge for the damage done to the lands of his northern supporters. He identified those responsible as the elder sons of the march wardens who had committed 'great and horrible outrages ... through the making of war'. Over the next quarter century, Archibald, master of Douglas, the chief target of Henry's accusations, would be a maker of wars. The English invasion which took shape in 1400 was only the first of his major campaigns in a career dominated by war. Until his death, almost inevitably on the field of battle, the fourth earl's search for the spoils of conflict would lead him into warfare in three kingdoms backed by a huge following drawn from across the south.[1]

The determined maintenance of the link between war and lordship by the master of Douglas was a lesson learned from the 1390s, when the family's hold on its border adherents appeared to weaken. During the same period, the Black Douglases were also exposed by their failure to retain the Lothian adherents of the Douglas earls. Instead these men had supported the enemies of Earl Archibald in attacks on his allies. Lothian, with its rich and powerful barons, had been the springboard for Douglas dominance in the marches in the 1330s and 1340s and, from 1400, the fourth earl of Douglas would make the recovery of his family's leadership in the province a major goal. With the end of the Douglas feud in 1399, this process was already underway. Men who had opposed Black Douglas claims for ten years quickly made their peace, drawn by ties to the Douglas earls established before 1388. Alan Lauder, William Sinclair, John Haliburton

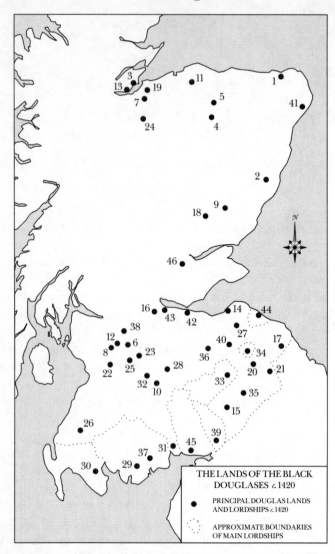

KEY TO MAP 3: THE LANDS OF THE BLACK DOUGLASES *c.*1420

Main Estates Acquired by the Family (Principal tenant in 1420 in *italics*)

A) The Murray Inheritance (1362)

1 Aberdour	*James Douglas of Balvenie*
2 Arburthnott	
3 Avoch	*James Douglas of Balvenie*
4 Balvenie	*James Douglas of Balvenie*
5 Boharm	*James Douglas of Balvenie*
6 Bothwell	
7 Brachlie	*James Douglas of Balvenie*
8 Carmunnock	

9	Cortachy	*Walter Stewart Earl of Atholl*
10	Crawfordjohn	*James Douglas of Balvenie*
11	Duffus (third)	*James Douglas of Balvenie*
12	Drumsergard	
13	Eddirdovar	*James Douglas of Balvenie*
14	Gosford	
15	Hawick	*William Douglas of Drumlanrig*
16	Herbertshire	*Willaim Sinclair Earl of Orkney*
17	Hutton*	
18	Lintrathen	*Walter Ogilvy of Lintrathen*
19	Petty	*James Douglas of Balvenie*
20	Smailholm*	
21	Sprouston	
22	Stewarton	*John Stewart Earl of Buchan*
23	Stonehouse (half)	*James Douglas of Balvenie*
24	Strathdearn	*James Douglas of Balvenie*
25	Strathaven	*James Douglas of Balvenie*
26	Traboyack*	*John Stewart Earl of Buchan*

B) **Other lands acquired by Archibald the Grim and family before 1388**

27	Clerkington	(1369)
28	Culter (half)	(1369)
29	Galloway	(1369)
30	Wigtown	(1372)
31	Nithsdale	(granted to William *William Sinclair Earl of Orkney* Douglas, Archibald's bastard son, 1387)

C) **The Douglas Inheritance (1388)**

32	Douglas	
33	Selkirk Forest	
34	Lauderdale	
35	Bedrule	
36	Romanno	
37	Buittle	*James Douglas of Dalkeith*
38	Fermes of Rutherglen	
39	Eskdale	
40	Heriot	
41	Rattray	*James Douglas of Balvenie*

D) **Lands acquired by Third and Fourth Earls of Douglas (1388–1420)**

42	Abercorn Castle (pre-1400)	*James Douglas of Balvenie*
43	Halls of Airth (pre-1400)	
44	Dunbar (1401)	Restored to George Earl of March (1409)
45	Annandale (1409)	
46	Dunbarney and Pitkeathly (1404 × 1406)	*James Dundas*

* Hutton, Smailholm and Traboyack were probably, though not certainly, part of Archibald's lands from his marriage to Joanna Murray

TABLE 4. THE BLACK DOUGLASES

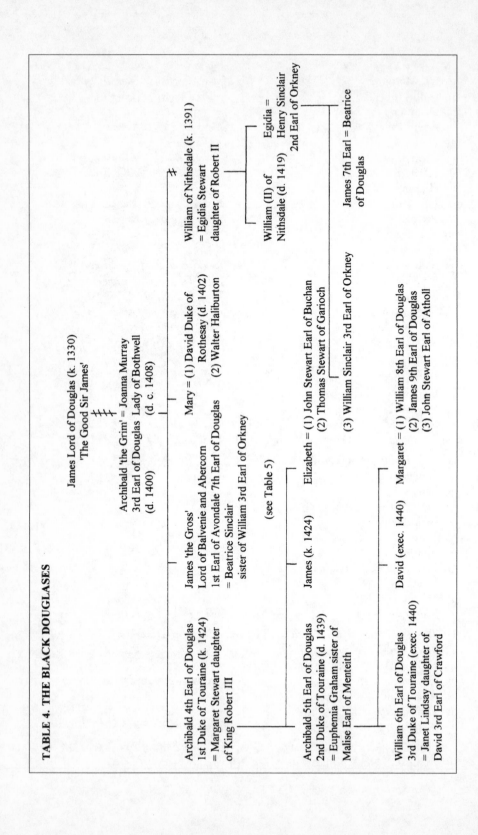

James Lord of Douglas (k. 1330)
'The Good Sir James'

Archibald 'the Grim' = Joanna Murray
3rd Earl of Douglas Lady of Bothwell
(d. 1400) (d. c. 1408)

Archibald 4th Earl of Douglas
1st Duke of Touraine (k. 1424)
= Margaret Stewart daughter
of King Robert III

James 'the Gross'
Lord of Balvenie and Abercorn
1st Earl of Avondale 7th Earl of Douglas
= Beatrice Sinclair
sister of William 3rd Earl of Orkney

(see Table 5)

Mary = (1) David Duke of
Rothesay (d. 1402)
(2) Walter Haliburton

William of Nithsdale (k. 1391)
= Egidia Stewart
daughter of Robert II

William (II) of
Nithsdale (d. 1419)

Egidia =
Henry Sinclair
2nd Earl of Orkney

James 7th Earl = Beatrice
of Douglas

Archibald 5th Earl of Douglas
2nd Duke of Touraine (d. 1439)
= Euphemia Graham sister of
Malise Earl of Menteith

James (k. 1424)

Elizabeth = (1) John Stewart Earl of Buchan
(2) Thomas Stewart of Garioch
(3) William Sinclair 3rd Earl of Orkney

William 6th Earl of Douglas
3rd Duke of Touraine (exec. 1440)
= Janet Lindsay daughter of
David 3rd Earl of Crawford

David (exec. 1440)

Margaret = (1) William 8th Earl of Douglas
(2) James 9th Earl of Douglas
(3) John Stewart Earl of Atholl

and William Borthwick had all served Earls William and James and had their service rewarded with grants of lands, in particular in the earls' border lordship of Lauderdale. By the 1390s, lands in Lauderdale were probably recovering their value. To obtain their profits, tenants needed the goodwill of their new lords, Archibald the Grim and his son.[2] Likewise, George earl of Angus needed his cousins' support to make good his rights in Jedworth and Liddesdale. Renewal of ties between lord and tenant generated new bonds of loyalty illustrated by the case of Richard Hangingside, a close supporter of Angus in the previous decade. In a bequest to Kelso abbey from his lands in Lauderdale, Hangingside remembered 'my lords of good memory, William and James earls of Douglas and ... my lord Archibald earl of Douglas'. After a decade of dispute, the continuity of Douglas lordship was being re-established. When, in late May 1400, the earl of Angus attended Archibald the Grim at Bothwell, the dispute seemed finally at an end.[3]

However, if Lothian lords initially looked to the Black Douglases in hope of gaining the profits of peace in their border lands, the dominance of the earls in the province would be established through the demands of warfare. The meeting of Archibald and Angus was set against a growing crisis in southern Scotland, tied up with the ambitions of the house of Douglas. The spark for this crisis came from the alliance between David duke of Rothesay and the Black Douglases. While it ended the struggle over the Douglas inheritance, the duke's marriage to Mary Douglas at Bothwell also brought an end to his liaison with Elizabeth, daughter of George Dunbar, earl of March, and created conflict in the marches and Lothian. Though bound up with issues of national war and allegiance, this conflict was driven by the ambitions of the two great noble houses of southern Scotland. Equally, though contemporaries saw the struggle as the product of the young duke's lust, George Dunbar's furious reaction to the news of the wedding at Bothwell was not simply anger for his 'defowled' daughter. Instead, deep-seated grievances over the decline in his family's fortunes, which had paralleled the rise of the Douglases, pushed the earl into drastic action.[4]

The Dunbars were the only southern magnates whose power stretched back to the thirteenth century. For them, sustained Anglo-Scottish war meant, not an opportunity for power, but the devastation of their lands in East Lothian and Berwickshire, the easiest cross-border route. Their allegiance to the Bruce cause was late and uncertain and, though in the 'Scottis fay' from 1334, Earl Patrick Dunbar left the defence of his lands to adherents like Alexander Ramsay. The construction of Tantallon castle by Earl William Douglas in the 1350s had further impact. The greatest lord in the south was now a close neighbour, based only seven miles from Dunbar itself, but despite this, rivalry between the two families was limited. Earl Patrick allied with William in war and politics after 1350. His successor, George Dunbar, was the protégé of David II, but, after 1371, established good relations with the Douglas earls and fought alongside Earl James at Otterburn. However, these Douglas allies increased the Dunbars' problems in maintaining their influence in the earldom of March. The earls' lands in the Merse had only recently been recovered from the English by the strenuous efforts of George Dunbar, while in East Lothian their tenants, the Lauders, Hepburns, Haliburtons and Towers, were increasingly Douglas servants and vassals. It was the Douglas earls who exercised greater pull as lords in Lothian between 1350 and 1388. The Dunbars, less wealthy and influential, and still maintaining connections with England, appeared an anachronism. These Anglo-Scottish links, which were a product of traditions stretching back beyond the wars, contrasted with the Douglases, who presented themselves and their servants as an unfailing defence against England.[5] Earl George Dunbar sought to alter his family's image and fortunes. From 1369 he led repeated attacks in the marches aimed at returning his lands to Scottish allegiance but, despite his efforts, while Douglas status in the kingdom was signalled by six marriages linking all the branches of the family to the royal house, after 1371 Dunbar obtained no similar alliance. The chaos after Otterburn was an opportunity for George earl of March to change this. He supported Robert of Fife's lieutenancy and in 1395, with the backing of this powerful ally, arranged, and paid for, a match between his daughter and the heir to the throne. Though the king delayed the marriage, in early 1400 Dunbar still antici-pated the place due to the father of a future queen and, with Rothesay's support, new influence in Lothian.[6]

Instead, he saw both marriage and influence diverted to the Black Douglases, a blow to his pride which he refused to tolerate. The master of Douglas's warmongering in the marches provided Dunbar with an obvious ally. He wrote to Henry IV of England asking for his help and in March 1400 crossed the border for direct discussions.[7] For the English, Dunbar's support for their planned invasion promised easy access to Lothian and perhaps the return of south-east Scotland to English allegiance. For Dunbar, English support added weight to his search for redress in Scotland. The earl threatened to take 'unheard of' action against his enemies and made plans which revolved around his possession of Dunbar castle and his influence in East Lothian. On 8 May he met with the key men of the district including Hepburn of Hailes, Lauder of the Bass, and his own nephew, Robert Maitland. Maitland was left to guard Dunbar castle while the earl went south to meet King Henry. With an English army of nearly 20,000 men preparing to invade the east march, the Scottish king and lieutenant would be more sensitive to the earl's demands.[8]

Though a similar strategy had worked only four years earlier, in 1400 the prospect of facing not just Dunbar's Scottish enemies but a supposedly friendly English army did not appeal to the men of East Lothian. Even before the earl went south a second time in June, Lothian landowners were looking to the Black Douglases for leadership. When Angus visited the old earl of Douglas in May he was accompanied by William Hay, William Borthwick and John Edmonston. All four had lands or interests in eastern Lothian. Their presence at Bothwell was connected with the defence of Lothian against Dunbar and the English. In future, all would be key supporters of the master of Douglas in war and politics. Rothesay also looked to Douglas in the growing crisis, granting the master the keeping of Edinburgh castle for life. While the duke, as lieutenant, was providing for the defence of the kingdom in the face of invasion, the master was exploiting the situation to his long-term advantage. He had ensured lasting custody of the stronghold which had been a base for both the knight of Liddesdale and Earl William, and which had been denied to the Douglases since 1360. From 1400 Edinburgh castle would be the younger Archibald's chief residence, the centre of his influence in Lothian and the south.[9]

The arrival of the master in Edinburgh had immediate effect. When March went to meet King Henry in late June, Douglas quickly made contact with Robert Maitland, the keeper of Dunbar. This contact was hardly surprising. Though Maitland was both nephew and vassal of March, his chief estates were at Thirlestane in Lauderdale.[10] In July, Maitland's other lord, the master of Douglas, appeared before Dunbar's gates. Without a show of resistance, Maitland handed over the castle. The loss of his stronghold dealt March's hopes a serious blow. He now pinned his hopes on the English invasion, negotiating an agreement with Henry IV which promised estates in England and military help in Scotland. In return, when the English army marched on Edinburgh in August, March was present to do homage to Henry as his true lord. Both Henry and his new vassal hoped to win support and limited the damage done to the lands of potential recruits. The effort was not entirely fruitless. According to Bower, 'the earldom of March and its inhabitants favoured the earl as their born lord', and lesser men, especially from the Merse, remained with their earl, who still held the castles of Cockburnspath and Fast in Berwickshire. With these men and bases, March would wage local war for his ancestral lands.[11]

However, George Dunbar had failed to retain the loyalty of the knights of Lothian. As the English army withdrew, Archibald master of Douglas

emerged from behind the walls of Edinburgh castle to offer his leadership
to a local community whose 'born lord' had shown himself a traitor. Though
contemporary chroniclers sympathised with the exiled earl, Archibald was
surely quick to present himself, like his forerunners, as the guardian of
Scottish liberties in contrast to March. Any appeal to national loyalties by
Douglas, though it clearly failed to win over all March's tenants, was also
backed by the more concrete benefits of patronage and physical protection.
The English invasion represented the escalation of conflict. In the borders
war continued unabated with English raids on Jedforest and Scottish attacks,
with mixed results, in all three marches. Lothian was now a second theatre
of war. In August the retreating English had attacked Dalhousie castle. In
February 1401, March and Henry Percy, son of the earl of Northumberland,
known as 'Hotspur', raided lands in East Lothian. The specific targets of
these attacks were Alexander Ramsay and Patrick Hepburn, men who had
failed to support March. The earl and his followers 'inflicted many evils
on Scotland' because they 'knew the weak places in the country'. Exposed
to 'malicious attack' by Percy, guided by men from the Merse, Lothian
barons looked to Archibald, now fourth earl of Douglas, for help. Leading
his men from Edinburgh castle, Douglas caught and routed March and
Percy's raiding force in early 1401 and, when war resumed in October after
a brief truce, 'the nobles of Lothian' sought 'the advice and support of the
earl of Douglas'. Significantly, this support was indirect. According to
Bower, the barons themselves organised and led raids during early 1402,
choosing leaders from their own community to avenge the attacks on their
lands.[12]

Douglas's role in the defence of Lothian was mirrored by his growing
political predominance in the south-east. The key to this predominance
was the patronage, direct and indirect, which, from early 1401, Douglas
delivered to local men to secure their loyalty. Three Lothian lords, William
Sinclair, John Edmonstone and Robert Maitland, were rewarded for past
and future services, in Maitland's case including the surrender of his uncle's
castle. Most importantly, in October 1401, as Douglas prepared to return
to war, he granted lands of strategic value at Pitcox near Dunbar and
Cranshaws on Lammermuir to John Swinton. The grant of lands on the
routes into Lothian signalled Swinton's importance in the earl's defence of
the province, a role to which, after thirty years of war in France and the
borders, he was well suited.[13] However, Douglas wanted wider support than
could be provided by his own patronage. The reward sought by many
tenants of Dunbar was their release from March's lordship. Vassals like
Patrick Hepburn, Robert Maitland and, from Berwickshire, Adam lord of
Gordon, saw March's absence as an opportunity to confirm their status

and independence by becoming tenants-in-chief of the crown. As in war leadership, the lesser nobles of Lothian and the east march were reducing their dependence on great regional lords. This ambition held a clear threat to the future power of the Douglas earls, but in 1401 Earl Archibald backed March's vassals as a means of winning their support and ensuring his own continued custody of Dunbar.[14]

To achieve these aims, Douglas needed Rothesay's continued acquiescence. In 1400 the duke had handed over the defence of the south, and probably custody of all three marches, to Douglas, but by the following year he was less keen to support the earl's military and political goals. In March 1401 the royal dukes, Rothesay and Albany, seeking to negotiate a truce with England, faced the open hostility of Douglas, who prevented them from entering the west march. When a truce was agreed in May, it was Douglas who set its terms and made it clear that George Dunbar was not included in it. Douglas used the truce to campaign against Dunbar's lands in the south-west. It was probably in June 1401 that Douglas led a retinue which included the earls of Angus and Orkney, John Swinton and John Edmonstone, to Lincluden near Dumfries, where he was joined by local lords. These leaders could bring major forces to bear on Annandale, and the lordship and its chief castle of Lochmaben were probably lost to Dunbar by the autumn, leaving Douglas ready to resume war on the English. While the earl prepared for war, though, both the English king and the Scottish dukes sought a truce which would extend to Dunbar.[15]

Further friction was caused between Douglas and Rothesay over the future of the earldom of March. Although, in February, parliament had passed a law allowing the crown to make permanent grants from earldoms in its temporary custody, no such grants were made from March in early 1401. Rothesay had his own plans for the earldom and by late summer was probably using the title himself. A royal earl claiming Dunbar was unacceptable to Douglas and unwelcome to the tenants of the earldom and when, in the autumn, Rothesay was taken captive by Albany, who had also been alienated by the lieutenant, the young duke would find no allies south of Forth. Rothesay's fate, and the political structure of the kingdom, was decided in a private council between Albany and Douglas at Culross in Fife. The greatest magnates from north and south of Forth agreed on the death of the heir to the throne and the neutralisation of the royal leadership which he represented.[16] Douglas's price for this support quickly became clear. In early October his grants to Swinton and Maitland were ratified and Maitland received the barony of Tibbers, no longer as Dunbar's vassal but as tenant-in-chief. Douglas himself was made lord of Dunbar, formal recognition of his leadership in the east. Less than a fortnight later,

Douglas restarted the war. In talks on the border, the Scottish negotiators, all the earl's men, blocked every English offer and, on 20 October, Douglas arrived with his 'whole army ... arrayed ... as if for war'. He broke off discussion and with his 'diplomats', Swinton and William Stewart, the earl crossed the Tweed 'with banner displayed' and raided east to Bamburgh.[17]

In war, as in the disribution of patronage in the south, Douglas's regional ambitions were now dictating the internal and external policy of the crown. Neither Rothesay, the lieutenant, nor Albany, his successor, had wanted Douglas confirmed in control of Dunbar or the war with England renewed. By late 1401 Douglas had achieved both aims. Even before Rothesay's arrest allowed Douglas to exploit Stewart rivalries, the earl had forced the pace of the attack on March and obstructed negotiations with England. What made this possible was Black Douglas leadership of lords and local communities across the south in the interlinked practice of war and politics. This now extended to Lothian, where Red Douglas and Dunbar adherents now looked to Earl Archibald as a source of protection and reward. The resumption of war in late 1401 allowed Douglas to strengthen his hold on this following, easing any doubts held by George earl of Angus about Rothesay's removal with demonstrations of patronage. In the early months of 1402 Douglas moved from Dunbar to the Forest, preparing the defence of his lands before returning to Edinburgh, surrounded by the veteran soldiers in his household, Swinton, Edmonstone and William Stewart. From Edinburgh, Douglas gave 'advice and support' to the Lothian lords as they began their raids on England. After initial success, the second of these raids was caught as it returned across Nisbet Muir near Duns by a mixed force of English and Scots led by the earl of March. The death of Patrick Hepburn, the heir of the lord of Hailes, and the capture of other former tenants and councillors must have been some revenge for March's humiliations of previous years.[18]

For March's greatest enemy, the earl of Douglas, Nisbet Muir was not complete disaster. The nobles of Lothian were now even more reliant on the protection of the earl and his massive following and, in the late summer of 1402, Douglas called out his men from Galloway, the west and middle marches, Clydesdale and Lothian to join him in arms. The men who came to his banner included March's brother, Patrick Dunbar of Biel, and two of the exiled earl's principal Berwickshire vassals, Adam Gordon and Alexander Hume, alongside George earl of Angus and his kinsmen. The service of these men to Douglas confirmed his successes in winning much of the south-east to his control since 1400. Finally, Douglas called in the last part of his deal with Albany. He sought the lieutenant's 'advice and backing' in the form of 'a band of knights and brave men' led by the duke's heir, Murdoch Stewart. From May onwards, the English had expected the attack of a huge Scottish host. By September the force, numbering perhaps 10,000 men, was ready and, under the command of the '*campi ductor*', the leader of the army, Archibald earl of Douglas, it crossed the Tweed into England.[19]

THE KING'S PRISONER

The waging of local war against Scottish rivals and the English enemy was a path to regional power well-trodden by Douglas warlords. It was a means to power which depended on military success. The price of defeat, experienced after Halidon and Neville's Cross, or even of a victory like Otterburn, was high. In 1402 Archibald earl of Douglas discovered the risks of military lordship at first hand. The huge Scottish army raided as far as the Tyne, but on its return was caught and trapped by a force led by Henry Percy 'Hotspur' and the earl of March. The Scots took up a position on Humbleton Hill, but found themselves cut down like 'fallow deer' by English archery. An attempt to break out led by John Swinton was crushed and the collapse of the Scottish host followed quickly. The ordinary men fled. The armoured leaders on foot were captured. Some eighty lords were taken prisoner, amongst them Douglas and his chief adherents. William Stewart was singled out for execution, accused of deserting English allegiance in the 1380s, the rest were led off for ransom. Many would be released over the next two years but not Douglas. He would spend the next six years as a prisoner, his lands and followers deprived of their lord. When, in 1346, Douglas of Liddesdale had faced similar captivity, he watched the gradual decline of his power at the hands of both English and Scottish rivals. Over the winter of 1402–3 Douglas, blinded in one eye by an English arrow and held prisoner by the Percy family, must have feared a similar collapse of Black Douglas lordship.[20]

The English threat came from Douglas's captors, the house of Percy. In March 1403 Henry Percy earl of Northumberland was granted the earldom of Douglas and all the lands of Archibald and his mother. In May the Percys began their effort to realise these claims with an attack on Cocklaws castle in Teviotdale. Though many Scots feared a collapse of allegiance in the marches, the crisis proved short-lived. The scale of Henry IV's patronage to the Percys was a mark of desperation in his dealings with a family which proved impossible to satisfy. By the summer 'Hotspur' was gathering forces for a revolt. Part of this army was provided by his captive, Archibald Douglas, who returned to Scotland to gather his followers. His contingent, including veteran soldiers like John Edmonstone and Robert Stewart of

Durisdeer who had been at Humbleton, joined Percy in Cheshire and was in the vanguard of his army when it met the royal host at Shrewsbury on 21 July. In return for his support, Douglas must have gained concessions about Percy claims to his estates and his own liberty. It would not be the last occasion on which Douglas would charge a price from a foreign lord for the support of his military retinue, nor would it be the last time such a plan ended in defeat. Though the 'giant' Douglas wreaked 'much slaughter ... with his great mace' among the royal household, when Percy was killed the rebel army fled. For the second time in a year, Douglas, who had lost a testicle in the fight, was captured by an army which included his enemy George Dunbar. Under heavy guard, 'the king's prisoner' was led off to renewed captivity.[21]

Despite his wounds, Douglas could at least take comfort from the disruption caused by the Percy rebellion to English ambitions in the middle march. In Scotland, similar comfort could be derived from the very completeness of the defeat at Humbleton. 'The flower of Scottish knighthood was captured', including the man most capable of challenging Black Douglas dominance in the marches, George, earl of Angus. At liberty, George Douglas could have revived his ambitions of 1398 for leadership of the family, as his father, Earl William, had done after Neville's Cross. The subsequent death of Angus from plague, leaving only an infant son by Princess Mary Stewart, freed the earl of Douglas from a major rival. Instead, Red Douglas adherents were forced into continued and not necessarily beneficial dependence on Earl Archibald and his agents. The death of Angus prevented renewed conflict over the Douglas inheritance. From 1403 what was at stake was Black Douglas leadership of Lothian. Although the lieutenant, Robert duke of Albany, had accepted this leadership in 1402, his attitude reflected short-term need, not long-term support, and he would show little sympathy for Douglas dominance of the south. Once the crisis after Humbleton was over, though, the duke faced his own problems in the north. Up to 1406 the Black Douglases faced rivals from south of Forth, from men returning from English captivity and seeking to challenge the captive earl.[22]

Henry Sinclair, earl of Orkney, and David Fleming of Biggar were unlikely rivals of the Douglases. Though Sinclair was from a Lothian family, neither he nor Fleming possessed the lands and formal power to rank with Douglas. In 1404, however, they operated in a regional society deprived of magnate leadership to an extent unparalleled since 1333. Just as after Halidon local knights had provided effective leadership, so after Humbleton Orkney and Fleming led in war and diplomacy. Fleming negotiated the truce with England in 1404, while the next year Orkney led a force to support a second Percy rebellion against Henry IV. This leadership was sanctioned by the aged king, Robert III, and by late 1405 Fleming and Orkney effectively controlled royal policy and powers. They held custody of Prince James, the young heir to the throne, and were seeking to establish the king's rights to act as tutor to his grandson, William, son of George Douglas and Mary Stewart. Control of the Red Douglas heir meant control of Tantallon and a base for leadership in Lothian. Already Sinclair was seeking adherents from his Lothian kinsmen and neighbours, and as he and Fleming prepared to march on Tantallon, Douglas leadership of the province seemed to be evaporating fast.[23]

In late 1405 Orkney and Fleming were directly attacking Douglas interests, blocking an attempt to exchange the earl for the head of the Percy family and showing ambitions in the marches as well as Lothian. The earl had to deal with these enemies from a distance, relying, as since Shrewsbury, on his family, servants and allies to remain loyal and represent him in his absence. In 1404, for example, William Muirhead, a knight in his parents' household, shadowed Fleming during negotiations with England, while the commission which discussed the truce included his father's friends, Douglas of Dalkeith and Bishop Glendinning, and his own adherents, John Edmonstone and William Borthwick. However, the key to the maintenance of the earl's power in the local warfare which erupted in late 1405 was the group of men left to run the earl's lands and following in his absence. As March's experience with Robert Maitland had shown, the failure of ties of blood and service with such agents spelled disaster. Douglas proved more fortunate in the custodians of his castles of Dunbar and Edinburgh, the centres of his power in Lothian. Dunbar was held by Richard Hangingside, an old Douglas retainer, and Edinburgh by William Crawford, whose loyalty in 'keeping the castle' in his master's absence was specifically rewarded by the earl on his return. Above these deputies stood the earl's younger brother, James, a figure with an enormous role to play in the fortunes of his family over the next forty years. James Douglas was left as warden of the marches but took on wider powers as active leader of the whole Douglas affinity.[24] When, in early 1406, Orkney, Fleming and Prince James moved on Tantallon with 'the leading men of Lothian', they found the castle's gates closed and James Douglas backed by a force from Edinburgh castle approaching rapidly. With the chief strongholds of Lothian held against them, Orkney and the prince escaped to the Bass Rock in the Forth, while Fleming sought to flee Lothian. Caught by James Douglas, Fleming's force was destroyed and its captain killed on Long Hermiston Moor west of Edinburgh. This running fight marked the nadir of Stewart kingship, condemning Prince James, soon to be James I, to English captivity, but for the Black Douglases success in local war had once again secured regional power. Like men from across the south for a century, the Lothian knights captured at Long Hermiston submitted to Douglas lordship imposed by the sword. Though still a prisoner himself, Earl Archibald could now turn to the consolidation of his hold on the south.[25]

He began by seeking reconciliation with the earl of Orkney, now also an English prisoner. A marriage between Orkney and Douglas's niece, Egidia, was probably agreed by the two men in London in 1406 and was designed to bring a rival for leadership in Lothian into the family circle. From early 1407, Douglas was increasingly allowed to exert direct influence in Scotland. Between March and November 1407, and after April 1408, Douglas was back in Scotland by permission of his captors. His temporary release was allowed in exchange for hostages given in his place. During 1407 and 1408 nineteen Scottish lords served spells in England for Douglas, principally motivated by personal obligations to their lord. Though the earl's sons, Archibald and James, were frequent sureties for their father, the bulk of the hostages were Douglas's councillors and adherents, like William Hay, John Edmonstone, Simon Glendinning and the son and grandson of that old family friend, Douglas of Dalkeith. These men performed duties for the earl in recognition or hope of his favour; and the readiness of a powerful group of lords to enter captivity for Douglas was a certain sign of the earl's continued lordship over these men.[26]

Douglas's parole began in early 1407 as a response to events in Scotland. The death of Robert III and the captivity of James I secured Albany's formal leadership of the kingdom as governor until the king's return. The governor's ambitions in the south remained limited. He did not attempt to impose his authority on the region or dismantle Douglas lordship. Instead he sought good relations with a magnate of such local power. He sent his second son, John, to act as a surety for Douglas's release in 1407–8 and recognised the earl's brother, James, as 'our lieutenant'. However, Albany also sought his own power-base and allies in the south. The absence of the king left the Stewart lands in the south-west without a lord, and in late 1406 the governor visited Ayr to establish his rights. Albany also sought the return of George Dunbar, earl of March. Despite their successes at Humbleton and Shrewsbury, by late 1406 the Dunbars were in dire straits. Attacked by Scots and northern English, the latter hating them for their part in Hotspur's death, the family were loaded in debt and unable to escape to their last refuge, Cockburnspath castle. Albany saw a chance to restore his former ally during Douglas's absence and in late 1406 probably established contact with the exile.[27]

Douglas may have been alerted to these discussions early in 1407, when a large group of his adherents, led by his brother James, attended him in London. Just as in 1402, the earl refused to accept the return of Dunbar, a man he had continued to harass in England as well as Scotland. While Albany was seeking characteristically to limit Douglas's predominance by careful diplomacy, the earl's, equally characteristic, response was to threaten the governor with war. On 14 March 1407 Douglas sealed two indentures with Henry of England. The first specified the terms of Douglas's release for thirteen weeks but also obtained Douglas's guarantee that he would keep a truce with England in 'all his bowndis'. The second agreement went further. Until he returned to prison, Douglas became King Henry's man 'ageynis al men' except King James. As James was in Henry's custody and the governor was not mentioned, the indenture was a potential alliance against Albany, who was still styling James 'the son of the late king'.[28] When Douglas returned to Scotland in 1407, Albany must have feared an open challenge to his authority by King James's brother-in-law, and Douglas may have initiated such a challenge in Ayrshire. The earl had interests in the region, where the old king's adherents resented Albany's takeover of a royal principality. Over the next three years, there was considerable local violence in the south-west and it was only in 1408, after the killing of James Kennedy, the principal royal adherent in Carrick, that Albany secured his control, binding the remaining Kennedys to his service.[29]

In the meantime, Douglas cemented his hold on the marches. By 1406 he was acting as keeper of the extensive lands of Coldingham priory in Berwickshire, employing his 'lufit squier and allie' Alexander Hume as his local agent and continuing to call himself lord of Dunbar. In 1407–8 Douglas also made grants of land or cash to two leading west march men, William Johnstone and Robert Maxwell of Caerlaverock, and employed Gilbert Grierson, an Annandale laird, as his bailie.[30] All these acts confirmed Douglas's lordship over former servants and neighbours of Dunbar. Whatever the governor's plans, Douglas was ensuring the survival of his regional leadership against Dunbar. The Red Douglases also experienced the earl's determination to deal with potential rivals. In 1406 the earl had probably gained the '*beneficio*' of Alexander Stewart, earl of Mar, for a further distribution of the old Douglas inheritance. Mar was the second husband of Isabella, elderly, childless heiress to the unentailed Douglas lands, and although George earl of Angus and his offspring had been named as heir to her border estates, in 1407 most of these lands were confirmed in the possession of other Douglas lords. Cavers, Drumlanrig and Selkirk burgh were now securely held by Archibald and William, bastard sons of Earl James Douglas, and Douglas of Dalkeith was granted regality powers in the barony of Buittle. These acts were all sponsored by the earl of Douglas to reward loyal kinsmen and to obscure his own quiet occupation of Jedworth forest. When Isabella died in 1408, the guardians of young William earl of Angus, his grandmother, Margaret Stewart, and kinsman, William Sinclair, faced the loss of almost the whole of their ward's inheritance. Fresh tensions were opened between Red and Black Douglas. From 1400 to 1406, Sinclair had supported the earl of Douglas but in 1407 his service ended abruptly as Douglas's protection turned to exploitation. Though Sinclair protested to Albany about the rejection of his ward's rights, the only part of Isabella's lands that passed to the Red Douglases was Liddesdale, already obtained by George earl of Angus before his death.[31]

When, in late March 1409, Albany instructed the sheriff of Roxburgh to invest George's son with the lordship of Liddesdale, it was a mark of an approaching settlement between the governor and Douglas. Albany had secured control of the Stewartry and in June 1408 had received George Dunbar and his family back into Scottish allegiance. However, Douglas, though still formally a prisoner, now had no intention of returning to England and remained firmly in possession of Dunbar castle.[32] Albany sought to resolve this stand-off without further conflict and in June 1409 the two magnates met at Inverkeithing. The result of this private council was an indenture between the 'mychty prince' and the 'mychty lord' with huge implications for the government of the whole kingdom. Douglas accepted the return of March as part of a territorial settlement to be confirmed later in the year, but the indenture dealt with wider issues. It was no bond of manrent. Though Albany was named as governor, the indenture made no reference to Douglas's obligations, or even his inferiority, to the duke, a situation which would only change should Albany become king. Instead the agreement specified mutual and identical help and support. Its principal purpose was to provide terms to settle 'discord' between the two lords and 'their men'. This would be achieved by personal arbitration of lords and councils rather than the public judicial powers of the governor. It was left to the two magnates to manage their own followings and resolve disputes arising from 'slaughter' or service. The power of Albany, like Douglas, rested on a huge network of personal bonds and relationships, and the indenture essentially recognised the separate and equal lordship exercised by the two magnates and sought to regulate future disputes after a period of tension and conflict between their adherents.[33]

The indenture set the character of political society for a decade. The formal reconciliation of Douglas and March in the autumn showed the winners and losers. At Haddington in early October, the earl of March and his son resigned the lordship of Annandale to Douglas in return for the lordship of Dunbar. A fortnight earlier Douglas had resigned the barony of Cortachy in Angus to Albany, but his concession was limited by the fact that, since 1402, the lands had been held from him by his Red Douglas cousins. From 1409 both Dunbar and Red Douglas would find lordship and politics in the south dominated by Earl Archibald, and this dominance accepted by the governor. The final settlement between duke and earl occurred in July 1410 when the marriage of Albany's second son, John earl of Buchan, to Douglas's daughter, Elizabeth, was agreed. The couple were to receive 200 marks (£166 sterling) worth of land from both families and the agreement specified that Douglas's portion should come from his Ayrshire estates. Though local conflict had been settled, Albany was keen to exclude the earl's influence from the Stewartry. He was still anxious about the power and ambition of the Black Douglas earl.[34]

GREAT GUARDIAN

During the whorlle-bourlle in Scottland ... the duke of Albanye governyd and ţoke uppone hym the reule of Scotland beyonde the Scottish See. And in the same wyesse dydde th'erlle Douglas bothe governe and reule over this syde the Scottische see.[35]

To an English contemporary, the hurley burley, the upheaval, in Scotland since 1400 had left Archibald earl of Douglas effective ruler of the south. The English had witnessed Douglas's direction of Scottish war and diplomacy and, in the agreement of 1407 between the earl and King Henry IV, accepted the 'boundis' of Douglas's lordship as stretching between 'Est see and West see and the Scottes see'. Though this English view ignored Albany's dominance in the Stewartry, by 1410 Earl Archibald possessed an unprecedented concentration of offices in southern Scotland. Justiciar south of Forth, keeper of Edinburgh castle, a post usually combined with the office of sheriff of Lothian, and warden of all three Scottish marches, Douglas had powers of justice and war leadership across the region. However, these official powers were not the key to Douglas's position in the south. As for most of the previous century, the exercise of power in Scotland depended on private resources, and offices were held by the men with the lands and followers to fulfill their duties. It was Douglas's military affinity which had allowed him to act as the defender of Lothian and the marches, and the same network of adherents forced Albany, who held the powers of the crown over justice and war, to accept the earl's predominance in the south. In two private councils, at Culross in 1402 and Inverkeithing in 1409, the two greatest Scottish lords, who were also the governor and march warden, determined the distribution of power in the kingdom in terms of private lordship and interests, not of the offices they held.[36]

The 1409 agreement left Douglas a free hand to strengthen the relationships on which his power rested, the ties of leadership and service between
magnate and those lesser lords with lands and influence at local level. As
had been the case for a century, regional lordship depended on Douglas's
acceptance as patron and protector by these locally important figures. By
attracting into his service men like John Seton and William Borthwick from
Lothian, Herbert Maxwell from Dumfriesshire and Alexander Hume from
the south-east, Douglas was forging ties of lordship with men who were
not simply significant local landowners, but themselves possessed adherents
and influence in their own communities. He did so not just by giving lands,
money and places on his council to these men but by making himself the
most effective source of lordship in the south.[37] Aristocratic rivals to Douglas,
magnates with the status and connections to challenge him for local support,
found their own interests subordinated or undermined by Earl Archibald.
Henry, earl of Orkney, whose influence in Lothian had threatened Douglas's leadership in the province in 1405, was drawn into the latter's following,
accompanying Archibald on his first visit to France in 1413. More serious
rivals were the Dunbars and Red Douglases. The settlement of 1409 left
both families exposed to the greater resources of the Black Douglas earl.
Without Albany's active support, the guardians of young William earl of
Angus were forced to accept Douglas's ascendancy. In December 1409 a
marriage was agreed between Angus and Margaret, daughter of William
Hay of Yester. As Hay was one of the earl of Douglas's closest councillors
and Angus's interests were represented in part by Orkney, the match was
clearly designed to link the young Red Douglas to his kinsman's following.
Earl William's grandmother, Margaret Stewart, whose determined quest
for the rights of her son and grandson had established them as major lords,
was forced, at the end of her life, to accept this sign of Black Douglas
primacy.[38]

Since 1400 the real rivalry in the south had been between Douglas and
the man whose military skill had twice led to his defeat and capture, George
earl of March. Although 1409 saw the end of open conflict, it also revealed
to Dunbar his political defeat, which far outweighed success in war. Dunbar
was restored to an earldom stripped of lands and vassals. Many of his
tenants, like young Adam Hepburn whose father had been killed at Nisbet
Muir, remained hostile to the family responsible for deaths and devastation
in the south-east. For such men, Douglas was still a potent master. His
resignation of the lordship of Dunbar did not end his local power. Douglas
continued as lord of those he had rewarded from the Dunbar lands for his
lifetime, ensuring their protection against March and maintaining his own
influence. The beneficiaries of his patronage, like the Hume and Swinton

families, continued to identify Douglas, and not March, as their principal lord. This influence in Berwickshire was increased further by Douglas's custody of the lands of Coldingham priory. Coldingham had a troubled place as a cell of English monks inside Scotland. Its traditional protectors had been the Dunbars, but, with the family's exile, the monks turned to Douglas. By 1406 he was 'keeper' of their estates and in 1414 was formally appointed as 'principal bailie' of the priory. His grip on the lands persisted despite the settlement of 1409 and despite the fact that March complained that Douglas held the office 'agayn my will'. Douglas's local power must have been stifling for March. Denied his rights in church and border defence, March saw his own vassals, like Alexander Hume and his brother David, raised in local status in Douglas's service and repaying this patronage with their loyalty and friendship. By contrast, March's own following consisted of family and household servants. His ambitions of 1400 had come to nothing.[39] The growing gap between the power of Douglas and March was demonstrated further in Annandale. Here Douglas's takeover occurred with barely a ripple, building on the local influence of his father, which stretched back to the 1360s and his own occupation of the lordship since 1401. In 1401 he had received support from the powerful lord of Caerlaverock, Robert Maxwell, and in 1410 appointed Robert's son, Herbert, as his steward of Annandale. As in the east, Douglas success raised the fortunes of ambitious local lords. For Maxwell and his neighbours, Carruthers, Grierson, Jardine and Carlyle, who attended their new superior at Lochmaben in 1411, there was no effective alternative to Douglas lordship. As march warden and lord of Galloway, Archibald's leadership was neither new nor subject to local opposition.[40]

The response of the Annandale men to their new lord applied across the south. The great religious houses south of Forth recognised Douglas's power to protect and support both inside and beyond the kingdom. The Coldingham monks were not alone in accepting Douglas as chief secular guardian and sponsor. After 1409 the earl was named 'special protector and defender' of Melrose abbey and 'principal protector' of Holyrood. He also maintained close ties with Dryburgh and Sweetheart abbeys, fulfilled his father's planned construction of a college at Lincluden and made his own plans to found Scotland's first Carthusian priory.[41] His concern for the church across the south from Lothian to Galloway emphasised the earl's physical power and the prestige and benefits of his lordship. In terms of scale and ambition, the fourth Douglas earl was looking beyond the regional power of his predeccessors. His visits to the French, English and Burgundian courts were not, like those of his father, as ambassador for the Scottish kingdom, but were to negotiate, on his own behalf, military alliances which provided him with employment and power. Abroad, Douglas played the part of a European prince. In Scotland, he sought to exercise lordship in settings which similarly displayed his princely aspirations. It was probably the fourth earl of Douglas who remodelled Bothwell castle. The construction of an impressive great hall and two attached tower houses created a palace complex within Archibald the Grim's fortress. Similarly, two free-standing halls were built on Threave island alongside the great grim tower built in the 1370s. These halls, to which may be added Newark castle in the Forest and the royal castle at Edinburgh, were residences for the head of a great household and backdrops to impress local communities with the wealth and prestige of their lord.[42]

If much of this wealth and prestige stemmed from Earl Archibald's great territorial possessions and even greater following from across the south, Douglas's power as principal magnate south of Forth was enhanced by his stranglehold on major offices in the region. In terms of finance the earl was almost certainly a bigger spender and more dependent on his cash income than his forerunners. The building at Threave, Bothwell and Newark was only one part of his expenditure. Douglas made several grants of annuities, either in connection with offices or in place of land, and also loaned or dispensed money to his close companions. Douglas's spells in England and on the continent also drew on his finances. In 1407 John Edmonston was given a nineteen-year lease on Tullialan in return for his gift of 240 marks given to Douglas in 'his great need', during his spell as an English prisoner; and the earl left a trail of creditors behind him in both England and France. Despite continued English raids, Douglas lands in the south probably returned increased revenues after 1409. To this income was added his £100 annuity as bailie of Coldingham and similar pensions or profits from other religious houses. However, these funds did not satisfy Douglas, and from 1408 he repeatedly exploited his offices of justiciar and keeper of Edinburgh castle to tap into crown revenues. The customs of Edinburgh were used as a private source of income by the earl. Initially this was justified as the pension due to his sister, the widowed duchess of Rothesay, as expenses for Douglas's duties in the marches or by order of the governor. Increasingly, though, Douglas simply took the balance of the customs without explanation. At least £500 per year, and in three years over £1100, were paid to or taken by Douglas from the customs revenue, while the profits of the earl's justiciary courts were also directed into his coffers.[43]

The powers of justice which the earl possessed as march warden and justiciar further increased his hold on his following. His followers could feel secure in his support in any legal dispute. The remission given by Douglas as justiciar to James Dundas and his accomplices in the 'cruel rapine of the husbandmen of Bathgate' was the verdict of a lord on his close follower. The earl similarly turned a blind eye to abuses of the customs by kin and councillors, especially his brother James, who treated Linlithgow as the earl treated Edinburgh. By contrast the earl's enemies had little recourse to the structures of law, making service to Douglas even more attractive. The earl's domination of local government extended beyond his own offices. In 1407, by re-issuing royal letters of fifty years before, Douglas appointed 'his very dear squire', William Hay, as sheriff of Peebles. The authority of a dead king was used to legitimise an appointment which should have been made by the governor. Sheriffs like Hay, Thomas Kirkpatrick in Dumfries and Uhtred MacDowell in Wigtown could be counted on to further Douglas's exploitation of the rights of the crown to sustain his own lordship.[44]

The governors of Scotland, Robert duke of Albany and, after 1420, his son Murdoch, made no serious challenge to this exploitation and usurpation of royal rights. Both accepted Douglas's dominance of local government and political society south of Forth as the management of his following agreed in the 1409 indenture. They were content to maintain a largely formal recognition of their governance in the region by holding general councils in Edinburgh, by ratifying land grants from the south with their great seal and, most impressively, by denying Douglas a monopoly over diplomacy and war with England, balancing his followers with their own and the Dunbars. Beyond this, though, there was no effort to undermine the earl's ascendancy. The talking to, which Albany delivered to Douglas in 1410 in response to the latter's plundering of the Edinburgh customs, was followed in the next account by grants of over £2000 to the earl. As in 1388 and 1402, the dukes lacked the motivation and resources for a major assault on the Black Douglases. Their hands were full with the maintenance and defence of their own power-base in the lands north of Forth from Fife to Argyll, in particular against the great Gaelic power of the west, the Lordship of the Isles. To this end they sanctioned the power of a second great regional lord, Alexander Stewart, earl of Mar. Mar's dominance of the north-east as justiciar and private lord matched that of Douglas in the south, although, unlike Douglas, Mar formally acknowledged his place as the governor's local lieutenant.[45]

By 1420 Scotland was dominated by structures of regional lordship which had grown up in the previous century of war and disorder. Albany, Douglas, Mar and the lord of the Isles held power, not by the design of kings and governors, but as a result of the demand for local defence and protection which they could provide to greater effect than the crown. In the meantime, the heir of Bruce as head of Scottish political society had been left in English captivity for fourteen years. James I was not, however, forgotten. Douglas in particular maintained contact with his king. Stewart family rivalries in 1384, 1388 and 1402 had been to the benefit of Douglas earls and in 1407 the fourth earl had been ready to pose as champion of his brother-in-law, the young king. A decade later, Douglas and the adult James I combined again to embarrass the governor. The two men led opposition to the aged Albany's refusal to abandon the anti-Pope, Benedict XIII. As the rest of Europe had settled the Great Schism in the Papacy, Douglas and the king made great international capital out of the issue, securing advancement for their ecclesiastical adherents and prestige for themselves. In the 1420s Douglas's relations with the king would again be bound up with his growing European ambitions. The decisions taken by the earl from 1420 onwards would form a crucial turning point for the Black Douglas dynasty and the Scottish kingdom.[46]

While these decisions will be considered in detail later, the ambitions of Douglas in the politics of western Europe still depended, in part, on the maintenance of his following in Scotland. By 1420, cracks were starting to appear within the earl's affinity. The very scale of the earl's lands and power was creating problems. The number and geographical spread of Douglas's adherents placed strains on lordship based on the personal bonds between magnate and follower. Some powerful Lothian barons had never fully accepted Douglas's leadership. Walter Haliburton, although linked to the earl by marriage to his sister, Mary, widowed duke of Rothesay, and by bonds of landholding, maintained equally significant connections with Albany and March. Elsewhere resentment at the direction of the earl's patronage created local friction. The elevation of Herbert Maxwell in Annandale may not have been universally popular with the men of the lordship, while the earl's grant of lands in Galloway to his close ally, William Hay, provoked the devastation of the estates in question. Douglas had to deal with such defiance himself, in this case ordering his local officers to punish those responsible and levy the rents of the estates from them. More seriously, in 1416 Edinburgh castle itself was held against the earl, possibly by Douglas's deputy, William Crawford, and the earl only regained the stronghold by a siege ended through negotiation.[47] Crawford's action suggests a serious deterioration of the bond between him and his lord, so valuable to Douglas when in English captivity, which may have resulted from the earl's adoption of the castle as his regular residence. The uncertainty of personal lordship also dealt a serious blow to Douglas's relations with the Hays of Yester. William Hay was amongst Douglas's closest associates, but his death and the succession of his younger son Thomas around 1420 broke the earl's personal connection with the family. The breach may have been caused by a dispute between Hay and William Borthwick which led to local violence in the 1420s. Borthwick continued to be one of Douglas's main councillors and, instead, Thomas Hay looked for help to his brother-in-law, William earl of Angus. By 1424, the marriage of Angus to Margaret Hay, designed to link the earl to the Douglas affinity, had backfired. Angus was ambitious for the inheritance of his father in the south and, by 1424, was building his own bonds with the knights of Lothian. With the death of George, earl of March, in 1420, Earl Archibald also faced more effective rivalry from the Dunbar family. The new earl of March began the process of recovering influence in the south-east, symbolised by the marriage of his daughter to John Swinton, son of Douglas's military companion. The re-emergence of effective rivals for lordship in Lothian, and evidence of local disputes involving Borthwicks, Hays, Abernethys, Douglases of Dalkeith and Dunbars in 1419–20, suggest limits to Black Douglas control in the province.[48]

Douglas may well have recognised the strains on his lordship, caused both by conflicts within his affinity and by the earl's own frequent absences in England and France. From 1419 his elder son, Archibald, was given an active role in the management of the lands and adherents of the family. The title of earl of Wigtown, revived for the Douglas heir on his father's initiative, gave him the status to lead Douglas retainers to France. Similarly, Wigtown gave his consent to grants from Lothian, Eskdale, Lauderdale and Bothwell. Douglas was grooming his heir to provide lordship in his own absence. Nor was Wigtown to be left alone in this task. When Douglas departed for France in 1424, he left his countess, Margaret Stewart, as mistress of Galloway. With James, Douglas's brother, left as counsellor to Wigtown, the earl was, as after Humbleton, using his family as substitutes for his personal lordship.[49] Throughout his career, however, Douglas had built connections with lesser men based on service in warfare. Continued raids on English territories in 1411, 1417 and 1420 maintained bonds established in the 1390s and 1400s. They may also have shown Douglas the limited rewards of such warfare for him and his men. From 1414 his ambitions centred, instead, on the opportunities of continental conflict. His negotiations with the rulers of England, France and Burgundy were designed to win money and power from Anglo-French war. The role of a mercenary prince with an army for hire was one which promised to satisfy the ambitions of Douglas and bind his retainers to his service.[50]

What would prove to be a brief career on the stage of European warfare represented the high point of Black Douglas fortunes. The army which Douglas and Wigtown took to France demonstrated the continued strength of the family's military and political lordship. Despite tensions in his following, Douglas's ability to assemble such forces marked him as still the greatest lord in the south. Thus, although he was son-in-law to the earl of March, the second John Swinton kept faith with his father's lord and followed Douglas's banner to France, accompanied by lords and knights from across the region. Douglas himself admitted no reduction in the scale or strength of his lordship. Never lacking in ambition or self-promotion, from 1419 the earl adopted a new title, used in dealings with foreign rulers. Earl Archibald was now 'great guardian of the marches of Scotland', a rank designed to give formal expression to his exceptional control of justice and war on Scotland's southern frontier. This control was the culmination of the efforts of Douglas magnates over the previous century. Lands and powers 'won in war' and earned from kings allowed the creation of a network of adherents from the Forth to the Rhinns of Galloway and built the reputation of Archibald fourth earl of Douglas as ruler of the south and great guardian of Scotland's marches.[51]

NOTES

1. *Scotichronicon*, ed. Watt, viii, 58–59; A. MacDonald, thesis, 129–36; *Cal. Docs. Scot.*, iv, no. 542; Walsingham, *Historia Anglicana*, ii, 242; *Rot. Parl.*, iii, 427–28; *The Royal and Historical Letters of Henry IV*, ed. F. C. Hingeston, 2 vols (London, 1860), 11–14.
2. *R.M.S.*, i, nos 638, 928; Fraser, *Douglas*, iii, nos 43, 49, 334; *A.P.S.*, vii, 139, 142–43, 159.
3. *Kelso Liber*, ii, 410–12; Fraser, *Douglas*, iii, no. 342.
4. Fraser, *Douglas*, iv, no. 53; *Scotichronicon*, ed. Watt, viii, 30–31.
5. *Wyntoun*, vi, 2–3, 53–57; *Cal. Docs. Scot.*, iii, no. 1033; Fraser, *Carlaverock*, ii, no. 15; *R.R.S.*, vi, no. 379; *R.M.S.*, i, no. 187, 265, appendix 2, nos 853–55. For a full study of George Dunbar's recovery of the Merse, see A. MacDonald, thesis. The Dunbars continued to patronise English ecclesiastical foundations and to maintain personal contacts with Englishmen throughout the fourteenth century (*Cal. Docs. Scot.*, iv, no. 451; *Rot. Scot.*, i, 877–78, 955, 975; ii, 136; S.R.O., GD 157/368/5; *The Correspondance ... of the Priory of Coldingham* ed. J. Raine, Surtees Society (London, 1841), nos lxii, lxv, lxix, lxx).
6. In 1371 George's sister was David II's intended queen, while his brother, John earl of Moray, had seduced and married Robert Stewart's daughter, Marjorie, in the 1360s against the future king's wishes. Neither match won any favours from the royal dynasty (Boardman, *Early Stewart Kings*, 23–25). In 1389 March had been with Fife in Tantallon, while Wyntoun suggests Duke Robert's longer-term support (S.R.O., GD 90/1/30; Andrew of Wyntoun, *The Orygynale Cronykil of Scotland*, ed. D. Laing, 3 vols (Edinburgh, 1872–79) iii, 78–79; Boardman, *Early Stewart Kings*, 203–4).
7. Fraser, *Douglas*, iv, no. 53; *Foedera*, viii, 131–32. Dunbar appealed to the English king on grounds of kinship. Their grandmothers were sisters and married into two of the main 'disinherited' families.
8. *Melrose Liber*, ii, no. 506; A. L. Brown, 'The English Campaign in Scotland, 1400', in H. Hearder and H. R. Loyn (eds), *British Government and Administration. Studies presented to S. B. Chrimes* (Cardiff, 1974), 40–54.
9. Fraser, *Douglas*, iii, no. 342; *E.R.*, iii, 515; Boardman, *Early Stewart Kings*, 203. With Angus and the others present, the third earl granted the Douglas fermes of Rutherglen to William Crawford. As Crawford would hold Edinburgh as the master's deputy, this gift may have been part of the preparations for the Douglas acquisition of the castle.

10. *A.P.S.*, vii, 159. In December and January 1399–1400, Maitland was also granted lands by William Hay, one of those with the earl at Bothwell (*A.P.S.*, vii, 136).
11. *Scotichronicon*, ed. Watt, viii, 32–37, 42–43; Wyntoun, *Chronicle*, ed. Laing, iii, 77–78; *Foedera*, viii, 153–54, 156–57; *Cal. Docs. Scot.*, iv, nos 551–52; Boardman, *Early Stewart Kings*, 229–32.
12. A. MacDonald, thesis, 139; *Cal. Docs. Scot.*, iv, nos 568, 585; *Foedera*, viii, 162; Wyntoun, *Chronicle*, ed. Laing, iii, 77; *Scotichronicon*, viii, 33–34, 42–43. Bower, a native of Haddington, may have experienced the events of 1401 at first hand.
13. *A.P.S.*, vii, 159; Fraser, *Douglas*, iii, nos 343, 345; S.R.O., GD 12/16; G. S. C. Swinton, 'John of Swinton: a border fighter in the middle ages', in *S.H.R.*, 16 (1919), 261–79. Cranshaws had been exchanged by Douglas with his mother, the dowager countess, Joanna Murray, in February 1401, indicating the start of the new earl's planned patronage (S.R.O., GD 12/14, 15).
14. *R.M.S.*, i, appendix 2, no. 1769; *H.M.C.*, Drumlanrig, i, 33.
15. *R.M.S.*, i, appendix 2, no. 1900; *Proceedings of the Privy Council*, ed. H. Nicholas, Rolls Series, 7 vols (London, 1834–37) i, 127; *Henry IV, Letters*, 52–56; Fraser, *Carlaverock*, ii, no. 21; *Cal. Docs. Scot.*, iv, no. 589. In June 1401 there was a sudden flurry of English activity with regard to the defence of Lochmaben suggesting preparations for an assault (*Cal. Docs. Scot.*, v, no. 900; *P.P.C.*, i, 135).
16. *A.P.S.*, i, 575–76; *E.R.*, vi, 55; Boardman, *Early Stewart Kings*, 232–47; *Scotichronicon*, viii, 38–41. Both James I and James II regarded Rothesay as having held the earldom of March after Dunbar's forfeiture (M. Brown, *James I* (Edinburgh, 1994), 155).
17. S.R.O. GD 12/16; *H.M.C.*, Drumlanrig, i, 33; *A.P.S.*, i, 159; Fraser, *Douglas*, iv, no. 54; E. L. G. Stones (ed.), *Anglo-Scottish Relations 1174–1328, Some selected documents* (Oxford, 1965), 173–82. Adam Gordon also had his lands regranted directly by the crown (*R.M.S.*, i, appendix 2, no. 1769).
18. *Yester Writs*, no. 56; Fraser, *Douglas*, iii, no. 346; iv, no. 54; *H.M.C.* Hamilton, no. 128; *Scotichronicon*, viii, 42–45. The earl of Angus was granted the barony of Cortachy in Angus in July while his uncle, Sinclair of Herdmanston, was confirmed in lands in Lauderdale in January. Angus had been in Rothesay's service the previous year and may have needed persuading about the duke's removal (*E.R.*, iii, 542).
19. *Scotichronicon*, viii, 44–49; *H.M.C.*, 10, appendix 6, 77–78; *Cal. Docs. Scot.*, iv, 402–403.
20. *Scotichronicon*, ed. Watt, viii, 44–49; Walsingham, *Historia Anglicana*, ii, 251–52; *H.M.C.*, x, appendix 6, 77–78.
21. P. McNiven, 'The Scottish Policy of the Percies and the strategy of the rebellion of 1403', in *Bulletin of the John Rylands Library*, 62 (1979), 498–530; J. W. M. Bean, 'Henry IV and the Percies', in *History*, 44 (1959), 212–27; Boardman, *Early Stewart Kings*, 267–73; *Scotichronicon*, ed. Watt, viii, 48–59; *Cal. Docs. Scot.*, iv, nos 633, 646; S.R.O., GD 15/333; *Foedera*, viii, 289, 313.
22. *Scotichronicon*, viii, 48–49. George was certainly dead by early 1406, when Mary Stewart was married to John Kennedy (S.R.O., GD 25/1/28; *R.M.S.*, ii, 378–80).
23. *Cal. Docs. Scot.*, iv, nos 660, 664; *Henry IV, Letters*, ii, 61–62, 73–74. Sinclair was lord of Roslin and Pentland in Lothian and related by marriage to the Haliburtons, Cockburns and Drummonds. Fleming was brother-in-law to the lord of Seton (*R.M.S.*, i, appendix 2, nos 1931, 1935). For a full analysis of the events of 1405–6, see Boardman, *Early Stewart Kings*, 278–97.
24. *Cal. Docs. Scot.*, iv, nos 660, 664; *E.R.*, iii, 618; Fraser, *Douglas*, iii, nos 350, 356; iv, no. 55; *Henry IV, Letters*, ii, 73–76. James Douglas was clearly exasperated by Orkney's attack on Berwick in 1405 which not only damaged the prospects for his brother's release but also prompted an English attack on the Douglas lordships of the middle march.
25. *Scotichronicon*, ed. Watt, viii, 60–63; Walsingham, *Historia Anglicana*, ii, 273. Orkney and Fleming's force certainly included members of the Towers, Cockburn, Gifford and Seton families who accompanied the prince to the Bass (M. Brown, *James I*, 18; *Cal. Docs. Scot.*, iv, no. 727).

26. Fraser, *Douglas*, iii, no. 351; *Rot. Scot.*, ii, 177, 180; *Cal. Docs. Scot.*, iv, nos 705–707, 729, 736–37, 752, 755, 762; A. Grant, 'The Higher Nobility in Scotland and their estates', unpublished D.Phil thesis, University of Oxford (1975), 336. Douglas made at least one grant, to Alexander Gordon of Stichill, which was clearly connected to hostage service (Fraser, *Douglas*, iii, nos 354, 357).

27. *Cal. Docs. Scot.*, iv, nos 676, 701, 734, 736, 762; *Rot. Scot.*, ii, 177; *R.M.S.*, ii, no. 27; *Foedera*, viii, 410; Fraser, *Douglas*, iv, no. 56. Fast castle had been handed over to an English keeper by 1404 (*Foedera*, viii, 370). Both Wyntoun and Bower identified Albany as the man behind Dunbar's return (Wyntoun, *Chronicles*, ed. Laing, iii, 78–79; *Scotichronicon*, ed. Bower, viii, 74–75).

28. *Rot. Scot.*, ii, 181; Fraser, *Douglas*, iii, no. 52; *Foedera*, viii, 478; *Morton Reg.*, ii, nos 215–16; *E.R.*, iv, 39, 55, 102.

29. S.R.O., GD 8/1; GD 25/31; Boardman, *Early Stewart Kings*, 282, 294. The Stewart lands in the west had been erected into a regality for Prince James in 1404. In early 1407 two Ayrshire men, John Montgomery and Hugh Campbell, were among those who attended Douglas in London. Douglas was lord of Stewarton in Cunningham (*Rot. Scot.*, ii, 181; S.R.O. GD 124/1/129; *H.M.C.*, Mar and Kellie, i, 7; Fraser, *Douglas*, iii, no. 356).

30. Fraser, *Douglas*, iii, no. 298; Fraser, *Carlaverock*, ii, no. 22; *Morton Reg.*, ii, no. 216.

31. Fraser, *Douglas*, iii, nos 43–46, 53; *H.M.C.*, seventh report, 727; Drumlanrig, i, no. 4; *Morton Reg.*, ii, no. 215; *Cal. Docs. Scot.*, iv, no. 736. Mar was with Douglas in 1406 and in 1410 received a £40 annuity for his '*beneficio*' from Archibald (*H.M.C.*, Mar and Kellie, i, 15). The earl of Douglas granted Hawick to Douglas of Drumlanrig, a frequent witness and hostage for the earl, in early 1407, while Cavers had been in dispute between Archibald and the Red Douglases since 1402 (Fraser, *Buccleuch*, ii, no. 22; Boardman, *Early Stewart Kings*, 288–89). In late 1408 Countess Margaret had her son's claim to Liddesdale recorded in a copy of the 1400 grant from Malcolm Drummond (Fraser, *Douglas*, iii, no. 51).

32. *Scotichronicon*, ed. Watt, viii, 74–75. Douglas left his younger son James, and perhaps also Simon Glendinning and Alexander Gordon, in custody when he failed to return to England (*Cal. Docs. Scot.*, iv, nos 762, 768, 770; *Rot. Scot.*, ii, 223).

33. Fraser, *Menteith*, ii, 277–80; Fraser, *Douglas*, iii, no. 300.

34. *R.M.S.*, i, nos 910, 920; *H.M.C.*, Hamilton, no. 128; Fraser, *Douglas*, iii, no. 359.

35. M. Connolly, '*The Dethe of the Kynge of Scotis*: A new edition', in *S.H.R.*, 71 (1992), 46–69, 49–50.

36. Fraser, *Douglas*, iii, no. 52; *E.R.*, iv, 19, 42, 80, 114, 133; N.L.S., Adv. Mss., 80.4.15, no. 1.

37. *R.M.S.*, ii, nos 70, 112, 119, 145, 242, 254, 255, 364; *Coldingham Corr.*, no. xcix; S.R.O., GD 157/371.

38. *Scotichronicon*, ed. Watt, viii, 80–83; Fraser, *Douglas*, iii, nos 351, 367; *R.M.S.*, ii, no. 112; *Yester Writs*, nos 45, 46, 50.

39. Fraser, *Douglas*, iii, no. 367; *R.M.S.*, ii, nos 119, 254, 255; *H.M.C.*, Milne-Hume, no. 1; *Coldingham Corr.*, nos xcviii, xcix, cii; S.R.O. GD 12/18, 19. After 1409, Douglas continued as superior lord of the lands of Wedderburn and Cranshaws granted by him to David Hume and John Swinton from the earldom of March.

40. Fraser, *Carlaverock*, ii, no. 21; *H.M.C.*, Drumlanrig, i, no. 110; *R.M.S.*, ii, no. 242.

41. *Calendar of Scottish Supplications to Rome*, vol. i, eds A. I. Cameron and E. R. Lindsay, Scottish History Society (Edinburgh, 1934), i, 68, 69, 106, 131, 142, 197, 230; Fraser, *Carlaverock*, ii, no. 21.

42. R.C.A.H.M.S., *Selkirk* (Edinburgh, 1957), 61–65; W. D. Simpson, D. J. Breeze, J. R. Hume, *Bothwell Castle*, H.M.S.O. (Edinburgh, 1985), 17–29; P. Yeoman, *Medieval Scotland, an archaeological perspective* (Edinburgh, 1995), 102–103; Francisque-Michel, *Les Ecossais en France: Les Français en Ecosse*, 2 vols (London, 1862), i, 112–115; *Cal. Docs. Scot.*, iv, no. 905.

43. S.R.O. GD 15/335; *Coldingham Corr.*, no. xcviii; *E.R.*, iv, 80, 116, 117, 130, 133, 143, 163, 175, 190, 224, 253, 277, 300, 309, 322, 324, 341, 368; *C.S.S.R.*, i, 77.

44. N.L.S., Adv. Mss. 80.4.15., no. 1; *E.R.*, iv, 113, 144, 193, 216, 244, 270, 296; *Yester Writs*, no. 44; Fraser, *Douglas*, iii, no. 368; *The Lag Charters*, Scottish Record Society (1958), no. 1. Douglas's elder son and adherents like William Borthwick, James Dundas and John Seton also exploited the earl's support to avoid or plunder the customs (*E.R.*, iv, 144, 224, 275, 278, 296, 301, 320, 322).

45. *R.M.S.*, i, nos 900, 919, 922, 928, 931; Fraser, *Carlaverock*, ii, no. 26; *Cal. Docs. Scot.*, iv, nos 804, 805, 833; *E.R.*, iv, 118, 163; S. Boardman, *Early Stewart Kingship*, 302–13; M. Brown, 'Alexander earl of Mar', in *Northern Scotland*, 16 (1996), 1–54.

46. Fraser, *Douglas*, iii, no. 52; *Foedera*, viii, 478; *Copiale Prioratus Sancti Andree*, ed. J. H. Baxter (St. Andrews, 1930), 27–28; *C.S.S.R.*, i, 224. For further discussion of the earl's relations with the Papacy see Chapter 9.

47. *Scotichronicon*, ed. Watt, viii, 75, 87; Frsaer, *Douglas*, iii, no. 371; *R.M.S.*, i, nos 897, 898, 900, 934;ii, no. 242.

48. *Yester Writs*, no. 50; *Calendar of the Laing Charters*, ed. J. Anderson (Edinburgh, 1899), no. 98; *C.S.S.R.*, i, 224–25, 236, 278; S.R.O., GD 12/20; *Melrose Liber*, ii, no. 512; Fraser, *Douglas*, iii, no. 373, 374; *R.M.S.*, ii, no. 364.

49. Fraser, *Douglas*, iii, nos 57, 62, 374, 380; *Melrose Liber*, ii, no. 507; *R.M.S.*, ii, nos 12, 13, 143.

50. *Scotichronicon*, ed. Watt, viii, 80–85, 86–87, 116–17.

51. *Scotichronicon*, ed. Watt, viii, 120–21, 124–25; Archives Nationales, J677, no. 20; *C.S.S.R.*, i, 8.

CHAPTER SIX

The Bloody Heart:
The Rise of the Douglases and the
Kingdom of Scotland

'The Bludy Hart', the red heart at the centre of the Douglas arms, was a symbol of power and prestige to the lords and earls of that family. It stood for the heart of Robert Bruce, carried on crusade by James Douglas in obedience to his royal lord's last command, and was probably adopted as the principal badge and mark of the family by William, the future first earl of Douglas, in the late 1340s. From then on, the heart appeared not just in the family arms, but carved into the stonework of their buildings and religious foundations, carried on the standard of Earl James at Otterburn, and even stamped into a wooden bowl found at the Black Douglas stronghold of Threave. However, the heart was not just a heraldic badge of the kind being adopted across fourteenth-century England and France, like the Prince of Wales's feathers or the bear and ragged staff of Warwick. Although these devices had significance for the lords who bore them, the bloody heart had symbolic and physical prominence in Douglas propaganda which they lacked. The heart itself was brought back to Scotland and buried at Melrose abbey, a house especially favoured by the king and which he had entrusted to the protection of James Douglas. For much of the next century, James's heirs continued as guardians of the heart's resting place and, although James himself was buried at St. Bride's Church in Douglas, his nephew, Earl William, who made the heart the symbol of his dynasty, chose Melrose as the site of his tomb.[1]

From the later fourteenth century, the bloody heart was well known in Scottish historical literature. In Barbour's *Bruce* and in the two surviving works of poetry from the Douglas household, Barry's Otterburn verses and Holland's *Howlat*, the heart was used to illustrate the qualities of the house of Douglas and to explain and justify their power in the kingdom. For a family of minor Clydesdale barons which had risen, in three generations, to be the greatest lords south of Forth, such an explanation was vital. While the two other rising powers of fourteenth-century Scotland, the house of

Stewart and the Clan Donald lords of the Isles, traced the roots of their power back through real or imagined ancestors to a distant past, the Douglases looked back only to the 'Good Sir James' as the source of their standing in the kingdom. The earliest indication of James's reputation was provided by Barbour's *Bruce* in the 1370s. The *Bruce* confirmed the heroic status of King Robert as the saviour of his people, drawing connections between his royal leadership and the freedom of Scots from English lordship, themes which were present in the king's own propaganda. In addition, though, Barbour depicted the reign as the heroic age of late medieval Scotland for the gratification of the heirs of Bruce's noble supporters. These heirs were the poet's patrons and audience and, while Barbour appealed to such 'ofspring' to follow the example set by their 'nobill eldrys', the description of the collective achievements of Douglas, Stewart and Randolph in Robert's service confirmed the pride and status of their families. For the Douglas lords, Earl William and Archibald the Grim, Barbour's account had special resonance. James Douglas's efforts in the Bruce cause received greater praise than those of his comrades, and his 'frendschip' with the king was depicted as a model of good lordship. Douglas 'servit ay lelely' and his master 'rewardyt hym weil his service'.[2] The rewards of King Robert had provided the house of Douglas with its network of estates in the middle marches and his trust had provided the family with its first powers of leadership in the marches. The tradition of service created by this personal bond was not simply a literary device. The relationship between Robert's son, David II, and James's son, Archibald the Grim, was, in part, a conscious renewal of paternal roles and duties. Archibald, like James, acted in the 1360s as the king's loyal lieutenant and David, like Robert, repaid service with lands, the titles to the Murray inheritance and eastern Galloway. For King David, it was by such 'faithful deeds ... in defence of our realm' that the legal heirs of James Douglas had received their lands and status too, and he sought to remind them of these obligations.[3]

However, the deeds ascribed to Sir James Douglas in general, and the tale of the bloody heart in particular, had a significiance for his successors which went beyond any simple tradition of loyalty and service to the Scottish crown. The relationship between Bruce and James Douglas, which led to grants of lands and authority to the knight, was a personal bond, described by Barbour as lasting 'quhill thai lyffand war'. With the deaths of the king and Douglas in 1329–30, and later with the death of David II in 1371, personal bonds of trust and loyalty between crown and magnate were broken. For most of the century from Robert's death royal lordship of the kind he had demonstrated, based on active command and generous reward, was absent from Scotland. Regardless of minority, captivity or incapacity among his heirs, none of Bruce's successors could afford to distribute the scale of patronage he had given his close allies. The reservoir of forfeited lands and of royal estates which Robert possessed was used, out of necessity, to create a nobility which would uphold Bruce claims to the kingship. The lords of Douglas who succeeded James were bound to the Bruce cause as the source of their own lands and power. However, for James's heir in the 1350s, Earl William, and Archibald Douglas himself after 1371, upholding the aristocratic establishment which Bruce patronage had created did not simply imply obedience to royal authority. In 1352, William could act as the defender of the Bruce cause whilst defying the king and his agent, Douglas of Liddesdale. Douglas magnates exploited the reputation of the 'Good Sir James' as a warrant for the special place of the house of Douglas in the kingdom, justifying their status as great regional lords in the south after 1330. Indications of how this was achieved are present in the *Bruce*. Barbour described how James Douglas was chosen to carry the dying Bruce's heart, first by his peers, and then by the king himself as 'best schapyn for that trawaill'. Douglas was the best knight for the task, a view repeated seventy-five years later in *The Buke of the Howlat*, where Holland gave pride of place to the tale of the bloody heart, emphasising its continued importance to his Douglas patrons. Both Barbour and Holland stressed the honour done to Douglas by the king's last command, which outweighed the 'mony larges and gret bounte that yhe haff done me ... sen fyrst I come to your service'. While the king had given extensive estates and rights to many of his noble adherents, the honour he showed to Douglas was worth 'mair ... than ony lordschipe or land' and led to the claim of the Douglases that 'the Bruse ... blissit that blud'. As James had served Bruce beyond the king's death, so Robert's blessing passed on his favour to James's descendants, the lords and earls of Douglas . The Douglases bore the bloody heart symbolically on their shields and banners as a reminder of Sir James's exploit in bearing the heart of his dead master and of the bond this gave James's heirs with the hero king. Rather than royal servants, the heirs of James Douglas drew from their founder's reputation the glamour and status to match their ambitions.[4]

Such claims had special significance after the end of the Bruce dynasty in 1371. Although the Douglases made no challenge to the rights of Robert Stewart as the Bruce's heir, their own claims to be the heirs of the greatest and best-loved of Bruce's supporters assumed a new importance. The Stewarts did not inherit the prestige of the Bruce kings, and to Earl William Douglas and Archibald the Grim, Robert Stewart was a magnate of similar origins and experience, building on the legacy of Bruce's patronage to establish his family's leadership in the disturbed local communities of the kingdom. From 1371 Earl William and his successors did not accord Stewart the respect received by his predecessors. The gap in status between the new royal line and its greatest southern vassals was clouded still further by seven marriage alliances between Robert II's family and the Douglases. While these family ties were intended to bring the Douglases into an extended royal house, from 1384 membership of the Stewart dynasty made the earls of Douglas an automatic participant in Stewart conflict. Instead of the convincing management of great lords by the crown, magnates like the earls of Douglas and the lord of Galloway were given a place in the management of the royal house and its policies. Between 1384 and 1424, Douglas magnates played a crucial part in the changes of five Scottish kings or lieutenants, exploiting these opportunities to extract fresh concessions and freedoms from the crown. By the end of the century the Douglases, if not kingmakers, were arbiters of the fate of a series of Stewart regimes.

The boasts of the house of Douglas about the origins and nature of their power were, however, the product of, rather than the spur to, their aggressive pursuit of greater power and independence in the high politics of the Scottish kingdom. However aided by the lordship of the Bruce kings, the roots of Douglas power were to be found in the border dales, not the king's chancery. The success of a whole series of lords of the Douglas name in placing themselves at the head of local communities across the south was behind the role of the earls as the makers and breakers of royal governments from the 1380s onwards. Even King Robert's 'gret bownte' to James Douglas was not simply his chosen means of rewarding the service of a loyal partisan. The grants of Selkirk and Jedworth forests and other lands in the middle marches marked the formal handover of title to lands and communities which had been under James's leadership for a decade. James's value to Bruce was as a man who through his own leadership and resources had won the men of the region to the king's cause. An identical process lay behind the grant of Galloway by David II to Archibald Douglas in 1369. David granted his adherent a lordship whose 'pacification' had been achieved in the previous five years with Archibald's own men and money. Grants of this kind were neither new nor unusual and harnessed the aims of crown and regional magnate, but they depended principally on aristocratic rather than royal resources, and the Douglases themselves claimed the leading role in the winning of their southern lands to 'Scottis fay' in the verses of *The Howlat*.[5]

For Robert or David to dispose of lands won by the Black Douglases to rival lords, or even to retain them, was bad lordship and bad politics, risking tension and conflict. From such a perspective, the gap between the service of James and Archibald to their Bruce masters and the careers of Earl William and the knight of Liddesdale seems far narrower. Like James and Archibald, these lords claimed to be winning or retaining regions of the south to the Bruce cause in the years from 1333 to 1358, when royal leadership was absent but the threat of English overlordship in the south remained very real. These men also expected rewards for what they presented as their services to Scotland and, in the series of royal grants to the knight of Liddesdale in the early 1340s, and by creating the knight's killer earl of Douglas and confirming his rights, not just to lands but to the leadership of the men of the middle march, David II was dispensing patronage. However, it was a process over which the crown had very limited control. In his rewards to the earl and the knight, David was acknowledging his inability to do more than confirm the existing regional power of these lords. For much of the next century, David's successors enjoyed little more in the way of control over the marches, as the fate of the frontier lordship of Liddesdale clearly illustrated. In 1342, 1354, 1389, 1397 and 1409 kings or lieutenants issued documents granting, or confirming, custody of Liddesdale to a series of Douglas lords. Rather than royal generosity, these grants marked royal acceptance of the physical control of the lordship by these lords. Since the 1330s there had been two rival Douglas claims to Liddesdale, those of Sir William Douglas and his heir James of Dalkeith, and of the first earl of Douglas and his Red Douglas descendants. Both claims rested, not on legal grant, but on the recovery of the lordship from the English by the heads of these rival branches of the Douglas family. During the fourteenth century, the crown alternately accepted both claims as genuine, dismissing the other on each occasion as lacking any legal right. The effect of such verdicts was not to settle the feud in the royal court but simply to ratify the *status quo* until the next turn of fortune. These turns, Halidon Hill, the murder of the knight of Liddesdale, the death of Earl James at Otterburn, were outside royal control and their consequences were worked out, not at court, but within the regional society of the south. The role of the crown in the control of this major southern lordship was to accept and legitimise the, often violent, verdict of Douglas family rivalries.[6]

Between 1329 and 1424 the only alternative to this policy for the crown was to risk conflict in a vulnerable region of the kingdom by challenging the power of these Douglas magnates. In the century from 1329 a series of such challenges was made, either directly by the king or his lieutenant, or by royal backing for a rival magnate in the region. With only one exception, these efforts met with disaster. In the 1360s David II, through his patronage and personal lordship, won many Lothian knights from the earl of Douglas's service to his own following, and in the marches he established his own adherents as rivals to the earl. Yet, in the long term, this exception proved the general rule. His greatest creation, the power of Archibald the Grim in Galloway, owed much to the methods of Douglas lordship elsewhere in the marches, and after David's death eastern Galloway served as a springboard for the extension of Archibald's predominance into Wigtown and Dumfriesshire. From the perspective of the crown, David II had created a monster, a second Douglas rival to royal authority in the marches. After 1389 this Black Douglas connection would represent an accumulation of lands and manpower in the south even greater than that of Earl William. All other royal attempts to weaken the leadership of Douglas lords in the marches ended in failure. The murders of Alexander Ramsay in Hermitage castle, and of William Douglas of Liddesdale in the Forest, removed local magnates who were acting, at the time of their deaths, as royal agents or allies in an attempt to weaken the power of leading Douglas magnates. Similarly, Malcolm Drummond, designated heir to the Douglas earldom and its estates by the lieutenant, John earl of Carrick, was denied his rights and forced into exile in fear of his life by the actions of Archibald the Grim and his allies in winning physical possession of these lordships. Finally David Fleming, appointed sheriff of Roxburgh and guardian of the heir to the throne, was killed on Long Hermiston Moor to end his royal-sponsored efforts to undermine Douglas lordship in Lothian. The crown, even working indirectly through powerful local lords, proved unable to alter the balance of power significantly in much of the south. The good relations between Carrick and the great southern magnates in the 1370s and 1380s were almost entirely dependent on his complete acceptance of the *status quo* in regional political society and his diversion to these magnates of renewed patronage, unalloyed by any efforts to increase his authority. The first crisis of this Stewart lordship in the south, the death of Earl James Douglas at Otterburn, caused the immediate collapse of Carrick's regional authority and led to his political eclipse in the kingdom. The stability of the region in the years before 1388 was an illusion which owed far more to the concerns and interests of the southern magnates than to the authority or good management of Carrick. In the crisis of 1388, as in 1342, powers granted by the crown carried little weight with Douglas magnates. Instead, by 1424, the earl of Douglas was making appointments to royal offices in the south, was using royal resources to finance his own lordship and was seeking European power and status by offering his services to the kings of England and France and the duke of Burgundy.

It was in the pursuit of greater influence in the local communities of southern Scotland that Douglas lords played most heavily on the image surrounding their family. The link between the Douglases and Robert Bruce, signified by the bloody heart, allowed the Douglases to identify acceptance of their lordship by men from Lothian and the marches with allegiance to the Bruce cause. The success of a series of Douglas leaders espousing this cause gave a propaganda value to the name of Douglas itself. As early as 1327 the cry of 'Douglas' was used to effect in warfare on the borders and beyond. The fame associated with their name was part of the reason for the rise of the house of Douglas and the eclipse of other southern magnates by the lords and earls of Douglas in the fourteenth century. Despite successful careers in war and politics, which earned them earldoms before the lords of Douglas, men like Thomas and John Randolph, Malcolm Fleming and Maurice Murray failed to create dynasties which compared with the Douglases. For these men, death or capture on the field of battle ended their dynastic ambitions and regional lordship. Their lands and power passed to heiresses or to heirs incapable of defending them in the competitive environment of fourteenth-century Scotland. The only exceptions to this process in the south, the Dunbars, produced no comparable reputation as upholders of the Bruce settlement, despite active careers in war and, from 1372, their place as the heirs of the Randolphs. Instead the continued cross-border traditions and contacts of this Anglo-Scottish dynasty won no special prestige for Earls Patrick and George beyond their own Berwickshire heartlands. By contrast, through the century there was a succession of effective leaders bearing the Douglas name, who modelled their activities on the roles and reputations of their forerunners in the Bruce cause. In the 1330s, for example, William Douglas took the place of his senior kinsmen left vacant by the deaths at Halidon, while in 1347, with William in captivity, his godson and namesake returned from France to maintain a Douglas presence in the Scottish leadership. These men based their leadership on the same men and communities of the south. This continuity of Douglas leadership was no accident. Unlike the Stewarts, whose rise to regional dominance in central Scotland owed much to the survival of Robert II through all the crises from 1332 to 1390, the Douglases enjoyed no freedom from death or capture. As the dying Earl James reportedly said at Otterburn, 'God be praised, not many of my ancestors have died in their beds'. Between 1330 and 1424 four lords of Douglas met death in battle and many more men of the name came to violent ends, but the family's identity survived. This survival was not, however, the result of a greater sense of family unity or cohesion than their neighbours.

To examine the rise of the Douglases from 1330 is to examine a series of prolonged and often violent family rivalries. Douglas of Liddesdale's efforts to appropriate the principal lordships of the senior line ended finally in his death in an ambush orchestrated by his godson, William lord, and then earl of Douglas. The feud did not end with the ambush in the Forest. Liddesdale's nephew, James Douglas of Dalkeith, inherited his ambitions and claims and found an ally and patron in Archibald Douglas. Archibald the Grim was not content to be the servant of his younger cousin, Earl William, and mistrust between the two men may have begun early. It was certainly fuelled by King David II who encouraged the ambitions of both Archibald and James at the earl's expense, and though the accession of the Stewarts cemented William's pre-eminence in the marches, the crisis which followed Otterburn revealed the survival of rivalries which went back to the 1330s and 1340s. These rivalries erupted into violent competition which lasted ten years and left the Black Douglas line of Archibald the Grim as earls. It also left the Red Douglases to seek the inheritance of Earl William, their ambitions producing conflict in the 1390s and remaining a threat to Black Douglas dominance into the next century. The rise of the Black Douglases to their place as leading magnates in the south was achieved in an interlinked, internecine feud as well as in wider conflicts.[7]

Ironically it was the very frequency and violence of these family rivalries which produced much of the conscious image-building undertaken by Douglas lords. The continued importance of the Douglas name was itself a by-product of internal wrangling. The entail of 1342, though an act with parallels in Scotland and elsewhere, was principally a device to give Douglas of Liddesdale a claim to the main lands of the senior dynasty. It was used, a half-century later, to prevent the succession of a Drummond earl of Douglas. Family name and family earldom were bound together, not just by this document but by the weight placed on the achievements of lords of the Douglas name in family and national propaganda. Much of this propaganda stemmed from Earl William Douglas and, in particular, his efforts to secure loyalties strained by his conflict with Douglas of Liddesdale. William's use of the bloody heart in his arms, his attempt to follow in the footsteps of the Good Sir James's Spanish expedition and his renewal of special links with Melrose, which included the burial and honouring of his, now safely dead, godfather in the abbey, all mark out the first Douglas earl as the architect of the family's image.[8]

The value to be derived from such image-building was not lost on William's cousin, Archibald the Grim. In 1388 Archibald probably played the same cards to deny Earl William's son-in-law, Malcolm Drummond, any claims on the Douglas earldom and the main border lordships of the family. His own links to his famous father were given concrete form by the tomb Archibald had built for Sir James in Douglas kirk, and with royal backing he had been given a role in the management of Melrose in the

1360s. This unshakeable Douglas pedigree gave Archibald a clear advantage in the struggle for the lands and lordship of the family after Otterburn and, as third earl of Douglas, he also drew a conscious link with his predeccessor, Earl James. The verses composed by Thomas Barry, provost of Archibald's foundation, Bothwell college, were bad poetry and poor history. However, they made the point that Earl James died a heroic death beneath the Douglas banner, 'the stars that embellish, the heart that gives heart', surrounded by his knights, John Towers, Simon Glendinning, William Gledstone and others. Earl James was 'the true light of the Scots', a 'precious champion for his homeland', who 'suffered martyrdom to protect liberty'. He was also 'the noble scion' of his line, who 'inherited his father's courage': phrases designed to pass praise to James's own kin. Such words, which contrast with the criticism of the earl as 'rekles' in works sponsored by the lieutenant, Robert earl of Fife, were designed to add Earl James to the list of Douglas war-heroes. By encouraging the portrayal of his predecessor as a martyr for Scotland, Archibald both enhanced the prestige of his dynasty and honoured the men who fought with James, and were also named in the poem. As in the 1350s, the head of the Douglas line disguised family conflict by appealing to the past heroics and common imagery of the dynasty. For Earls William and Archibald, such continuity of image encouraged continuity of support from lesser lords.[9]

The crisis of 1388 revealed much about the character of Douglas lordship and the methods of succession to the leadership of the family. As in 1334, 1347, 1353 and in the 1390s and 1400s, control of the Douglas lands and followers was won and maintained by the magnate best able to provide the kind of lordship required in the south. This lordship centred on war. Archibald the Grim won support in the marches, not just as the son of the Good Sir James, but as a veteran leader in border warfare, a man able to protect and lead his adherents. Such a reputation had less effect in Lothian in 1388–89, but in 1400, when war returned to the province, Lothian's knights and barons looked to the earls of Douglas for 'advice and support'.

The image and power of the house of Douglas was built in war. The claims of the family to be the upholders of King Robert's legacy were given credibility by the role of James Douglas and his heirs in the physical defence of Bruce's realm. Their actions staved off English overlordship and annexation and maintained adherence to the 'Scottis fay', support of Bruce dynasty and land settlement. In their own propaganda, *The Howlat*, the Douglases were the 'werwall', the war wall, an unfailing military defence for those of 'Scottis blud'.[10] The individual magnates who built Douglas power and fame were, above all, war leaders, captains and warriors who presented their military achievements as being in the Scottish cause and earned reputations accordingly. James Douglas was 'the hammer of the English'. The knight of Liddesdale was called 'wall of the Scots', and was a lord who 'accomplished many deeds for the liberty of the realm'. Earl William 'waged fierce war ... to maintain his marches in liberty', while his son, James, was 'always hostile to the English'. Finally, Archibald the Grim's bastard, William Douglas of Nithsdale, a dark-skinned giant like his father and grandfather, was a knight who was 'indefatigable in harrying the English', and his deeds were described with pride by contemporary chroniclers. For such Scottish writers, the Douglases were a dynasty of warriors in the Scottish cause, the bearers of the bloody heart, but their martial reputations were not just for 'national' consumption. Douglas fame was earned by unceasing military activity in the marches. The fearsome reputations of Douglas lords produced a dynasty which was recognised and supported as a powerful protector and a dangerous enemy by border kindreds and communities.[11]

NOTES

1. H. Laing, *Impressions from Ancient Scottish Seals*, 44; G. H. Johnston, *The Heraldry of the Douglases* (Edinburgh, 1907), 14–15; P. Yeoman, *Medieval Scotland*, 103–104; *Scotichronicon*, ed. Goodall, ii, 400; *The Bruce*, XX, lines 575–600, 607–11.
2. *Wyntoun*, v, 256; *The Bruce*, XX, lines 624–29; S. Väthjunker, thesis, *passim*; S. Boardman, *Early Stewart Kings*, 59–61; A. Grant, 'Aspects of National Consciousness in Late Medieval Scotland', in C. Bjørn, A. Grant and K. Stringer (eds), *Nations, Nationalism and Patriotism in the European Past* (Copenhagen, 1994), 68–95; R. J. Lyall, 'The Lost Literature of Medieval Scotland', in J. D. McClure and M. R. G. Spiller (eds,) *Bryght Lanternis: Essays on the Literature of Medieval and Renaissance Scotland* (Aberdeen, 1989), 33–47; S. Boardman, 'Chronicle Propaganda in Fourteenth Century Scotland: Robert the Steward, John of Fordun and the 'Anonymous Chronicle', in *S.H.R.*, 76 (1997), 23–43; M. Brown, 'Rejoice to Hear of Douglas': The House of Douglas and the Presentation of Magnate Power in Late Medieval Scotland', in *S.H.R.*, forthcoming.
3. *R.R.S.*, vi, no. 51.
4. *The Bruce*, XX, lines 190–248; *Longer Scottish Poems*, 63, lines 456–57.
5. *R.R.S.*, vi, no. 451; *The Bruce*, XX, line 234.
6. A. Grant, 'The Otterburn War', 63.
7. Froissart, *Chroniques*, iv, 15. The Randolph line died with Earl John at Neville's Cross. The Steward made clear his ambitions on their northern lands by marrying John's widow. Maurice Murray also died at Neville's Cross, most of his lands passing ultimately to Archibald the Grim. The fate of the Flemings has already been noted above, see Chapter 3.
8. Fraser, *Douglas*, iii, no. 23; *Scalachronica*, 175; *Scotichronicon*, ed. Goodall, ii, 357; *Scotichronicon*, ed. Watt, vii, 297–301; *R.R.S.*, vi, no. 51.
9. *Scotichronicon*, ed. Goodall, ii, 406–13; *Scotichronicon*, ed. Watt, vii, 420–43; D. E. R. Watt, *Scottish Graduates*, 31–32; *The Bruce*, XX, lines 595–600; *R.R.S.*, vi, no. 450.
10. *Longer Scottish Poems*, i, 60, line 382.
11. *Scotichronicon*, ed. Goodall, ii, 329, 347, 400, 403; *Scotichronicon*, ed. Watt, vii, 68–69, 138–39, 270–71, 274–75, 402–403, 410–11; *Chron. Fordun*, i, 346.

CHAPTER SEVEN

The War Wall:
The Douglases at War

'WITH DINT OF DERF SWERD'

The fame of the Douglases, their lands and followers was won 'with dint of ... derf swerd'.[1] Blows from Douglas swords established the family's place in Scottish political society. The rise of James Douglas from dispossessed baron to magnate and royal lieutenant was based on his leadership in war, winning lands for himself as well as his royal lord. Similarly, returning from exile, the two William Douglases, the knight of Liddesdale and the lord of Douglas, built their power on the leadership of the war against England in the front line of conflict. Even from the 1380s, when most of the south had been returned to the 'Scottis fay', Black Douglas leadership in continuing warfare remained vital to their exercise of lordship. Through almost all of the fourteenth century, Scotland south of Forth was a land of war. The patterns of local power, of royal authority and of political society in general in southern Scotland were linked to the course of the wars with England. An event like the capture of Roxburgh castle by Alexander Ramsay in 1342 was not just a military success in the struggle against the English. It also altered the balance of power in the Scottish marches, signalling a challenge to the leadership of William Douglas of Liddesdale and precipitating a murderous conflict. Conversely, Scottish political rivalries could affect allegiance in the marches. The feud between Douglas of Liddesdale and his godson, the lord of Douglas, led to the former seeking the English king's lordship. Malcolm Drummond also turned to England in 1389 and George Dunbar became an English liegeman following the master of Douglas's seizure of his main castle. These events produced or threatened conflicts which combined internal Scottish disputes with Anglo-Scottish warfare.

The link between national warfare and southern politics was demonstrated most dramatically by the impact of military disaster on the power of Douglas lords. The pursuit of regional dominance through leadership in war risked the disintegration of that dominance in the aftermath of defeat. In 1333 after the death of the Tyneman at Halidon, in 1346 after the capture of the knight of Liddesdale at Neville's Cross, and in 1388 following Earl James's death at Otterburn, their leaderless followings in the marches and Lothian collapsed in the face of the English advance or fragmented between rivals for the family's power. In 1402 at Humbleton, the Black Douglas line, which had so successfully exploited the aftermath of Otterburn, only narrowly escaped the high price charged for the link between war and Douglas lordship. The readiness of a dynasty of Douglas lords to lead from the front in the war against England, displaying their skills as commanders and fighters, brought these lords great rewards for their efforts but, as many of them discovered, they risked capture or death and the collapse of their hard-won regional lordship.

The wars in which these Douglas magnates fought were part of a sporadic conflict with England which had significance at international, national and local levels. Scotland was one of a number of theatres in the escalating political and military rivalry between the French kings and the house of Plantagenet, kings of England and dukes of Gascony. Like Gascony, Scotland was a flashpoint in Anglo-French relations from the 1290s, and provided the impetus for sustained conflict in the late 1330s. From 1337 onwards direct and indirect Franco-Scottish military co-operation was an occasional but vital element in the Hundred Years' War in both kingdoms. It drew English resources away from the north and saw English ambitions in Scotland tempered by diplomatic efforts to win the Scots from their support of France. As will be discussed later, this 'auld alliance' held a special significance for the house of Douglas, which was keenly aware of this European dimension to Anglo-Scottish conflict.[2]

In the eyes of Scottish historical writers of the late fourteenth century, though, the wars in which the kingdom was involved were a clash of national communities. The Scots defended the 'liberties' of an independent realm against the English king. English aims were the absorption of Scotland under direct control or the recognition of the English king as feudal overlord of Scotland, to be achieved between 1333 and 1356 by the support given to Edward Balliol. Royal-led expeditions to Scotland between 1296 and 1400 were aimed at forcing the Scots to submit to these aims. In return, major Scottish armies crossed the border, not to challenge the national allegiance of northern England, but to disrupt the region and display Scottish military power. In both attack and defence, the Douglases played a major role. In 1333, 1388 and 1402, Douglas magnates headed large Scottish armies in battle and, as has been stressed, it was in terms of national defence that the family presented its achievements, fighting to win back men to the 'Scottis fay' and the king's peace.[3]

This large-scale, national warfare had a clear impact on the course of Anglo-Scottish conflict and it was certainly the part of that conflict which attracted European attention. Destructive campaigns by armies numbering 10,000 men and, still more, victory in a major battle could change the course of the war. A success like Bannockburn had a huge moral and political impact, winning power and prestige for Bruce. Defeats like those of Halidon, Neville's Cross or Humbleton deprived the Scots of leadership and brought a collapse of resistance in their wake, especially in the south. The English were given an opportunity to capture strongholds and force the submission of local men to the English king. They also gained prisoners, like Douglas of Liddesdale, the fourth earl of Douglas or even David II, as a source of ransoms and, more importantly, a means to exert political leverage on Scotland. Despite these risks, in defence or attack the Scots felt it to be vital to produce armies of a size to meet major English forces in the field. The ability to recruit such forces was a mark of the kingdom's ability to stand as a European realm and have an impact on European warfare. From the battle of the Standard to Flodden, Scots leaders, like their counterparts across the west, saw battle as the ultimate test of military power and political strength.[4]

However, as in the rest of medieval Europe, campaigns by such armies were rare and battles even rarer. Instead warfare was waged, most frequently, with forces numbering tens and hundreds not thousands, and at the level of local communities, not of kingdoms. It was in this type of warfare that the Douglases won their fame and built their power, and in this type of warfare too that Scots achieved their real successes in the fourteenth century. Bannockburn was an exception as a complete battlefield victory. From Dunbar to Humbleton, the Scots' experience on the stage of major warfare was disastrous. The ground lost on the fields of defeats like Halidon and Neville's Cross was recovered in sustained local warfare which chipped away at English bases and adherents in Scotland, grinding down English resources and morale over a number of years. The masters of this style of warfare, James Douglas, the knight of Liddesdale, and the first earl of Douglas were the makers of Douglas power in the marches, and their success in war and politics was closely intertwined.

Scotland south of Forth had a longer and more intensive experience of war with England than the rest of the kingdom. From the fall of Perth in 1339, the war was chiefly waged in Lothian, the marches and Galloway. These lands contained the principal remaining English garrisons and the remaining communities in English allegiance. Such garrisons and communities were those most easily supported from England, and from 1334 the six sheriffdoms of southern Scotland formed what the English regarded as a pale, a defended zone ruled by the crown through its own officials and noble vassals. The aim of the Scottish war effort in Lothian and the marches was the recovery of these communities, which the English held, wholly or in part, not just in the 1330s but before 1314 and after 1346. Areas which had been part of the thirteenth-century kingdom were to be restored to their allegiance to the heir of Bruce as Scottish king, and protected against attempts to force them back into English allegiance.[5]

Yet, although expressed in terms of national loyalties and objectives, the localised warfare waged across the south for much of the century from 1296 was dominated by men like the Douglases. The successful return of southern lords and lands to the political and social structures of the kingdom depended on the ambitions, skills and resources of local magnates. Their methods and goals essentially determined the character of war in the region. In some areas, for example, the Douglas name alone was capable of winning local support. The earliest indication of this was in Douglasdale itself. In 1307, James Douglas won his first followers from among the tenants and neighbours of his ancestral lands. Forty years later, James's nephew and heir, William, the future first earl, repeated the feat. On his arrival in Scotland after fourteen years in exile he made first for 'his native land' of Douglasdale, where he also received support. Of even greater importance to the Douglases' military role in the south, throughout the fourteenth century, was the support and secure base provided by the Forest, the uplands of the middle marches. James Douglas's occupation of the district in 1307 itself exploited the support which earlier leaders like William Wallace and Simon Fraser had received from local men. James's use of the region as a military base was turned into formal control by royal grants of lands from Eskdale to Jedworth forest and, from the 1320s, the Douglas name had its own effect. William Douglas of Liddesdale directed his initial military efforts to the recovery of the Forest, and by the early 1340s his 'company' included men of the region. When William lord of Douglas returned to the marches in 1347, he too headed for the heartland of the Forest, where he was greeted by 'luffande folk', men who recognised and welcomed the credentials of the Douglas name and were ready to renew service to such a lord. By the 1380s, the same effect extended across the central borders. The succession of Archibald the Grim to the Douglas lands of the middle marches owed much to his reputation and experience as a border lord and something to his fame as the son of Sir James Douglas. These communities were responsive to the appeals of Douglas lords for support, identifying in them magnates able to provide protection from attack and to lead them in war. It was as the means of winning such support from border communities that the house of Douglas presented itself as a dynasty of warlords in a patriotic cause. For the men of the Forest, as early as the 1340s, the return of Douglas leadership was seen as both the alternative to English lordship and as the opportunity to end allegiance to England.[6]

The readiness of the Forest men to back a series of Douglas war leaders was a vital element in the war in the marches. The Forest was a large upland region of difficult terrain, which the English found impossible to subdue effectively. Attempts to base forces in the Forest in the 1300s made little lasting impact and led, in 1309, to the ambush and defeat of such a force by James Douglas. After this the English were reduced to launching sporadic raids on the Forest and the surrounding districts. Such attacks did not affect the value of the area as a refuge for men and herds or as the base for Scottish offensives into Lothian, Galloway, lower Teviotdale and England itself. Writing in the early 1340s, an English correspondent informed Edward III's council that, since the Scots had, for once, vacated the Forest, a 'certain secret matter which was discussed in the king's presence' might be attempted. This enterprise 'if it could be accomplished ... may have as great an effect on the war as a battle'. The goal of such a plan may well have been control of the Forest, which would have been a major blow to Scottish efforts in the marches. In the winter of 1341–2, Edward III based himself at Melrose and made an attack on the Forest at a time when the Scots, who lived off the land, were traditionally rendered vulnerable by lack of supplies. If this was the secret plan, it failed dismally. Efforts to disrupt the control of the Scots, and especially of Sir William Douglas, over the Forest were thwarted, not least by the weather, and King Edward returned to England 'in a melancholy with them that movid hym to that jornay'.[7] The security of the Forest allowed its inhabitants to give ready support to a series of Douglas lords without fear of full English retribution in 1342 and later in 1356. In most other regions of the south, choices of allegiance were more constrained by fear and force, English and Scottish. The only equivalent to the Forest was Galloway, where there were also upland refuges from attack. For the Galwegians, however, adherence to their Balliol lords aligned them with the English. In provinces like Galloway, where there was sympathy with English war aims, or in regions like Annandale, Teviotdale and Lothian, while they were under effective English control, Scottish allegiance, often represented by Douglas lordship, had to be imposed by 'force and negotiation'.

Scottish success was achieved by disrupting English military and political control in these districts. The cornerstone of this control was not the large armies brought north by English kings and lieutenants. These large paid forces were expensive and therefore could only be fielded infrequently and for limited periods. Instead a permanent military presence was provided by garrison troops. The spread of garrisons reflected the fortunes of English authority in Scotland. In the 1300s and after 1333 the English garrisoned the south at will, and in the years after Halidon, Edward III and his chief lieutenants refortified the castles at Edinburgh, Roxburgh, Jedburgh and Lochmaben. These strongpoints and Berwick were intended to protect the English pale, but despite the reverses caused by Neville's Cross, the Scots gradually picked off these castles from 1341 onwards, as they had between 1311 and 1314. By 1409 the English retained only Roxburgh and Berwick.[8] This process was the key to Scottish military recovery in the south. English garrisons were not passive, defensive blockhouses but bases for military and political control of the surrounding regions. Garrisons of about one hundred and thirty soldiers, comprising mounted men at arms, archers and hobilars, lightly equipped horsemen ideal for local raiding, were stationed at Edinburgh, Berwick and Roxburgh in the 1330s. These 'castle-troops' were professionals, English, Scots and even Germans. In 1341 the men of Roxburgh were described as '*compagnons*', a term used on the continent for mercenary bands who fought for profit. In the local warfare of southern Scotland, forces of this size and character could have a major impact, defending Scots in English allegiance and the English borders from attack and leading raids against the neighbouring Scottish communities. In these roles they would provide a professional core to stiffen locally raised forces of Scots or borderers. In 1347, for example, the constable of Roxburgh, John Copeland, led 'many soldiers' of his garrison 'into Teviotdale, hoping to confirm the men of Teviotdale in English allegiance'. The force of local Scots and English garrison-troops was routed by William lord of Douglas, but it was this type of operation which was essential to maintain, not just the military, but the political influence of the English crown in southern Scotland.[9]

English garrisons also had a crucial political function. Castles formed the centre of English attempts to manage the surrounding sheriffdom, binding Scots to English allegiance. This function was not simply carried out by alien military rule. In the 1330s the sheriff and constable of Edinburgh was John Stirling, a Scottish knight. His garrison in Edinburgh was about a third Scottish and included knights from Lothian, like William Ramsay and John Crichton, men from the Borders and even Edinburgh burgesses. Stirling himself lived outside the castle, suggesting an, ultimately misplaced, sense of security in the loyalty of the burgh. John Copeland, the Northumbrian constable of Roxburgh for much of the period from 1346 to 1363, also sought to maintain the loyalty of the local Scots by acts of lordship. In 1358 he granted lands in Attonburn to John Kerr, and, despite the efforts of William lord of Douglas over the previous decade, a significant group of local landowners, Kerr, William Rutherford, James de Lorraine, William Gledstanes, Robert Colville and others, remained in English allegiance. In part, this reflected English military control. Change of loyalty would mean the loss of lands for these men. However, efforts were made to maintain standards of rule and prevent the acts of one constable in the 1360s, who 'oppressed the people under colour of his office', retaining 'Scottish grooms and other unfit persons' as his garrison. From this perspective, Edward III's confirmation of the ancient rights of Teviotdale in 1356 attempted to foster a sense of English lordship which would survive, in a limited sense, until the fall of Roxburgh in 1460.[10]

The aim of Scottish war leaders in the fourteenth century was to break down the bonds of allegiance to England and show that they and not the English could guarantee the security of lands and livestock. Much of the small-scale warfare in the south was designed to weaken the power of English garrisons, either as a prelude to an assault on the garrison's castle or to reduce English control in the surrounding region. Barbour's account of James Douglas luring the garrison of Douglas castle from behind its walls and into an ambush by displaying only part of his force described a tactic repeated in the 1350s by William lord of Douglas to trap the garrison of Norham. With greater significance, in 1337 Douglas of Liddesdale caught John Stirling outside Edinburgh with a force from the garrison. After 'mekyl peyne', Stirling was captured and his retinue killed or taken. Although Douglas's threat to hang his prisoner if Edinburgh castle was not surrendered did not work, defeat in a skirmish involving less than a hundred men dealt a major blow to English influence in Lothian. A similar skirmish was crucial in the fall of Hermitage in the 1330s. Douglas of Liddesdale ambushed and captured the supplies being sent to the castle from Melrose, and the isolated and hungry garrison quickly surrendered. For troops in the frequently plundered marches, the maintenance of supplies and communications was vital. In 1398 the master of Douglas also targeted the food, fodder and water supplies of Roxburgh. His destruction of the town and bridge of Roxburgh, only twenty years after George earl of March had done the same, impoverished, isolated and demoralised the English and demonstrated their inability to defend themselves or their local allies against attack.[11]

Garrisons in this situation were more vulnerable to direct assault. The attack on Lochmaben, launched by Archibald Douglas and George earl of March in 1384 at the expiry of the formal Anglo-Scottish truce, had been prepared by local attacks by the two magnates and their local adherents over the previous fifteen years. Besieged by Archibald, with the earls of Douglas and March, the garrison surrendered after eight days. The success of the siege, interpreted by the English as the result of treachery, was caused by the garrison's 'shortage of supplies'. Despite the successful captures of Berwick in 1319, Perth in 1339 and, finally, Roxburgh in 1460 after formal sieges, most such ventures, like the attempts on Roxburgh in 1385, 1417 and 1436, met with failure. Instead the Scots most frequently took English strongholds by surprise attack. The links between English garrisons and local Scots often proved decisive. In 1341, after several failed attempts, William Douglas took Edinburgh castle with the help of a number of burgesses, one of whom, William Fairley, was also a member of the garrison. In the same way, Alexander Ramsay captured Roxburgh the next year at 'the council and instigation' of a local man, 'Hude of Ednam'. Similar local help had aided the capture of the same castles by Thomas Randolph and James Douglas in 1314. All these attacks were surprise assaults on unwary garrisons. Although the capture of Jedburgh castle in 1409 by local lords was probably also achieved by sudden escalade, the Scots' failure, elsewhere, to repeat the achievements of 1314 and 1341–42 after 1346 suggests greater English precautions against these tactics.[12] The capture of English-held strongholds had a major impact on the surrounding regions. The fall of Roxburgh to James Douglas and, later, to Alexander Ramsay brought the men of Teviotdale into Scottish allegiance, while William Douglas's capture of Edinburgh effectively ended English influence in Lothian. Except for large-scale and short-term English invasions, as in 1356, 1385 and 1400, Lothian was relatively secure for the rest of the century. Fears concerning the possible delivery of Dunbar to the English in 1400 related to a renewal of war and the dispossession or defection of Lothian men in the face of a local English garrison. Even forces based in fortalices like Fast and Cockburnspath committed 'many evil deeds in Lothian'.[13]

However, while the recapture of English-held castles was a goal of the Scots into the fifteenth century, this was not the only means of winning the war in the south. Much fighting in the region took the form of local raiding, attacks by small, mobile forces, whose aims, like the Scottish raid on Galloway in 1332, which 'burned, spoiled and led off the cattle but killed few men', were to impoverish as much as kill the enemy. Such attacks had parallels with warfare in the lordship of Ireland and much of the fighting in Gascony. As in these realms, warfare in the Scottish marches had political as well as purely economic or military goals. In 1347, after defeating local Scots and English troops, William lord of Douglas kept his 'gret company' in lower Teviotdale, occupying the district until local men submitted. The raid displayed Douglas's ability to deprive the men of Teviotdale of lands and goods, and to defeat the English, with the aim of enforcing local support. The same balance of fear and protection determined allegiance in many border communities, and Galloway, Annandale, Teviotdale and the Merse, in particular, experienced periods of raiding by both English and Scots. It was the Scots who won consistent and lasting successes in this war of raid and counter-raid. From 1350 onwards, the area of the marches acknowledging Scottish lordship steadily increased as a result of local conflict. In 1353 and 1354 the principal captains of Galloway submitted to William lord of Douglas in the wake of a major attack on the province. A decade later this 'pacification' was completed by Archibald the Grim, whose lordship bound Galloway into Scottish allegiance. The 1350s also saw men from Tweeddale submit to William Douglas, and the same lord recover Hermitage castle and extend his lordship over upper Liddesdale. The final period of Scottish success, from the late 1360s to the 1380s, witnessed sustained small-scale Scottish attacks on Annandale, lower Teviotdale and the Merse which eroded the number of men and settlements in English allegiance in these regions. In 1384, in the last campaign of his long and successful career, Earl William Douglas brought Teviotdale back to the 'Scottis fay', this time permanently. The flow of men into Douglas's service and the 'king's peace' marked the final success in this long struggle. Though Roxburgh held out and Teviotdale was subjected to English attacks during subsequent decades, its principal men remained in Scottish allegiance. Moreover the recovery of Scottish communities exposed the north of England to attack. The period 'when Teviotdale was English' was a time of relative peace and prosperity for the men of Redesdale. When in the 1380s both Teviotdale and Annandale were brought back under Scottish control, the neighbouring English communities lay open to sustained and ruinous Scottish raiding.[14]

For regions like Galloway and Teviotdale, entry into Scottish allegiance was marked by submission to the magnate who had led the attacks on them, and who often had formal rights of lordship granted by Bruce or his heirs. The men of such regions would hardly welcome this new lord as a simple liberator. For the Douglases, the submission of local men, though returning communities to Scottish allegiance, also represented the winning of 'lordschips sere', of many lordships, in warfare against the 'sonnis of Saxonis'. War and political power in the marches remained bound together even after the recovery of the principal border communities in the 1380s. In these decades and earlier, attacks on England and the defence of southern Scotland against attack had a political, as well as military, purpose for leaders like the Douglases. Raids on England, launched with greatest effect in the years of Scottish ascendancy between 1314 and 1327, and between 1374 and 1389, were designed to weaken the enemy's morale and resources by winning plunder and extracting blackmail and ransoms from English border communities. These plundering attacks on England included even major Scottish invasions, like those of 1385, 1388 and 1402, involving thousands of men. Although the sustained raiding of Robert I's reign pressurised northern English communities into seeking peace or even lordship from the Scottish king, like contemporary English warfare in France, the principal goal of such attacks was to impoverish enemy territory and enrich the raiders. The financial and political benefits of such an expedition for its leader had a value in maintaining support and winning prestige, and as the Black Douglases found in the 1390s, failure to provide such leadership could weaken the political power of a border magnate.[15]

The same was true in defence. In all medieval societies, lords had a duty to defend their own lands. To maintain their hard-won leadership in the south, Douglas magnates could not simply withdraw in the face of major English armies, the strategy of most Scottish leaders. Instead, although they avoided open conflict with large English forces, many Douglas lords maintained a military presence in the south. In 1317, James Douglas, with two hundred men, ambushed a small English force at Lintalee, near Jedburgh, and then harried the retreat of the main army. Following a similar strategy, Douglas of Liddesdale harassed English campaigns into Carrick and Clydesdale in 1336: 'He approached the English army secretly and killed many men'.[16] The greatest political value from this kind of warfare was probably extracted by William, lord and future earl of Douglas, in 1356. Edward III's winter invasion of Scotland, 'the burnt Candlemas', caused massive destruction and had political and military impact on the south. According to an English account, Edward received the submission of 'many Scottish magnates', probably including men from Teviotdale, while 'the people took their goods beyond the Scottish sea'. In this situation, William Douglas 'sent letters to the [English] king that none should come to the king's peace'. Through the rest of the campaign, Douglas harried English foragers, destroyed a force detached to raid Ettrick forest and, on Edward's withdrawal, 'restored the *Scotos Anglicanos* to the [Scottish] king's faith'. William had preserved and even extended his lordship in the south in the face of massive opposition. Lesser English attacks could be met directly. While he sat out the English invasion of 1400 safely behind the walls of Edinburgh castle, the next year Archibald fourth earl of Douglas led a force from the castle to rout a raid of Hotspur and the Dunbars on Lothian and secure his place as principal defender of the sheriffdom. Similarly, the active defence of the east march by the earl's father in 1389 formed a valuable display of his credentials as a man capable of protecting the marches, following the dispute over the Douglas succession during the previous winter.[17]

The depiction of the house of Douglas as the war-wall of Scotland was not just poetic propaganda from *The Howlat*. It was a claim which rested on the family's military achievements, crucial in forming the loyalties of southern Scotland, and also on their management of border defence. In the late 1440s, as *The Howlat* was being penned, the men of the west march met at Lincluden to record the statutes 'ordanit to be kepit in blak Archibald of Douglas dais and Archibald his sonnis dais in tyme of weirfar'. These military laws dealt with the raising and discipline of raiding forces, the crucial issue of the division of spoils and ransoms and the punishments for disobeying orders, especially the obviously unpopular command to dismount and fight. They also specified the maintenance of 'baills', beacons warning of attack, on peaks throughout the west and the duty of borderers to take up arms in response to their lighting. Such laws, which indicate the organisation behind the waging of war in the marches, were identified with the Black Douglas wardens. It was Archibald the Grim, warden of the west for nearly forty years, who was seen as the architect of these statutes and they reveal the authority established by his family in attack and defence in the front line of Anglo-Scottish conflict.[18]

WARDENS AND WARLORDS

Like Archibald the Grim, who 'had everywhere in his following a large company of knights and men of courage', the Douglases waged war at the head of a military retinue. It was the ability to raise and maintain men in arms which gave Douglas lords their place in the Scottish war effort. Archibald's own career had begun as a household knight in the service of Douglas of Liddesdale and later of Earl William in the warfare of the 1340s and 1350s. While such households had political, as well as military, functions, their origin and continuing character was as an armed force. The companies gathered by James Douglas in 1307 and by William Douglas of Liddesdale in 1334, which formed the basis for their subsequent retinues, were made up of desperate, dispossessed men looking for a leader capable of protecting and maintaining them. Though later Douglas retinues were based on more settled relationships, their military purpose was never far away. In 1356 William lord of Douglas led his 'knights' to France and, at the battle of Poitiers, they 'seized hold of their lord ... and removed him from there forcibly' at the cost of their lives or liberty. At Otterburn, Earl James also entered battle at the head of 'his knights'. The earl's death was, in part, due to his impetuousness in leaving the safety of this following. Though two knights fell in his defence, James's household, whose stated duty was to protect their lord, could only recover his dying body. In the local warfare of the south, these battlefield bodyguards provided the basic unit of military organisation and the basis of the Douglases' military power.[19]

The Douglas retinue in war reflected the structure of the family's political following. Service in war and service in council was performed by the same close groups of adherents. The family's military following included kinsmen, the brothers and nephews of Douglas of Liddesdale or Archibald Douglas, the Lindsays and Sinclairs, who served Earl William Douglas, for example. Tenants also served in a military capacity, though the vital early support which James Douglas received from men of his ancestral lands, like Thomas son of Richard and Sim of Leadhouse, came less from formal obligations than personal ties to the Douglas line. The bulk of the family's military household was built on such personal connections to friends and allies. Such relationships could, however, have an overtly military foundation and character, as shown by the service of John Swinton to a series of Douglas lords between 1378 and 1402. Swinton was recruited and maintained by these lords as an experienced soldier, a veteran of English campaigns in France. His employer, John of Gaunt, duke of Lancaster, had paid double the normal fee to retain his services, and Swinton returned to Scotland in 1378 with a following of sixty men. Earl William clearly sought Swinton's service, brokering the deals by which the knight recovered his family's Berwickshire lands, and, by the 1380s, Sir John was part of the inner circle of the Douglas affinity. After Earl William's death, John married his widow and, as 'our dearest father', accompanied his stepson, Earl James Douglas, in his attacks on England. Otterburn, where Swinton played a prominent part, severed the link of service to the Douglas earls for a decade. Significantly, the link was renewed with the resumption of major warfare by the fourth earl of Douglas in 1400. Swinton's appearance as a well-rewarded councillor of the new earl was related to his military experience. He was given lands of military significance and accompanied Douglas on campaign before leading his men to Humbleton, where, though killed, he showed himself the only Scottish captain with any initiative. Swinton was not the only experienced soldier on Earl Archibald's council or in his retinue. The presence of the crusader, John Edmonston, and the border warrior, William Stewart, both veterans of Otterburn, and of Robert Stewart of Durisdeer, a companion of the earl's half-brother, William, in wars in Ireland, England and Prussia, added to the military atmosphere of the household. It seems likely that, not just the second Archibald, but Earls William and James valued and actively sought out such veterans as their advisers and lieutenants.[20]

Though men like Swinton and Edmonston attended the earls of Douglas in military as well as political endeavours, there is no evidence that they received regular fees to retain their services. Swinton, who had been retained by Gaunt for war by paid contract, provided much the same service in Scotland for less structured returns. Both Edmonston and Swinton received a number of grants of property from Earl Archibald, but while these sealed the close relationship between magnate and followers, they were given for unspecified services. In Swinton's case, moreover, the new tenant paid for his lands with silver vessels worth 500 marks. As has been mentioned, the fourth earl's search for funds may indicate his use of money to maintain his massive political affinity, but there was no stated link between payments in land or cash and military service under the earl. Even a professional like Swinton served the earl in return for protection and the expectation of reward. In 1372, however, Earl William Douglas issued letters of retinence for life to James Douglas of Dalkeith, paying James 600 marks in return for his service, 'in peace and war', with eight men-at-arms and sixteen archers. This exceptional document was not purely designed to organise military service, however, but had political overtones. James formally agreed thereby to enter the retinue of his long-standing enemy after the death of his principal protector, David II. James wished to avoid the kind of mauling he would receive from William's son, George earl of Angus, in 1398. The terms of the agreement may have been borrowed from English practice as a means of expressing the desired relationship, but, if a political as much as a military agreement, the indenture also suggests the formalisation of the place and structure of armed retinues in Scotland. Only two years earlier James Douglas had been granted 'the leadership of all the men of his lands' by King David. The grant had military overtones, specifying James's right to lead his men on 'ridings in the marches' and denied the earl's rights as private lord and march warden. The 1372 indenture was designed to establish the earl's leadership over James and his men in both the politics and warfare of the south, and to counter the dead king's grant. There was no extensive system of paid military retinues in Scotland. In the 1380s Earl James Douglas granted Drumlanrig to his bastard son in return for 'the service of one knight in our army'. Military service rested on a variety of contacts tenurial, financial but mostly personal. Such variety related to the origins of Douglas retinues in the warfare of the 1300s and 1330s. Lords like the knight of Liddesdale lacked the regular resources to support retainers with cash or land, and instead followings, so-called 'schools of knighthood', consisted of volunteers serving an effective leader in hope of receiving the profits of war, booty and ransoms. The immediacy of war in southern Scotland throughout the century, which meant the need for protection and the profits of war were never far away, perpetuated this form of military service.[21]

The 1372 indenture is, however, useful in providing evidence of the scale of Douglas retinues. The twenty-four men promised represented a significant force in the context of local warfare. Douglas of Liddesdale had won his successes with a 'rowte', an armed following of between thirty and a hundred men, while his godson, Earl William, took a band of forty knights on the ill-fated pilgrimage which ended at Poitiers. In Scotland, the earl was probably accompanied by a larger force, and these organised companies of well-equipped but mobile horsemen were strong enough to win the victories of the Douglas family. This armed household played a similar role to the English garrisons and warden's retinues by maintaining a regular military and political presence in the south. It also allowed Douglas lords to react at once to crises or opportunities in the war. In February 1401, 'with his men from Edinburgh castle', Archibald earl of Douglas defeated a raid on Lothian by Hotspur and George Dunbar, while in October of the same year he broke up negotiations on the border and, with the 'whole army ... arrayed as if for war', led the force on an immediate raid across the border. This small army, including Swinton and William Stewart, was probably the earl's own retinue, performing its duty by escorting the march warden to the border negotiations.[22]

Most importantly, the retinues of the Douglas earls served as the core of much larger forces. The Douglas household was itself composed of smaller retinues. For example, William Ramsay led a force of ten knights in the company of Earl William Douglas in 1356, and men like Ramsay, James Lindsay of Crawford, who fought with the earl's knights at Otterburn, and Sir John Swinton, who had a company of sixty men on his return to Scotland, were capable of raising considerable forces. As has been stressed, the extension of political influence was closely bound up with that of military success and military service. In the 1350s, 1370s and 1380s, and again after 1400, Douglas political influence in Lothian was linked to their activities in war. The earls tapped into military structures which had grown up in the 1330s, when leadership came not from magnates but local knights. Alexander Ramsay led his neighbours, John Haliburton, John Heryng, Patrick Dunbar and Laurence Preston, with 'the young men of their households and kin ... and defeated the English'. In the 1400s in the face of renewed attacks on Lothian, local barons and knights again led their own retinues in war. The raids of 1402, commanded and carried out by the Haliburtons, Hepburns, Lauders and 'the flower of the fighting men of Lothian', displayed the continued power of such men in warfare. These knights followed the leadership of Earls William and James, and accepted the 'advice and support' of Earl Archibald. In the 1350s, John Haliburton, William Towers and William Ramsay all served Earl William in war, and in the 1380s Earl James continued to lead the men of Lothian. Ties of kinship, landholding and counsel between these knights and the Douglas earls were matched by service in war. In military terms, it was the harnessing of these Lothian knights to the pursuit of their war aims in the marches which was the principal achievement of the house of Douglas in warfare. The support obtained from a community, which, after 1341, was secure in Scottish allegiance, was a crucial factor in the recovery of the south for both kingdom and Douglas family, and bonds of landholding and loyalty were fostered to perpetuate this link.[23]

The marches themselves were a source of military manpower for the Douglases, though one where service was bound up with issues of loyalty and physical survival. Success in war brought Douglas lords the service of borderers who had submitted to them. The men of Teviotdale, who acted as Douglas of Liddesdale's followers in the early 1340s, defected from him and Scottish allegiance after Neville's Cross. Earl William's campaign of 1384 did not just force men like Robert Colville and William Stewart to submit, they also joined the earl's service in war. Border lords were more than victims of the war which was fought over their lands and loyalty. They were active participants, 'robbers and freebooters', launching attacks regardless of formal allegiance. Already in the 1340s the Kerr brothers had earned the reputation of 'the greatest enemies' of the English in the Forest, and by the 1370s lords like John Johnstone in the west and John Gordon in the Merse were leading their own men in attacks on English-held lands as well as following the wardens. Such attacks were not simply border lawlessness; English evidence shows border knights planning '*chivachie*' together and with magnate encouragement. As the Lincluden statutes show, the Scots leaders, who enjoyed greater success in 'ridings' in the later fourteenth century, were also well organised and in the marches, even more than in Lothian, structures of lordship revolved around leadership in war. The regional power of magnates like the earls of Douglas and March depended on their relations with local captains and the protection and profit they could provide such men in warfare. The importance of William Stewart of Teviotdale in particular was a result, not of his landed power, but of his connections with Teviotdale kindreds such as the Rutherfords, Colvilles and Turnbulls. These connections, probably based on his abilities as a warleader, made Stewart a vital member of the Douglas affinity. His death, executed by the English on the grounds that he had deserted his native allegiance, stemmed from a desire to eliminate an enemy whose influence had prompted the renewal of Douglas aggression in the marches in 1398.[24]

This accumulation of adherents and allies in war and politics was the basis of the military power of the Douglases. As early as 1335, William

Douglas, Earl Patrick Dunbar and Alexander Ramsay led 800 of their men, 'the floure of the south half the Scottis see' to Culblean in Aberdeenshire. In 1347, William lord of Douglas assembled a 'gret company' for his campaign in Teviotdale which seems to have included his own retinue from France and Douglasdale, the men of David Lindsay, his uncle, and probably of other Lothian lords, a company of Edinburgh burgesses and the supporters he had just won in the Forest. Similarly in 1401, Archibald fourth earl of Douglas, setting off with his retinue from Edinburgh, gathered enough support from the 'country' to rout a raiding party of 2,000 men. Such armies were built on the structures of lordship fostered by the Douglases in politics as well as war. They were given recognition by the crown in the 1350s when William Douglas was not just made earl but also given the right to lead the men of upper Clydesdale, Roxburgh, Selkirk and Peebles.[25] Powers like these complemented and reinforced existing military relationships and confirmed the place of Earls William and James as the embodiment of the Bruce cause in the disputed communities of the south.

Powers of leadership also raise questions about the organisation of Scotland for war in the fourteenth century. The ideal, as presented by Barbour, was the royal leadership of Robert Bruce, which fulfilled the king's duty to defend the community in person. The king organised and directed the war, summoning his subjects to his army from the whole kingdom or from specific regions. Though this 'common army' could only be maintained in the field for a short time, it represented the effort and support of the community in war. As Barbour stressed, the king depended heavily on his noble lieutenants, who upheld his cause in local war, but Robert maintained control by his personal leadership in Scotland, England and Ireland. Delegation in war was unavoidable. The English recognised this fact, appointing wardens in east and west marches in 1315 to oppose Scottish attacks and, soon afterwards, James Douglas was given an equivalent post. In Scotland, however, this office did not reappear until the 1340s, and recognisable powers in war and justice only gradually became associated with it.[26]

The revival of the Scottish wardenships in 1340 showed that, as in political society, the 'ideal' of Robert's reign was impossible to maintain without the king. Between 1329 and 1341 and from 1346 to 1357, war was directed by a series of lieutenants, including Archibald Douglas the Tyneman, Andrew Murray, John Randolph and Robert Stewart. These lieutenants could raise armies from across Scotland, as at Dupplin, Halidon and the Steward's siege of Perth. They also led campaigns south of Forth, like the raids on Galloway and Annandale and the attacks of 1335 and 1337 on Edinburgh. However, except for the Tyneman, the interests of these men were centred north of Forth or west of the Clyde, and their intervention in the south was infrequent and inconclusive.[27] The favourable accounts of later chroniclers cannot conceal the facts of the Steward's guardianship after 1346. His brief visit to Berwick after its capture in 1355 and his son's campaign in Annandale hardly represent active leadership in war. Instead Robert's concerns were with the entrenchment of his family in the vacant earldoms north of Forth. Even in the 1330s the war in the south had been run by local leaders. Though men like William Douglas and Alexander Ramsay recognised the authority of the lieutenants, leading their men to Culblean in 1335 and Perth in 1339, this help was delivered north of Forth and, in 1340, the role of southern leaders in their own areas was recognised when the march wardenships were revived. This revival reflected the Steward's lack of direct interest in the region and was possibly encouraged by David II on the eve of his return to place formal control of war in the marches in the hands of his allies. By the naming of three wardens, Ramsay in the east, Douglas in the middle and John Randolph in the west, the border was divided 'as it had been held in the past'. Bower's phrase implies the formalisation of earlier distinctions between Randolph, Douglas and Dunbar spheres in the marches, perhaps in force from the 1320s, with Ramsay acting for Earl Patrick and Douglas for Hugh the Dull. Private magnate power and independence from central authority determined the structure and role of the Scottish march wardenships from their origins. Thus although there was no English middle march until later, Teviotdale and the Forest formed a natural military zone in which Douglas showed a refusal to accept external leadership.[28] This distinction remained. From the 1350s onwards, though it extended to the central borders, the eastern march was divided into two parts, specified as the bounds of the earl of March and those of the earl of Douglas. The two earls acted as joint wardens, recognising respective zones of influence in the marches, their authority in war and justice stemming principally from their private lordship in these lands. The association of Earl William Douglas and Patrick earl of March was the basis of military organisation in the eastern borders in the 1350s. This collaborative approach to war could lead to 'discord between the magnates', as in 1355, when March, supported by Thomas earl of Angus, captured Berwick against Douglas's advice. Despite the English invasion which this provoked, this discord was a difference of approach, not of objectives or ambitions, and caused no conflict like that of Ramsay and William Douglas, when the former encroached on the latter's march. Angus continued to be a close ally of Douglas, and William Towers, also involved in the attack, soon became Earl William's brother-in-law. Most importantly, when a truce was agreed between Earl William and the English in April 1356, Douglas guaranteed the protection of March's lands. This truce indicated the character of war and politics in the borders. It resulted principally from Douglas's own interests, his plan to depart on pilgrimage. The terms of the truce protected Earl William 'and his people' until the autumn, and Douglas agreed in return 'not to molest English forays' except those in the 'bounds' of March. Douglas would not oppose English attacks elsewhere, perhaps indicating the west where other lords were leading Scottish efforts. Though praised as the defender of 'Scottis fay', Earl William's activities in war were dictated by personal interests. He was not prepared to risk his own security in the defence of regions under the lordship of other Scottish leaders, except for those of his ally, March. In the years of lieutenancy, the aims and prosecution of Anglo-Scottish conflict were determined by the goals of individual southern magnates.[29]

The return of King David in 1341, and again in 1357, affected this situation. David II expected to renew his father's personal leadership and the overall control of war against England. In the early 1340s the king initiated a series of raids on England, though the first was made 'under of Murrawe the baner', under the formal leadership of John Randolph earl of Moray, one of the march wardens. David's attempts to alter the balance between these wardens by promoting Ramsay were disastrous, and so, ultimately, was the return of active royal leadership of the Scottish community in war. For his campaign of 1346 the king summoned the national host, but his call produced an army riddled with rivalries. The defeat of this force at Neville's Cross was as much to do with these conflicts as with military failure. David's experience in 1346 coloured his attitude to warfare after his release from English captivity.[30] In the 1360s, David generally sought peace, and attempts to change the management of the marches were indirect. In 1367 he employed a commission of seven to act as wardens of the east march. Though Douglas and March were included, they were associated with five of the king's closest councillors. By contrast, in the west march, David replaced a minor lord with an increasingly powerful magnate, Archibald the Grim. The king's attitude was based on his own sponsorship of Archibald, and from 1368 also extended to the new earl of March, George Dunbar. This tolerance included renewed warfare. David seems to have accepted Dunbar's seizure of English-held lands in the Merse and raids by both magnates on Annandale. Rather than an anticipation of renewed Anglo-French conflict from 1369, these attacks were the work of new border lords, keen to win profits and followers by leading in war, taking the initiative from Earl William.[31]

As in the management of politics, the Stewarts brought a change in atmosphere from David II in war and diplomacy. The reign of Robert II saw an intensification of the local war begun in the late 1360s. By 1378 the

English had clearly identified their principal enemies, claiming that 'the earl of March and Douglas and the latter's cousin, Sir Archibald, are harassing the English borderers by imprisonment, ransoms and other-wise'. As in the 1350s, attacks were chiefly led by the march wardens in a collaborative attack on English interests. The targets of these attacks, by March in the Merse, by March and Archibald in Annandale, by Earl William in Teviotdale and by all three, alone or together, across the border, principally reflected their own interests.[32] Earl William's aggression in Teviotdale was overtly dealt with in this light, as a private conflict with the Percy family over rival claims to Jedworth. The pursuit of such rights was hardly new, stemming as it did from the conflicting grants of English and Scottish crowns fifty years earlier. The presentation of local war in this way reflected the attempts of the overlords of both Douglas and Percy to separate national from private conflict and control local warfare. In the 1370s and 1380s, the Stewarts took a more active role in war than in the years before 1357. John earl of Carrick sought a connection in the south by forging alliances with Archibald the Grim, George Dunbar and, most importantly, Earl William and Earl James Douglas. Management of rela-tions with England was part of this, and from 1381, as 'lieutenant of the marches', Carrick acted as leading Scottish negotiator. When full-scale war did start in 1384, the lieutenant had powers to lead national war. The names of those involved in the campaigns of 1385 and 1388, and the leading role played by Carrick's brother, Robert of Fife, in these efforts suggest that, for the first time since 1346 and encouraged by the active support of France, war was being directed and led by the crown.[33]

However, the extent and significance of this Stewart leadership can be exaggerated. Unlike Robert I and David II, Carrick never took the field himself, and the role ascribed to Fife came from the earl's own propaganda in the 1390s. Even ignoring the error-ridden account of Jean Froissart, which made James earl of Douglas its hero, the war in the 1380s continued to be dominated by the leadership and interests of southern magnates. Earl William's Teviotdale campaign of 1384, although officially sanctioned by the crown 'for the common weal', marked the imposition of Douglas lordship on a community which had been a target of the family for most of the century. Similarly, the capture of Lochmaben, a natural target for both Dunbar and Archibald the Grim, was launched by Archibald 'out of regard for his men of Galloway who had suffered countless losses at the hands of the garrison'. Even in 1388, the attack of William Douglas of Nithsdale on the coast of Ireland was not simply part of a co-ordinated three-pronged Scottish offensive, but a raid in the centuries-old tradition of Galloway and a venture repeated by Galwegians in 1405. In war, as in

Hermitage Castle in Liddesdale. The grim border keep taken by William Douglas 'the knight of Liddesdale' in about 1337. Within its walls Douglas starved his rival, Alexander Ramsay, to death.

Tantallon Castle. Built in the 1350s by William first earl of Douglas, this massive red stone fortress became his principal base in Lothian and, from 1388. was the stronghold of his Red Douglas descendants.

Home of the Heart. Melrose Abbey was the last resting place of Bruce's 'bludy hert' and, despite tense relations between the monks and their Douglas neighbours, the burial place of three Douglas magnates in the fourteenth century.

House of Ill-repute? Lincluden Priory near Dumfries, where, at the urging of Archibald the Grim, the community of nuns were expelled for alleged immoral behaviour. Their departure cleared the way for Archibald's foundation of a collegiate church. Lincluden was the final resting place of Margaret, duchess of Touraine and countess of Douglas, whose tomb survives in the church.

The Green Tree. The imprint of the seal of William. first earl of Douglas (d. 1384) shows the bloody heart and three stars of the Douglases quartered with the arms of the earldom of Mar which he held in right of his wife. Behind the shield stand two trees, a favourite symbol on the Douglas arms.

The Wild Men. The seal of Archibald the Grim as lord of Galloway also bears the Douglas arms. To distinguish them from his cousin's the background to the heart is of ermine. The arms are supported by wild men, a motif designed to portray Archibald's subjugation of Galloway and also used by his allies, the Douglases of Liddesdale and Dalkeith.

Threave Castle on its island in the River Dee. The massive tower formed the centre of Douglas power in Galloway for eighty years, from its construction by Archibald the Grim in the 1370s until its surrender to royal forces after a long siege in the summer of 1455.

Found at Threave, this seal matrix bears the arms of Margaret, duchess of Touraine, the widow of the fourth earl of Douglas and sister of King James 1. To the left is the royal lion rampant. To the right are the Douglas hearts and stars and the lions of Galloway. From the 1420s to 1440s Margaret was lady of Galloway, ruling her province from Threave.

Also found at Threave, these everyday wooden objects include a bowl stamped with the ubiquitous heart, the badge of the Douglases.

From Warlords to Princes. Bothwell Castle above the Clyde shows the growth of Black Douglas power. While Archibald the Grim created a crude fortified base from the ruins of the earlier castle, his son, the fourth earl, added new buildings to the east wall (nearest the camera) which included a great hall, a chapel, a square tower-house and stylish round tower in the French style, which made Bothwell a palatial residence.

The tomb of Archibald, fifth earl of Douglas and lieutenant-general of Scotland, in St. Bride's Kirk at Douglas. The earl was the first lord of Douglas to be buried in the church since his great-grandfather the 'Good Sir James'. The return to family roots could not hide the damage which the earl's premature death in 1439 caused the dynasty.

Château-Gaillard above the Seine in Normandy had a special significance in the rise of the Douglas dynasty. It was here that William Douglas, the future first earl, and perhaps also his cousin, Archibald the Grim, spent their childhoods in the company of David II. The experience would make them and their offspring valued friends of France.

The lure of the Loire. The donjon at Loches in Touraine. Part of the rich duchy given to Archibald fourth earl of Douglas in 1424. The new duke of Touraine stayed in Loches the night before his spectacular entry into Tours on 7 May to take possession of his ducal capital.

The magnificent tomb of James the Gross, seventh earl of Douglas, at St. Bride's. Douglas was a worthy monument to a career dedicated to the accumulation of power. The immensely fat earl was a true political heavyweight: the survivor of numerous changes of regime since 1400 and the veteran of a series of coups the last of which brought him the earldom of Douglas.

The magnificence of Earl James's tomb did more than commemorate his past deeds. Built by his brood of sons, themselves represented on the base of the monument, it was a work designed to display the wealth, status and size of the Black Douglas dynasty. The family arms appear twice, supported both by the traditional wild men and by an angel. The construction of such a tomb at Douglas was designed to link the new line of earls with their famous ancestors and was part of a conscious image-building effort in the 1440s to lay claim to the traditional status and rights of the dynasty.

politics, Carrick depended on the support of southern magnates, and this was reflected by his unprecedented patronage of these men. Rewards were directed, overwhelmingly, to men of military significance, not just the three wardens, but aggressive captains like Henry Douglas, brother of James of Dalkeith, William Stewart and William, bastard son of Archibald Douglas. The latter William received the lordship of Nithsdale, an annuity worth £300 and marriage to Carrick's sister, exceptional rewards for 'skill in war'. The French shared these priorities when they sought to cultivate the Scottish nobility in 1385. Forty thousand *livres tournois* distributed to this end went overwhelmingly to southern magnates. Earl James Douglas received 7,500 *livres tournois*, the most given to any lord, Archibald Douglas got 5,500 *livres tournois* and March, 4,000, while Carrick, the lieutenant, was only paid 5,500 and Fife, 3,000. These sums reflect the value of various lords in French plans to wage war in the marches and suggest that the wardens with their retinues and experience remained the key figures in conflict with England. This extended to the organisation of defence and the making of truces. In the 1380s, as in the 1350s, it was the wardens who negotiated temporary periods of peace, and when, in 1385, Archibald Douglas agreed such a truce with the earl of Northumberland, the question of its extension to the whole border was decided, not by Carrick, but by Earl James Douglas. Looking back from the next century, it was from the statutes of 'blak Archibald' that the west march men drew their practices of local defence.[34]

Rather than a break in patterns of military leadership, the campaigns of 1384–1388 represented a greater level of support for the existing structures of Douglas and, to a lesser extent, Dunbar war lordship. The sources of this support, Carrick and the French, had no intention of reducing the effectiveness of these lords. Carrick's position as lieutenant was built on the backing of Earl James and his extensive connection, and it was Earl James, rather than Carrick or Fife, who was central to the Scottish war effort in the 1380s. With his death in 1388, the whole edifice collapsed. Stewart attempts, from 1393 to 1401, to challenge the new Black Douglas ascendancy in marcher war and diplomacy directly, met with complete failure. By 1401 Earl Archibald was in a position to exclude the lieutenant from the making of war and peace on the border. Albany's recovery of a limited role in both was bought at the price of his physical support of the earl's adventures, and as his purely temporary leadership in the crisis after Humbleton showed, marked the duke's own limited aspirations in the borders.[35]

The Douglases' prominence in the Scottish war effort was built on their ability to provide manpower for the war with England. By the 1380s this Douglas affinity in arms could produce impressive forces. Though apparently the smaller of two Scottish armies and part of a general host, the men led by James Douglas to Otterburn were not a national force but the Douglas affinity at war, 'our army' as the earl described it in one charter. The force, numbered at either 2,500 or 7,000, was dominated by the earl's adherents. The men of the earl's own company, the Lindsays and Sinclairs, were themselves powerful lords, the three Lindsays collecting £2,000 from the French war subsidy. Such kinsmen formed the core of an army which also included the earl's brother-in-law, Malcolm Drummond, his cousin, James Sandilands, his uncle, Patrick Hepburn, and his stepfather, John Swinton. Hepburn was also part of the wide Lothian following present with Earl James, and Ramsays, Lauders, Haliburtons and Abernethys, families associated with the Douglas earls since the 1350s, were also at Otterburn. The middle march provided similar indications of this link between lordship and military service. The Colvilles, Rutherfords, Gledstones, Glendinnings and William Stewart were borderers whose presence in Scottish allegiance was bound up with their adherence to the Douglas earls. To see this army as a regional division of a national host ignores the ties of Douglas lordship which bound it together and the control which Earl James exercised over the force. Even the presence of the Dunbars, Earl George and his brother, John earl of Moray, and of Carrick's retainers, Thomas Erskine and David Fleming, represented political allies of Douglas, who had brought their men to follow his banner at Otterburn, as on other occasions.[36]

The army which Archibald fourth earl of Douglas led to Humbleton was a further display of aristocratic control of warfare. Bower described Douglas's expedition as 'his army' and, though Albany contributed to the force, he sent men from his own following led by his son. Like that of Earl James, the core of Archibald's army was his household and kin. John Swinton, William Stewart, John Edmonston, Stewart of Durisdeer and William Hay, servants like William Crawford and William Muirhead, and kinsmen like Douglas of Nithsdale and three sons of Douglas of Dalkeith, all fought alongside the earl. Much of the army came from the earl's massive connection in the south, including not just the Lothian and middle march men who had fought at Otterburn, but also lords from the west march and Galloway and from the Merse. The presence of MacDowells, McCullochs and Thomas Kirkpatrick reflected, not a national summons, but the well-established leadership of the Black Douglases over these men. Similarly, the service of Berwickshire lords, Adam Gordon and Alexander Hume, was not based on appeals to national loyalty but on Earl Archibald's successes in the region. The southern earls, Angus and Orkney, who brought their own men to Humbleton, also responded to the call of

Douglas, whom they had supported in the war against the Dunbars and the English since 1400. Although, as justiciar and sole march warden, Douglas represented the merging of public and private power in the south, the host he raised in 1402 was built, primarily, on obligations of leadership, not on the duties of national service. It should be added, however, that the Humbleton campaign, resembling that of Neville's Cross, supports Bower's view of the latter battle that, regardless of royal or aristocratic leadership and recruitment, Scottish hosts could be huge in numbers but were unwieldy and vulnerable to better-organised English forces.[37]

The military power of the Douglases was not unique in early fifteenth-century Scotland. The 'force of knights and brave men' sent to Douglas in 1402 was made up of Albany's own retainers and their adherents from the areas of his influence, and Albany probably sent a similar force to aid Alexander earl of Mar at Harlaw in 1411. Mar himself had raised 'all those he could have' from the north-east to oppose the attack of the lord of the Isles. His army, numbered in thousands, was led by men who had provided Mar with his retinue in previous years or who would become his military lieutenants. Militarised society was not limited to the borders. North of Forth, as in Lothian and the marches, structures of lordship provided the basis of both small retinues and major armies.[38] The culmination of this militarised lordship came in the early 1420s, when armies of up to seven thousand Scots fought the English in northern France. These forces, serving overseas for sustained campaigns lasting several years, were not produced by summoning the national army, which obliged Scots to fight for a limited period and within Scotland or on its borders. Instead, Scottish armies in France were raised and led by their lords. The structures of military lordship, which had provided forces for the local warfare of the marches and which, by 1400, could produce full-scale armies, could also be used to provide mercenary contingents for service in continental warfare. Control of these structures, established hand-in-hand with the rise of the family in Scotland, now gave the Douglases a place in European war and politics.

NOTES

1. *Longer Scottish Poems*, i, line 379.
2. J. Campbell, 'England, Scotland and the Hundred Years War in the Fourteenth Century', in J. R. Hale, J. Highfield and B. Smalley (eds), *Europe in the Late Middle Ages* (London, 1965) 184 – 216; M. Prestwich, *The Three Edwards* (London, 1980), 42–78, 167; J. Sumption, *The Hundred Years War*; A. Goodman, 'Introduction', in Goodman and Tuck (eds), *War and Border Societies*, 1–29, 3, 14–16.
3. *Scotichronicon*, ed. Goodall, ii, 333, 348–49; *Scotichronicon*, ed. Watt, vii, 146–49, 274–75;

A. Grant, *Independence and Nationhood*, 3–57; R. Nicholson, *Edward III and the Scots*, 151–62; A. Grant, 'The Otterburn War', 30–34.

4. G. W. S. Barrow, *Robert Bruce*, 332, 381, 387; A. A. M. Duncan, '*Honi soit qui mal y pense*: David II and Edward III 1346–52', in *Scottish Historical Review*, lxvii (1988), 113–41.

5. *Cal. Docs. Scot.*, iii, no. 1127; *Wyntoun*, vi, 124–27.

6. *The Bruce*, V, lines 271–302; VIII, 425–27; *Scotichronicon*, ed. Goodall, ii, 346; *Wyntoun*, vi, 100–101, 194–95; *Scotichronicon*, ed. Watt, vii, 140–147; *Cal. Docs. Scot.*, v, no. 809; *R.R.S.*, v, 143, 166–67, 184, 267.

7. *Cal. Docs. Scot.*, v, no. 809; *Scalachronica*, 299.

8. *Cal. Docs. Scot.*, ii, no. 1241; iii, nos 218, 279, 336, 1122, 1283, 1323, 1382; R. Nicholson, *Edward III and the Scots*, 189, 224–6.

9. *Cal. Docs. Scot.*, iii, nos 1323, 1382, pages 360–63; v, no. 809; *Scotichronicon*, ed. Goodall, ii, 346–47; *Scotichronicon*, ed. Watt, vii, 270–71; *Wyntoun*, vi, 194–97.

10. *Chron. Lanercost*, 292–96; *Cal. Docs. Scot.*, iii, pages 360–63; iv, no. 64; *S.H.S. Miscellany*, v, nos 13, 16; *R.R.S.*, vi, no. 130. In the fifteenth century a, primarily economic, connection continued to exist between Roxburgh and local men.

11. *The Bruce*, VI, lines 375–453; *Scotichronicon*, ed. Goodall, ii, 350; *Scotichronicon*, ed. Watt, vii, 138–39, 278–81; *Chron. Lanercost*, 296; *Wyntoun*, vi, 120–21; *Foedera*, viii, 57–58.

12. *Scotichronicon*, ed. Goodall, ii, 332, 334, 397; *Cal. Docs. Scot.*, iii, pages 361, 363; iv, no. 331; *The Bruce*, X, lines lines 359–496; *Wyntoun*, vi, 138–147, 160–63; *Scotichronicon*, ed. Watt, vii, 144–47, 150–51, 394–97; viii, 72–73, 86–87, 296–97. For a full account of the raids launched on Annandale in the 1370s and 1380s, see A. MacDonald, thesis, 28, 33, 36, 76–77. Grant suggests Berwick was seen as impossible to take and hold against English attack after the experience of 1355, when its capture prompted the 'Burnt Candlemas'. Later seizures of the town by local Scots were opposed by the earl of March, whose lands were most at risk (Grant, 'The Otterburn War', in Goodman and Tuck, *War and Border Societies*, 36–37).

13. *Cal. Docs. Scot.*, iii, no. 894, page 313; *Wyntoun*, vi, 162–63; *Scotichronicon*, ed. Watt, viii, 74–75.

14. *Chron. Lanercost*, 269; *Wyntoun*, vi, 194–95, 222–23; *Scotichronicon*, ed. Goodall, ii, 356, 400; *Scotichronicon*, ed. Watt, vii, 296–97, 402–403; *R.R.S.*, vi, nos 137, 451; *Rot. Scot.*, i, 826; A. MacDonald, thesis, 59–60, 77–78; M. Vale, 'Seigneurial Fortification and Private War in Later Medieval Gascony', in M. C. E. Jones (ed.), *Gentry and Lesser Nobility in Later Medieval Europe* (Gloucester, 1986), 133–48; R. Frame, 'The Defence of the English Lordship', in T. Bartlett and K. Jeffery, *A Military History of Ireland* (Cambridge, 1996), 76–98; J. A. Tuck, 'War and Society in the Medieval North', in *Northern History*, 21 (1985), 31–52.

15. *Longer Scottish Poems*, i, 68, lines 574–77; *Wyntoun*, ed. Laing, iii, 29–30, 32–36; *Scotichronicon*, ed. Watt, vii, 368–73, 378–79; viii, 44–49; J. Scammell, 'Robert I and the North of England', in *English Historical Review*, 73 (1958), 385–403; J. A. Tuck, 'Richard II and the Border Magnates', 30–44.

16. *The Bruce*, XVI, lines 336–501; *Chron. Guisborough*, 397; *Chron. Lanercost*, 287, 288. Douglas of Liddesdale also harried the English army approaching Edinburgh in 1337 at Crichton (*Scalachronica*, 103; *Wyntoun*, vi, 118–19).

17. *Scotichronicon*, ed. Goodall, ii, 354, 356; *Scotichronicon*, vii, 296–97; *Chron. Knighton*, ii, 85; *Scotichronicon*, ed. Watt, viii, 32–33, 34–37; *Westminster Chron.* 396–97. Edward Balliol was probably included in those submitting to Edward, by the resignation of his royal title.

18. *A.P.S.*, i, 715.

19. *The Bruce*, V, lines 279–302; *Scotichronicon*, ed. Goodall, ii, 317, 357; *Scotichronicon*, vii, 106–109; 146–49; 298–301; viii, 34–35; Froissart, *Chroniques*, iv, 15.

20. *The Bruce*, V, lines 279–302; X, lines 352–505; G. S. C. Swinton, 'John of Swinton: A Border Fighter of the Middle Ages', in *Scottish Historical Review*, xvi (1919), 261–79; A. Goodman, *John of Gaunt, The Exercise of Princely Power in Fourteenth Century Europe*

(London, 1992), 216–17; S.R.O., GD 12/ 1, 14, 15, 16; AD 1/ 27; *Melrose Liber*, ii, no. 491; Fraser, *Douglas*, iii, nos 344, 346; Fraser, *Carlaverock*, ii, no. 21; *Scotichronicon*, ii, ed. Goodall, 404; *Scotichronicon*, ed. Watt, vii, 414–15; viii, 48–49.

21. S.R.O. GD 12/ 40; *Morton Reg.*, ii, nos 129, 154; *R.R.S.*, vi, no. 457; *H.M.C.*, Drumlanrig, i, no. 2; J. Wormald, *Lords and Men in Scotland* (Edinburgh, 1985), 42–45; A. Grant, *Independence and Nationhood*, 135.

22. *Wyntoun*, vi, 147, 150–51; *Scotichronicon*, ed. Goodall, ii, 316, 329–30; *Scotichronicon*, ed. Watt, vii, 106–109; 138–39; viii, 32–33; E. L. G. Stones, *Anglo-Scottish Relations*, 173–82.

23. *Rot. Scot.*, i, 793; *Cal. Docs. Scot.*, iv, no. 254; *Scotichronicon*, ed. Goodall, ii, 317, 333, 350, 357; *Scotichronicon*, ed. Watt, vii, 146–48, 278–81; viii, 42–43; Froissart, *Chroniques*, iv, 15.

24. *Rot. Scot.*, i, 685; A. Grant, 'The Otterburn War', 42; Froissart, *Chroniques*, iv, 15; *Cal. Docs. Scot.*, iii, no. 1564; v, no. 809; *Wyntoun*, ed. Laing, iii, 10–11, 13; *Foedera*, viii, 54–58; *R.M.S.*, i, no. 850. Stewart was executed for having left English allegiance, and this group of borderers who switched sides was seen as the main cause of disorder by the English in 1397 (*Foedera*, viii, 54–57).

25. *Scotichronicon*, ed. Goodall, ii, 346–47; *Wyntoun*, vi, 192–93; *Scotichronicon*, ed. Watt, vii, 270–71; viii, 32–33; *R.M.S.*, i, appendix 1, no. 123.

26. *The Bruce*, XI, lines 448- 469; XVII, lines 181–87; R. R. Reid, 'The Office of Warden of the Marches; its Origins and Early History', in *English Historical Review* (1917), 479–96; R. L. Storey, 'The Wardens of the Marches of England towards Scotland', in *ibid*.(1957), 593–615; T. I. Rae, *The Administration of the Scottish Frontier, 1513–1603* (Edinburgh, 1966); *Scotichronicon*, ed. Watt, vi, 882–83. The first official record of the title of march warden only occurred in 1355, referring to William lord of Douglas (*R.R.S.*, vi, no. 137).

27. P. F. Tytler, *A History of Scotland*, ii; R. Nicholson, *Edward III and the Scots*, 91–138; *Wyntoun*, vi, 53–55, 124–25; *Scotichronicon*, ed. Goodall, ii, 330–31; *Scotichronicon*, vii, 140–45; *Chron. Lanercost*, 269, 270–71, 278; *Scalachronica*, 103.

28. *Wyntoun*, vi, 152–55; *Scotichronicon*, ed. Goodall, ii, 333–34.

29. *Cal. Docs. Scot.*, iii, no. 1607; *Rot. Scot.*, i, 913–14; ii, 73, 85–86; *Scotichronicon*, ed. Goodall, ii, 351–54, 356–57; *Scotichronicon*, ed. Watt, vii, 278–83; *Chron. Fordun*, i, 373; A. Grant, 'The Otterburn War', 36; J. Campbell, 'England, Scotland and the Hundred Years War', 199–200.

30. *Wyntoun*, vi, 162–69; *Scotichronicon*, ed. Goodall, ii, 341–43; *Scotichronicon*, ed. Watt, vii, 252–63; *Scalachronica*, 115; *Chron. Knighton*, 32–33, 41–43.

31. *Rot. Scot.*, i, 913–14, 921; *Cal. Docs. Scot.*, iv, nos 47, 100; A. MacDonald, thesis, 18–20.

32. *Cal. Docs. Scot.*, iv, 223, 231, 242, 260, 315; *Wyntoun*, ed. Laing, iii, 12–15, 18–20, 24, 30–31; *Scotichronicon*, ed. Watt, vii, 368–73, 379–81, 394–97. The warfare of the years 1368–1389 has been the study of much recent examination, in particular by A. Mac-Donald in his thesis, by A. Grant in 'The Otterburn War' and S. Boardman in *The Early Stewart Kings*.

33. *Rot. Scot.*, i, 955, 965; ii, 3; S. Boardman, *Early Stewart Kings*, 114–15; A. Grant, 'The Otterburn War', 33–34, 45–49.

34. *Wyntoun*, ed. Laing, iii, 29–34; *A.P.S.*, i, 552, 715; *R.M.S.*, i, nos 752, 753, 770; *Morton Reg.*, ii, 158–59; *Scotichronicon*, ed. Goodall, ii, 403; *Scotichronicon*, ed. Watt, vii, 394–97; 412–15; S. Boardman, *Early Stewart Kings*, 286; Archives Nationales, J677, no. 15; *Rot. Scot.*, ii, 73, 85–86.

35. S. Boardman, *Early Stewart Kings*, 142–53; *P.P.C.*, i, 127.

36. Froissart, *Chroniques*, iv, 15; Wyntoun, *Chronicle*, ed. Laing, iii, 38. Though Froissart's accuracy on Scottish issues has been thrown into doubt, the overlap between the men (or those identifiable after their names had been mangled) in the battle and those found counselling Earls William and James is striking (Fraser, *Douglas*, iii, nos 22, 23, 25, 293, 323, 330, 332, 333, 334, 335; *H.M.C.*, Drumlanrig, i, no. 2; Hamilton, i, no. 126; Milne Hume, no. 582; *North Berwick*, xxxvii; Fraser, *Scotts of Buccleuch*, no. 7; S.R.O., GD 12/1; RH6/155, 191; *St. Giles*, 7–8; *Melrose Liber*, nos 490, 491).

37. *Scotichronicon*, ed. Watt, vii, 262–63; viii, 44–49; *Cal. Docs. Scot.*, iv, 402–403; *H.M.C.*, 15,

appendix 10, 77–78; Fraser, *Carlaverock*, ii, no. 21; S.R.O. GD 12/14, 15; Fraser, *Douglas*, iii, nos 298, 346, 351.

38. *Scotichronicon*, ed. Watt, viii, 44–49, 74–75. For the details of Mar's retinue and the development of militarised lordship in northern Scotland, see S. Boardman, 'Alexander Stewart Earl of Buchan' and M. Brown, 'Alexander Stewart Earl of Mar, in *Northern Scotland*, 16 (1996), 1–54.

CHAPTER EIGHT

Douglas Lordship

COUNCIL AND HOUSEHOLD

In Barbour's *Bruce* the power of the Douglases was traced to humble origins. Returning to Douglasdale from exile the young James Douglas sought out a local man, 'Thom Dicson',

> That wes of frendis rycht mychty
> And rich of mwbill [goods] and catell
> And had beyn till his fader lele

Dicson sheltered James in his house and secretly summoned the 'leill men of the land' who had been 'duelland' with James's father to come 'and mak him manrent':

> Douglas in hert gret blithnes had,
> That the gud men of his cuntre
> Wald sa gat bundin till him be.

Barbour's description must have rung true to his audience in the 1370s. It presented the basics of lordship, the establishment of bonds between great and lesser men symbolised by the homage and manrent done by Thomas and his 'frendis' to James. Like many others, these men bound themselves to James in circumstances of local warfare and, like his heirs, James wanted military service first and foremost from his men. But, even in 1307, the Douglases saw themselves as more than just leaders in war, heads of 'schools of knighthood' built around the quest for protection and the profits of conflict. Though the needs of war left a long and deep impact on the structures and character of the Douglas affinity, there were other factors at work.[1]

The first of these was land. Douglas was in his 'awn cuntre', where his claim to be his father's heir as lord of Douglas carried with it rights to land and the loyalty of local men. Despite the insecurity and damage caused by war, territorial lordship retained significance in southern Scotland. Douglas lords who had won lands in war still sought legal title to their holdings, as Douglas of Liddesdale's acquisition of estates in the marches from 1339 demonstrated. Formal possession of land brought with it claims to judicial

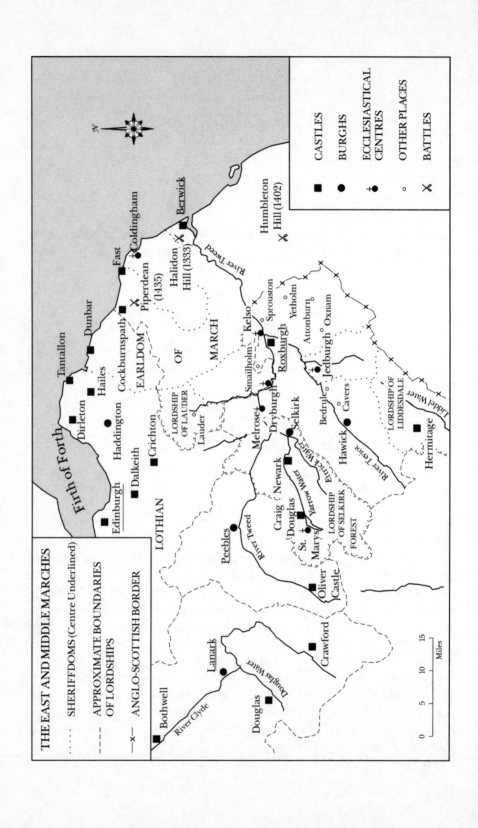

THE EAST AND MIDDLE MARCHES

.... SHERIFFDOMS (Centre Underlined)

– – – – APPROXIMATE BOUNDARIES
OF LORDSHIPS

–×– ANGLO-SCOTTISH BORDER

CASTLES   ■

BURGHS   ●

ECCLESIASTICAL
CENTRES   ✛●

OTHER PLACES   ○

BATTLES   ✕

Firth of Forth

Edinburgh

Tantallon
Dirleton
Hailes
Dunbar
Fast
Coldingham
Berwick
Haddington
Cockburnspath
Crichton
Dalkeith
Piperdean
(1435)
Halidon
Hill (1333)

EARLDOM

OF

MARCH

River Tweed

Humbleton
Hill (1402)

Sprouston
Yetholm
Attonburn
Oxnam
Jedburgh
Kelso
Roxburgh
Smailholm
Melrose
Dryburgh
Selkirk
Bedrule
Cavers
Hawick

LOTHIAN

LORDSHIP
OF LAUDER
Lauder

River Tweed

Peebles

Newark

Craig
Douglas
St.
Marys

Ettrick Water
Yarrow Water

LORDSHIP
OF SELKIRK
FOREST

River Teviot

LORDSHIP OF
LIDDESDALE

Liddel Water

Hermitage

Oliver
Castle

Crawford

Lanark

Bothwell

Douglas

River Clyde

Douglas Water

0   5   10   15
Miles

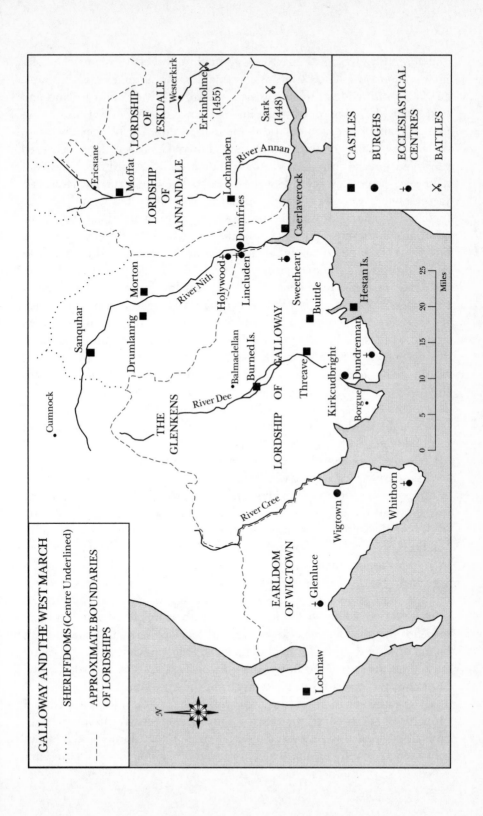

GALLOWAY AND THE WEST MARCH

SHERIFFDOMS (Centre Underlined)

APPROXIMATE BOUNDARIES
OF LORDSHIPS

CASTLES   BURGHS   ECCLESIASTICAL
                    CENTRES   BATTLES

LORDSHIP
OF
ESKDALE

Westerkirk

Erkinholme✗
(1455)

Sark ✗
(1448)

Moffat LORDSHIP
        OF
Ericstane ANNANDALE

Lochmaben

River Annan

Caerlaverock

Dumfries

Morton

Drumlanrig

Sanquhar

Cumnock

Holywood✝
Lincluden

River Nith

Sweetheart
Buittle

Hestan Is.

THE
GLENKENS

Balmaclellan
Burned Is.

River Dee

Threave

Kirkcudbright

Dundrennan

Borgue

LORDSHIP
OF
GALLOWAY

River Cree

Wigtown

Whithorn✝

EARLDOM
OF
WIGTOWN

Glenluce✝

Lochnaw

0   5   10   15   20   25
Miles

N

and administrative rights, like those granted by Robert I to James Douglas in the 'Emerald Charter', and land remained the normal basis of power. Anglo-Scottish conflict was expressed in terms of rival claims to land and lordship held by kings or by magnates like Douglas and Clifford, and secure and profitable possession of land in the marches was the principal war aim of Douglas magnates throughout the century. Control of land delivered rents and judicial profits. It was also the currency of lordship. Grants of land were the greatest reward given by lord to follower for service. However, these services were very rarely specified in military terms and, in late medieval Scotland, lordship did not rest solely or even principally on formal tenurial relationships, on 'feudalism' for want of a better term. The real link between landholding and lordship was as the focus for wider networks of bonds and connections resting on a range of contacts expressed, as in the *Bruce*, in terms of friendship or personal loyalty, not of tenantry. A personal relationship, like that between James Douglas and the local free-holder, Thomas Dicson, was the key to winning the support of lesser men from the 'cuntre'. From 1307 this bond served both lord and man well. Thomas received Symington from the king for his loyalty, and by the 1370s his successors had established themselves as keepers of Douglas castle for their Douglas lords. In return these lords gained the 'de Symington' family as servants and deputies in upper Clydesdale. Across the south, the Doug-lases sought similar relationships and offered similar rewards to men of local influence with 'frendis' of their own.[2]

Collectively, these individual relationships formed an affinity, a political, social and military grouping which, though built on informal or semi-formal ties, represented the fullest expression of Douglas power. The 1409 inden-ture between the fourth Douglas earl and the duke of Albany moved on from potential disputes between the two magnates to the management of conflicts between their followers. Lords were expected to defend 'thair men' in law, arbitration and, if necessary, in arms and Douglas and Albany's agreement to 'do thair power' to seek settlement in such disputes 'be trety in lufely manere' and, in any case, to avoid a direct confrontation stemmed from a real possibility that conflict between their followers would draw in the earl and the duke. Local justice and local politics in Scotland were bound up with the management of magnate followings and, by the opening decades of the fifteenth century, such ties of lordship were regarded as the most effective structures of government in the kingdom.[3]

These networks of lands and adherents were built around the personal property and personal relationships of the Douglas earl. Terms like 'friends' and 'allies' applied to individual nobles were not just a form of address but could reflect real friendship which, for example, Hume of Godscroft later claimed existed between his ancestor, Alexander Hume, and the fourth earl of Douglas. It was from such personal ties, not just his inherited power, that the earl gained the power to lead and arbitrate between his men. The death of an earl weakened these ties and if, as in 1388, there was no obvious heir, the lands and following of the Douglas earl fragmented not just according to legal rights but, as with Archibald the Grim and George Douglas, according to the ability to forge these personal ties of lordship in the local communities dominated by the dead earl. Even the absence of the fourth earl abroad from 1413 to 1414 placed pressures on those who looked to him for lordship. The earl of Douglas was called in to arbitrate in a 'discord' between Melrose abbey and a local squire, David Haig of Bemersyde. In 1416 he stated that he had hoped to settle the dispute earlier 'bot be cause of hee and grete besiness that we had appoun hande to do in sere countreis in the tyme of the rising of this discorde', he had been unable to make a 'finable ende' to the dispute. Strains in the earl's following may have been exacerbated by Douglas's absence in England and France. When he departed for France in 1424, Douglas sought to replace his own lordship by putting territorial and political power in the hands of his family. Earlier, in 1402, his younger brother, James, had provided leadership for the Douglas following in the marches and Lothian. In 1424, intending a longer absence, the earl formally appointed his heir, Wigtown, 'steadhaldand and governor' of Annandale and probably of the other family lands in the marches and Clydesdale. These formal commissions, allied to the leadership exercised by Wigtown in France since 1419, were intended to create a smooth passage of power from father to son, like that of the 1390s, when the handover began perhaps two years before Archibald the Grim's death.[4]

However, in their efforts to run their lands and connection the Douglas earls could look beyond their family. The 1409 indenture gave a formal role in the settlement of disputes to the two magnates' councils, and the Douglases, like most great lords, relied on the advice and support of their close adherents in justice and politics. Unlike the royal council of James I, though, it is hard to identify the council of the Douglases as a formally summoned and constituted body. Instead the witnesses to Douglas charters suggest that those around the earl of a certain status provided him with counsel. For example, the Dumfriesshire barons and Douglas officials, Robert and Herbert Maxwell, attended the fourth earl when in Annandale and Nithsdale, and came to Edinburgh when he was dealing with lands or issues of direct interest to them. Similarly, regular councillors of the earl in Edinburgh, like William Borthwick and John Seton, did not follow him to the south-west in 1418 and 1419. This variable composition does not indicate the council's lack of importance but rather suggests its value as the means of contact between lord and local adherents. When the fourth earl gave judgements in Galloway surrounded by men like Thomas McCulloch, Uhtred and Fergus MacDowell, John Herries and Alexander Gordon he involved the chief men of the province in their lord's judgements and patronage. Similarly in 1425 the fifth earl gave formal judgement in the dispute between Haig and Melrose in company with a panel of local Roxburghshire men. Such councils enhanced the significance of the earl's personal presence, maintained his connections with local adherents and increased the authority of his decisions. The earl's council, which could be military staff, the means of contact between an exile and his homeland or the legal and political focus of the affinity, was central to the structure of Douglas lordship.[5]

However, Douglas magnates did not rely solely on meetings of local adherents for council. They also employed ecclesiastical advisers in their entourage for their skills as scribes and lawyers. For example, in the 1340s and 1350s Richard Small, rector of Ratho, transferred from Liddesdale to Earl William's council, his continued employment suggesting his value to both lords. Earl William certainly employed a secretary from the 1350s, and in the 1400s a body of clerical servants was retained as the earl's writing office. By 1413 this was headed by a chancellor, first Alexander Cairns and then William Foulis, the previous secretary, both of whom had lengthy careers in Douglas's employ. The division of labour between chancellor and secretary perhaps reflected the fourth earl's need for a greater number and range of servants to match his greater power and pretensions. Men like Foulis and Cairns, and other clerks of the earl like Gilbert Cavan, James Fawside and Matthew Geddes, were used as messengers, ambassadors and councillors in Scotland and beyond. In short they were professional agents who built their careers in Douglas service.[6]

The fourth earl also employed a number of household servants. The steward of the household, John Livingston, was employed in taking large sums from the Edinburgh customs for his master – like stewards in English noble households, suggesting that he was the earl's chief financial official responsible for receiving and distributing funds. Although not a councillor, Livingston, a layman, was clearly an important servant able to count on Douglas's protection. A second household official, Patrick Cunningham, was 'gardrober' of the earl, responsible for Douglas's wardrobe, his jewels, clothes, moveable goods and, clearly, the earl's 'hôtel' in London. References to these officers suggest the existence of a group of intimate lay servants who did not witness the earl's acts but were vital to the running of his lands and following. There were similar clerical servants, most importantly the fourth earl's confessor, John Fogo, a highly influential monk of Melrose, and John Gray, former dean of the faculty of medicine at the university of Paris, who was Douglas's 'familiar' and probably his personal physician from 1406 to 1415. A final servant, Robert Storm, rector of Covington, was dean of the earl's private chapel, in charge of the daily mass attended by the whole household.[7]

The possession of a large, structured household establishment was an indication of status and power as well as an instrument of government. Formal households were probably established for the wives and adult heirs of the earls of Douglas. Certainly these institutions were given formal duties in the absence of the earl himself. Both Countess Margaret Stewart and the earl of Wigtown had their own officers to aid in the exercise of their local authority in 1424 when the fourth earl left for France. It was from their own household that the Douglas earls drew their armed escort. The size of this following, up to a hundred men, added to clerks and other non-combatants, suggests that the Douglases maintained a household comparable with those of English magnates. This retinue was designed to have a political impact. The efforts of kings like David II and James I to limit those who rode with a magnate to no more than 'may suffice ... till his estate' indicate the value of such retinues to their lords. With this backing Douglas magnates could deal with any local opposition.[8]

In the absence of anything approaching English livery rolls or household books the composition of this retinue is hard to establish. However, under the fourth earl a significant group of men were described as esquires of the earl. This was not simply a casual reference to non-knightly status. William Hay of Yester was a knight and sheriff of Peebles but on two occasions he was called '*notre trescher esquier*' by the earl. Similarly, Alexander Hume, an influential local figure, was named the earl's esquire four times. Both were intimate councillors of Douglas, and it was this service which was possibly indicated by their place as his squires. More than twenty others were given similar titles. Many were junior members of kindreds connected to the Douglases. David Hume was Alexander's brother, Henry Haliburton and Thomas Turnbull kinsmen of powerful local lords, William Borthwick, Alexander Stewart and Robert Gordon sons of Douglas tenants and neighbours, and Adam Hepburn the heir of the knight killed at Nisbet Muir. A second group of squires were minor figures, some of whom, like James Dundas and Michael Ramsay, received lands and money through the earl's patronage. Ramsay, Douglas's 'familiar', may also have gained reward through the dubious privilege of being the husband or father of the earl's mistress, Christian. Ramsay, Dundas, John Durant, Simon Dalgleish and others witnessed the earl's acts after major lords and ranked alongside

Douglas's secretary. It seems likely that these squires formed a group of household retainers giving the earl support as councillors, messengers and soldiers and reinforcing links with local families. These landless squires probably served in the expectation of landed rewards, like those given to Henry Haliburton, Michael Ramsay and David Hume by the fourth earl, and for a regular fee. According to James I's legislation, 'ful and redy payment' was the proper basis for retinue service and, though evidence of individual annuities is hard to find, the Douglases possessed the resources to maintain a core of paid servants as well as hangers on, living on the earl's food and hoping for such fees.[9]

The earl's household and retinue played a vital role in the management of Douglas lands and interests. The scale of these interests placed limits on the earls' influence and, like all medieval princes, to maintain their lordship the Douglases made regular progresses between their principal residences accompanied by the secretariat, financial offices and staff of advisers, messengers and guards included within their household. The movements of this entourage were essentially between the four principal regions in which the earls had major interests, Lothian, Clydesdale, the middle march and the south-west. Of these, the Douglas earls dwelt longest in Lothian. Although they had only minor estates in the province, Lothian's physical security, economic wealth and access to lands beyond Forth gave it an importance signified by Earl William's construction of a massive fortress at Tantallon which he used as the base for his influence in the region. From 1400 Edinburgh performed the same function for the fourth earl. Most of his acts were issued from the castle or burgh. By comparison, Clydesdale was of secondary importance. Douglas castle, small and remote, retained value as a staging post and must have had an emotional significance which led Earl William to go there to die in 1384. From 1388, however, the earls' main residence in Lanarkshire was Bothwell castle. Archibald the Grim's first and favourite castle, Bothwell, was, after Edinburgh, the most regular residence of both his son and grandson and was extensively improved by them.[10] There was no castle of comparable size in the middle march where such strongholds either lay in ruins or English hands. Courts were held at traditional sites like the manor of 'Edirbredshiels' or at Tinnis and Erncleuch deep in the Forest. The visits of Douglas lords to these houses were rare but probably of great significance to their tenants and neighbours. It was only in the early fifteenth century that a more impressive and defensible residence was built in the region. Even so, Newark castle near Selkirk was no Bothwell. It was a tower house, an easily guarded base from which to exert influence over the Forest and Teviotdale. The picture was similar in the south-west. Though the acquisition of Lochmaben in

1409 provided the Black Douglases with a large and easily accessible fortress in the region, their own construction had been Threave, a second massive tower. After 1424 Threave would provide the main residence of the fourth earl's widow as lady of Galloway, though both she and her husband would also hold court at Wigtown, the principal centre of west Galloway.[11]

Present or not, the earls of Douglas needed such bases across the south and relied on their followers to guard them. In 1421, for example, Michael Ramsay was made keeper of Lochmaben castle for life, promising 'to kepe the ... castel ... agayne all other'. In return he received a fee of £80 in goods and cash as well as assigned lands and rents. Such terms were not unique. The Symingtons held four vills in Douglasdale as the hereditary fee for keeping Douglas castle, while William Crawford, paid out of the fourth earl's £133 6s 8d pension as keeper of Edinburgh, also received lands in connection with his role as constable of the castle. These Douglas keepers formed a group with impressive credentials. Ramsay was the fourth earl's trusted 'familiar', William Borthwick, who succeeded Crawford in Edinburgh, was his most regular councillor and William Symington was a member of a family whose record of loyalty to their lords went back a century. Richard Hangingside, keeper of Dunbar castle in the 1400s, had served all four Douglas earls, and Alan Lauder, custodian of Tantallon for Earl William, was among his master's most favoured adherents.[12] The Douglases often relied on, and paid for, the service of the same men as their deputies in holding courts and exacting revenues in their local estates. As well as keeper of Lochmaben, Michael Ramsay was chamberlain and chancellor of Annandale, and John Turnbull combined duties as keeper of Newark and bailie of Sprouston for the fifth earl. Other local officials came from the earl's immediate retinue. Alexander Hume was the fourth earl of Douglas's deputy as keeper of the lands of Coldingham priory, while another squire, Alexander Mure, went on to become steward of Kirkcudbright. Payment for these posts took various forms. Hume received £20 from the £100 paid to his lord by Coldingham, Michael Ramsay was assigned rights to grain from local renders and to the customs charged on local waterways, while Herbert Maxwell, as steward of Annandale, was given licence to take £20 from the proceeds of justice in his court. For the Douglases such fees were well worth paying. It was with the backing of their deputies, keepers, stewards and bailies that the earls exerted influence across the south, both in terms of formal relationships and by maintaining more personal contacts. The nature and extent of this influence can best be understood by an examination of the workings of Douglas lordship in the various communities south of Forth.[13]

LAND AND LORDSHIP I: THE MIDDLE MARCH

For the century from the 1320s the lords of Douglas were by far the greatest landowners in the middle march. Earl William held the lordships of Eskdale, Liddesdale, Lauderdale and Jedworth Forest and the baronies of Cavers, Ednam and Bedrule in Teviotdale. Though 1388 saw Liddesdale, Jedworth and Cavers pass to junior Douglas kindreds, Archibald the Grim added Sprouston and Hawick baronies to the remaining lordships. However, in a period dominated by warfare, the key to Douglas power in the central borders was provided by the Forest of Ettrick and Selkirk. Although control of the Forest had been won through 'force and negotiation' by a series of Douglas lords, and although its importance owed much to its security in war, the value of the Forest to the house of Douglas should also be understood in territorial and political terms. Lordship over the Forest gave the Douglas earls the means to establish their predominance in the middle march. The grant of Ettrick Forest by Robert Bruce to James Douglas represented a unique piece of patronage. The Forest was a new lordship. Before the 1320s it had been the principal royal forest of the south, run by the kings to their own maximum advantage through special officers and courts. In 1324, the 'Emerald Charter' gave Douglas, who had previously been *officiarius*, chief royal official, in the Forest, hereditary control of this system. For the next hundred and thirty years, James's heirs seem to have maintained the structures of this royal forest. The lordship remained divided into three wards: Ettrick, between Ettrickdale and the Yarrow; Yarrow, from Yarrow to Tweed; and Tweed, from Tweed to Gala water. The running of these wards was committed to three pairs of officials, a master and a 'currour' or ranger, which in the fifteenth century were hereditary posts, held by local families in return for a fee of one steading, designated as 'le Maisterstede' and 'le currourstede'.[14]

However, apart from these holdings, what was striking about the Forest when recovered by the crown in the 1450s was its territorial integrity. The eighty-six steadings which comprised the lord's demesne formed a huge proportion of the Forest. Though the church of Glasgow and Melrose abbey had established local rights, the Douglases had no major tenants in the Forest. The inhabitants, even if of growing importance elsewhere, had only limited holdings within the Forest. The Douglases deliberately maintained this situation, making no major grants from the lordship and keeping the lands of forfeited inhabitants like Eustace de Lorraine in Ettrick ward in their own hands. The only example of Douglas patronage within the Forest comes from two charters to William Middlemast in the 1420s, which confirmed the local man as master of the ward of Yarrow and granted two steadings held by a traditional lease. This small-scale and temporary grant to a hereditary official reinforces the impression of deliberate seigneurial control in a period of eroding territorial lordship elsewhere. Such an approach had clear financial advantages. In 1456 the Forest rendered £519 13s 4d in rents, more than wealthy earldoms like Mar and Strathearn. While this must have risen since the end of major warfare, the Forest had always had a value as a source of timber and pasture. Each steading paid a traditional rent of one or two cattle and ten or twenty lambs to the lord, and the Douglases must have gained considerable revenue from these flocks and herds throughout their tenure.[15]

However, it was the men, not the sheep, of the Forest which were of greatest importance to the Douglases. The support they gave to Douglas lords between 1307 and the 1350s was crucial in the rise of the family. The Douglases may have been helped by the lack of any major landed rivals, which led freeholders like the Middlemasts and Robsons to follow them, but they were also aided by the backing they received from crucial local kindreds. Among these were the official families of the Forest. The Pringles, rangers of the ward of Tweed, who held Whitebank and their fee of Redhead near Clovenfords, were long-standing councillors and servants of their lords, while the Scotts, rangers of Ettrick, though less intimately connected to the Douglases, also followed the family in war and politics. It is possible to identify links between the Douglases, the Forest and a number of other rising kindreds in the middle march, such as the Kerrs and Turnbulls. The Kerrs were described as men of the Forest in the 1340s and 'of Ettrick Forest' a decade later, while the Turnbulls received lands at Philiphaugh by Ettrick water from Bruce, and possibly already possessed lands at Whitehope on the Yarrow. During the same period the two kindreds also had ties to Douglas lords, serving Douglas of Liddesdale in the 1340s and Earl William in the 1350s, when Henry Kerr acted as sheriff of Roxburgh. Their kindreds would remain closely bound up with Douglas lordship into the next century. John Turnbull of Fulton was keeper of Newark castle in the Forest, and Andrew Kerr proved to be a diehard partisan of the family in the 1450s and '60s.[16]

The Douglases extended these ties of lordship beyond Ettrick Forest. Earl William granted Richard Kerr lands in Lothian in 1350s, in the 1400s the Turnbulls received the barony of Bedrule in Teviotdale, in 1408 the fourth earl granted Robert Pringle the east demesne lands of Smailholm barony and in 1426 Walter Scott received Lempitlaw from the fifth Douglas earl. Douglas lordship probably also delivered indirect rewards. The Turnbulls' acquisition of Minto in the 1340s, and of Wauchopehead and Hassendean in the 1400s, represented the promotion of a kindred in Douglas service to a position of local significance in upper Teviotdale. Similarly the fifth earl's confirmation of Andrew Kerr's purchase of lands near Primside in 1431 represented the enlargement of the family's holdings, established by a variety of means, in the districts to the east of the Teviot. Though Scotts, Kerrs and Turnbulls were capable of building up their own interests, their rise also reflected the support available from the Douglases to kindreds with long-standing connections to the Forest.[17]

The Forest was not the only source of adherents drawn on by the Douglases to strengthen their position in the middle march. Throughout the century from the 1320s, and especially in the years of major warfare before 1389, Douglas lords sought more than purely military support from followers from beyond the marches. In the 1350s, for example, Earl William Douglas granted an Angus adherent, Thomas Restennet, lands in Roxburghshire, and in the 1380s a Lanarkshire tenant of the earl, Laurence Govan, received 100 shillings' worth of rents in the same sheriffdom. Another family of close Douglas adherents probably also originated in Lanarkshire. The Gladstones may have followed the Douglases from Clydesdale into the borders, and in the 1350s William Gladstone emerged from English captivity to become Earl William's principal local agent as bailie in both the lordship of Lauderdale and the barony of Cavers near Hawick. One kinsman, possibly a son, died at Otterburn, another was captured at Humbleton, while a third, James, defended Cocklaws castle in Teviotdale after the defeat. James Gladstone was also a local official of the Douglas earls in the middle march, and links of office-holding and counsel, rather than landholding, seem to have been the key to relations between lord and man. Again, though, the lands acquired by the Gladstones in Peeblesshire and upper Teviotdale may have been the indirect product of Douglas support.[18]

In the late 1350s and early 1360s, following the years of major Anglo-Scottish warfare, Earl William Douglas sponsored a series of transactions in his lordship of Lauderdale. At a time when the earl was probably keen to reward support and stabilise personal lordship, Douglas increased the holdings of a number of his close allies. For example Alan Lauder, the earl's constable of Tantallon, was allowed to buy Merton, Newbigging and half the mains of Lauder and was granted Wormiston in a series of deals between 1359 and 1366. Similarly, John Haliburton, whose father had died in Douglas's company in 1355, was granted Dalcove, the Maitland family received Thirlestane and other holdings, and in the same category was the grant of Ledgerwood by Douglas's local ally, John lord of Gordon, to William Borthwick, whose family held lands in the Forest. These grants, especially those to Lauder and Haliburton, had a specific political goal. By giving lands in Lauderdale to men whose connections with him had been based on Lothian, Earl William gave these powerful adherents a stake in the defence of his own interests. Haliburton, Lauder and other Lothian men would be more willing to bring their kin and followers from elsewhere to Douglas's support.[19]

It was to secure his own lordship over the local communities of the middle march that William sponsored the interests of men from the Forest, Lothian and elsewhere, in these districts where the principal challenge to the Douglases came from England. The two decades up to 1360 showed the strengths and limits of Douglas lordship in the dales of Tweed, Teviot and Leader. In the early 1340s William Douglas of Liddesdale, as guardian of the Douglas heir and keeper of Roxburgh, attracted followers from Eskdale, the Forest and Teviotdale. Neville's Cross saw the collapse of this following and, from 1347 to 1356, even men from the Forest like John Kerr, James de Lorraine, Walter Turnbull and William Gladstone were in English allegiance. In these circumstances of apparent defeat, the desire to protect lands which they had received in Teviotdale made them accept English lordship. However, in the longer term, previous ties of land and lordship to the Douglases and the effectiveness of Earl William's campaigning saw a return to his following. While John Kerr 'of Ettrick Forest' remained in English allegiance, his kinsman, Henry, supported Douglas and the 1356 campaign brought others, including the Turnbulls, back to Scottish allegiance. By the late 1350s those men of the Forest still in the English camp were being paid heavily for their loyalty by their lords. John Kerr received Attonburn and Mow and James de Lorraine obtained an annuity of £20 and Caverton. The 1356 campaign also brought men from Teviotdale into Douglas's following. Bernard of Hadden, who had witnessed a charter at English-held Roxburgh in 1354, was forfeited by the English as a rebel in 1357 and on Douglas's council soon after. His example was copied by others, including Henry Haig of Bemersyde and William Learmouth, but many Teviotdale men, William Rutherford, William Roulle and Robert Colville among them, remained in English allegiance.[20]

These men or their kin were finally won back during the local warfare of the 1370s and 1380s. Families like the Rutherfords, De Lorraines and Colvilles had relinquished lands elsewhere in Scotland by their adherence to England, and the possibility of regaining them, agreed by the king in 1384, allied to the pressure of Scottish attacks, had a gradual effect. The combination of Douglas lordship and royal patronage brought about the submission of Robert Colville of Oxnam in 1384. He recovered the barony of Ochiltree in Ayrshire forfeited by his ancestors. He also entered the circle of the earl of Douglas. His heir married the daughter of Douglas's

ally, James Lindsay of Crawford, and the Colvilles appeared as councillors
of Earl James and adherents of the Black Douglases in the 1380s, 1390s
and 1400s. The same process accompanied the recovery of Robert Roulle
of Primside, Richard Rutherford and William Stewart of Teviotdale to
Scotland, while the Kerrs were allowed to retain the lands of Attonburn,
granted by English owners but held in Scottish allegiance from the Black
Douglases. While the English passed sentences of forfeiture against these
defectors, they were forced to admit their inability to recover the lands
concerned. The contraction of English control to the immediate environs
of Roxburgh was both cause and consequence of these defections. The
ability of Douglas earls to lead, protect and punish local men and affect
their tenure of land determined loyalties in Teviotdale. While adherence
to England was signalled by attendance at the courts of the constable of
Roxburgh, from the 1340s to the 1420s return to Scottish allegiance was,
for many middle march men, accompanied by service to and reward from
Douglas lords.[21]

Douglas lordship and the goals of Douglas lords moulded political society
in the central borders in the fourteenth century. The family's leadership
in war, described in the previous chapter, was harnessed to bonds of land
and service to extend Douglas influence in Lauderdale, Teviotdale and
Tweeddale. Families like the Kerrs, Colvilles and Gladstones, which forged
close ties with the Douglas earls, received lands and offices. Those which
refused to come to terms with the magnates whose estates and influence
dominated the region declined. Most obviously, the de Lorraines, Auldtons
and Burrells, who persisted in English allegiance, lost lands and goods in
Scotland, but even within the Scottish faith local men lost out to those like
the Kerrs or Alan Lauder who had the support of the earls of Douglas.
The earls' deliberate patronage of such adherents in the region continued
into the fifteenth century. The Black Douglases granted away Hawick to
William Douglas of Drumlanrig in the 1400s and supported his appropriation of Selkirk burgh and the acquisition of Cavers by his brother,
Archibald, seeing them as manageable neighbours. The attitude of the
Douglas earls to their border estates in Teviotdale and Tweedale contrasted
consistently with their treatment of the Forest. Earl William granted extensive lands in Lauderdale and the whole barony of Bedrule while
Archibald the Grim and his heirs disposed of, not just Hawick, but half
the demesne of Smailholm and virtually the whole barony of Sprouston
between the Tweed and the Cheviots. Grants from this exposed estate were
made to Scotts, Kerrs and others who could be relied on to defend their
lands. While Ettrick Forest was preserved as an economic and political
base, the Douglas earls used their border lands further east as the means

of winning and maintaining service from crucial local kindreds. In the marches, land, lordship and war leadership were closely bound together.[22]

LAND AND LORDSHIP II: THE WEST MARCH AND GALLOWAY

Between 1369 and 1409 the Black Douglases established a stranglehold on the west march. The lordships of Galloway, Annandale, Nithsdale and Eskdale and the earldom of Wigtown were all acquired by the family and, as in the middle march, territorial resources and personal lordship were linked in the management of local communities. Galloway east of Cree and Wigtown were the first parts of the region in which Archibald the Grim was given lands. His authority in these estates was established by forceful means, and Archibald's early rule suggests a determination to quash any local opposition by increasing his and his supporters' lands in the province at the expense of native families. Before 1371 he extracted Borgue from the MacDowells and gave it to his ally James Douglas of Dalkeith who, in turn, granted the lands to his bellicose brother, Henry. A story exists that, at about this time, Archibald expelled the Agnews from their lands in Wigtownshire and, whether true or not, by the early fifteenth century one William Douglas, possibly another kinsman of Dalkeith, was lord of Leswalt in the same region. By the 1380s Archibald sought the same ends by financial means. Galloway, subjected to warfare since the 1290s, was still depopulated in the 1380s and the rents of Buittle and Preston in the 1360s were between a quarter and a half of pre-war levels, although they recovered under Archibald's rule. Despite increasing revenues Archibald, backed by his rich Clydesdale baronies, exploited the poverty of local lords. He bought the valuable lands of Balmaclellan from Gilbert MacLellan, a member of one of the main Galwegian kindreds, and obtained Stockarton near Kirkcudbright for ten years in return for a £40 loan to its lord.[23]

The Douglases also used their estates in Galloway to provide for their adherents. Probably in the 1390s Archibald's allies, the Glendinnings, received Parton in the Glenkens and, after 1400, the fourth earl gave lands in the province to a number of his men, including his chancellor, Alexander Cairns, and his familiars Michael Ramsay and William Hay. Even after thirty years as lords, the Douglases still chiefly relied on such men as their local officials. The stewards of Kirkcudbright, Thomas Hert and Alexander Mure, were both Douglas servants from elsewhere, and the sheriff of Wigtown in the early 1420s was William Douglas of Leswalt. These appointments suggest mistrust between the Douglases and the native kindreds, and in 1418 the earl ordered his officers to 'impose distress' on men refusing their rents to their new lord, the 'outsider', William Hay. Force remained an element in Douglas management of Galloway fifty years after Archibald the Grim had 'pacified' the province.[24]

However, neither bringing new men into Galloway nor the use of force against native kindreds marked the Douglases out as different from earlier lords. Edward Balliol had rewarded his partisans with lands in Galloway, and even the native lords of Galloway had maintained Anglo-French tenants and servants. Families from elsewhere in the south, like the Gordons of Stichill and the Maxwells of Caerlaverock, had estates in Galloway before 1300, and the fourth earl confirmed and extended these holdings in return for their support. More importantly, the Douglases saw themselves as heirs to the rights of previous lords. Although they had been granted Galloway in two parts, the first simply of crown lands between Nith and Cree, the second of the earldom of Wigtown without regality rights over justice and administration, the Douglases treated the province as a single lordship with its own legal customs. In 1384 Archibald the Grim defended these customs in a general council. In particular he secured Galloway's exemption from laws concerning the pursuit of criminals by citing the structures which existed under the *leges Galwidiensis*. These structures, however, revolved around the native kindreds and their captains. The captains maintained officers, sergeants, with powers of summary justice, and played a key role in the containment of local blood feuds. They could also demand *calp*, a levy of livestock, from lesser men in return for protection. Such powers lasted at least to the end of the fifteenth century, and the act of 1384 makes clear that the Douglases did not seek to challenge local custom. Archibald was probably not just defending the traditional rights of his tenants. It is likely that the lord of Galloway was protecting his own powers of justice and lordship. The rentals of Galloway from the 1450s record extensive renders in cattle and meal which may include traditional payments for the lord's protection. The harrying of those who refused their rents to William Hay in 1418 by the lord's officers probably represented the use of powers of coercion in the robust tradition of Galwegian lordship. The officers, Robert Crichton from Dumfriesshire and the 'fallow Macgyewe', perhaps a local man, were perhaps the lord's sergeants west of the Cree.[25]

Although the Black Douglases had taken over Galloway by cutting down the captains, as lords they had no desire to disrupt the structures of law and lordship in the province. The heads of kindreds like the MacDowells and MacLellans retained their importance, and after the 1380s there is no evidence of continued appropriation of their lands by the Douglases. Instead these captains were vital allies of the Douglas lords of Galloway, with the

McCullochs and MacDowalls providing men and ships for Douglas military expeditions and serving as the earls' councillors in the south-west; and by 1414 Uhtred MacDowell was sheriff of Wigtown, an office apparently created by the Douglases rather than the crown. In return the Douglas lords of Galloway gave these kindreds protection after decades of war. Archibald the Grim's attack on Lochmaben in 1384 was in response to the damage inflicted on the Galwegians by the garrison. Archibald's bastard son, William Douglas of Nithsdale, provided active military leadership. His raid on Carlingford in 1388 was launched with men from the south-west and returned to Loch Ryan, and William's local standing may have been inherited by his son and namesake who headed Earl Archibald's council in the south-west on several occasions. In 1424 on his departure to France, the earl could safely leave Galloway in the hands of his countess, and her rule in the province, though not without disturbance, was clearly acceptable, even welcome, to local men who wished for the benefits of lordship based within Galloway. Countess Margaret, for her part, took her secretary, Thomas McGuffoch, and chancellor, John McCulloch, from native kin-dreds and took council from MacLellans and MacDowells as well as Douglas of Leswalt and Alexander Mure, her husband's men.[26]

Black Douglas lordship in Galloway, based on traditional rights and occasional force, had similarities with the management of Gaelic kindreds by the Stewarts in Atholl and Moray. Such overlordship also had parallels with Douglas lordship in the upland regions of the marches. In Liddesdale and Eskdale, for example, magnate influence rested on the relations between lord and a handful of powerful local kindreds. In Eskdale, a lordship formed from the baronies of Westerkirk, Staplegordon and Kirkandrews, the Doug-las earls, though feudal superiors, had only minimal lands as demesne. Much of the lordship, including its legal centre at Dalblane, was held by the Douglases of Dalkeith as lords of Westerkirk and Staplegordon, but of even greater importance locally were kindreds like the Glendinnings. The Glendinnings were, probably hereditary, bailies of the Douglas earls in Eskdale and were responsible for exercising the lord's judicial and territorial rights in the lordship to mutual profit, and for bringing their kin and neighbours to serve the earl in war. In return the family was established at the head of local society. Kindreds like the Glendinnings and the Arm-strongs and Nixons in neighbouring Liddesdale were, in circumstances of semi-permanent war, increasingly assuming formal roles in the management of local justice and politics. The inclusion of members of all three families as guarantors of the Anglo-Scottish truce for Archibald the Grim in 1398 was a recognition by the earl of the importance of these kindreds in the militarised society of the marches. Both Black Douglases in Eskdale and, from 1400, Red Douglases in Liddesdale, sought to exercise their rights through such families.[27]

The final area of Douglas power in the south-west lay between Eskdale and the Nith, a region dominated by the lordships of Annandale and Nithsdale. Nithsdale retained little in the way of territorial significance. It was held by William, bastard son of Archibald the Grim, for only four years before his death. Resumed by David earl of Carrick, by 1406 it had been successfully recovered by William's son and namesake who acted as a councillor of his uncle, the fourth earl of Douglas, until his death in 1419. Nithsdale then passed to William's sister and her husband, Henry earl of Orkney, but the couple, whose holdings were surrounded by those of the Douglas earl, experienced trouble in obtaining their rights from local men who, despite Douglas's earlier friendship for Orkney, may have had Earl Archibald's support.

By contrast with Nithsdale, the lordship of Annandale remained a re-markably consistent and coherent local community. Despite a century of war which left the lordship's main castle of Lochmaben in English hands for fifty years, families like the Johnstones, Kirkpatricks, Jardines, Corrys and Herries, councillors of their Bruce lords in the early thirteenth century, attended a succession of lords from the 1320s, the Randolphs, David II, the Dunbars and, from 1409, the Black Douglases. Emerging families like the Carruthers and Carlyles probably had local origins, and only a few, the Crichtons and Murrays of Cockpool for example, were from outside. Although the Douglases only became leaders of this close-knit community in 1409, their influence as neighbours and march wardens had been felt in Annandale since the 1360s. Significantly, the family's closest connection in Annandale before 1409 had been with the Murrays of Ae and Cockpool, junior kinsmen of Countess Joanna, who were in the service of both Archibald the Grim and the fourth earl. After 1409, these kinsmen received lands in Annandale and elsewhere from the Douglas earls.[28]

For other local lords, the pull of Douglas lordship must have increased with the occupation of Lochmaben by the fourth earl, probably in 1401, and during the next eight years there were strong indications of Douglas influence over local men. Against this background, George Dunbar's resig-nation of Annandale to Earl Archibald was hardly a major upheaval. Despite this, the new lord's first formal court at Lochmaben in late 1411 was clearly a significant gathering. Grants and confirmations of land were issued by the earl to his new tenants, and the witnesses to these documents show Douglas surrounded by an impressive council of family and retainers and attended by local men, John Carlyle, Humphrey Jardine, Thomas Murray, Robert Herries and Simon Carruthers. These men were anxious to secure the good lordship of Douglas, his protection and patronage. For his part, the earl was keen to bind Annandale into his wider connection, granting at least two local landowners lands in Galloway, and bringing in adherents from elsehwere as new tenants and officials. Even in a lordship like Annandale, where tenurial bonds between lord and vassal remained strong, personal relationships were the key to the exercise of magnate power.[29]

The significance of such personal bonds is clear in the relations between the Douglases and the Maxwells of Caerlaverock. The Maxwells were not early adherents of the Black Douglases. Landowners in Roxburghshire as well as the south-west, they were in the front line of warfare and, as intermittent supporters of the Balliols in the south-west, the family had suffered during the decades of local war. They probably resented Archibald the Grim's growing power in the region and supported his rivals in the 1370s. Only the threat of renewed war with the defection of George Dunbar in 1400 induced Robert Maxwell of Caerlaverock to deal with the fourth earl of Douglas. There was a clear value for the earl in securing the services of a family which retained influence in Nithsdale and Annandale as well as a stronghold like Caerlaverock. In the next few years, Robert's son, Herbert, stood hostage for the earl in England and, in return, Douglas entered an agreement with him which promised Maxwell an annuity of 40 marks to be paid until he received lands worth £40. The bond promised more than just material reward. Douglas also guaranteed to 'supowelle and defende' Herbert in 'all his ryghtwys cause' as 'our man and our kosyn', linking the grant of cash and land with the promise of support and protection. For Douglas to state the terms of his lordship was unusual. It may reflect lingering mistrust, but was also a formal declaration of lordship between men who lacked established landed or political connections. There is certainly no indication of renewed tension between Douglas and Maxwell, and in 1410 the earl made his 'kosyn' his steward of Annandale. Though the appointment of Michael Ramsay to a number of lesser offices in the lordship may have been designed to balance Maxwell's local authority, the grant represented a major reward for the lord of Caerlaverock. A family which had lost out in competition with the Douglases established a basis for influence in Annandale as servants of their former rivals.[30]

LAND AND LORDSHIP III: LOTHIAN AND LANARKSHIRE

The support of Lothian knights and barons was crucial in the rise of the Douglases. The lords and earls of Douglas sought their services as councillors, agents and allies in war and politics. The backing of these allies, who were secure in Scottish allegiance from the 1340s, was a major element in Douglas dominance further south. The family's main cadet branch, the Douglases of Lothian, was one of these locally powerful dynasties. In the knight of Liddesdale and Douglas of Dalkeith it produced men with the resources to pursue ambitions well beyond Lothian. The Douglas earls harnessed the resources of similar families to their own interests in the marches and sought to become leading lords within Lothian itself. To this end, they acted as physical defenders of the province and offered landed rewards to their Lothian adherents. Their numerous grants of land to these men in Lauderdale and elsewhere have already been discussed as a means of strengthening the Douglas earls' hold on lordships threatened by the English. In addition, such patronage created bonds of land lordship between the Douglases and the Haliburtons, Sinclairs, Lauders and other Lothian nobles. The same goal lay behind the grants made by the fourth earl of Douglas to a number of Lothian men. James Dundas, William Cockburn, William Towers and Archibald Hepburn received lands in Douglas's baronies of Bothwell and Drumsergart. These Clydesdale baronies were not under physical threat. Instead they provided rich rewards for lesser men, securing their service to the earl.[31]

However, the importance of such bonds to the earls was a product of their limited resources within Lothian. The Douglases had none of the territorial power which they possessed in the marches. Robert I, so generous to James Douglas further south, followed his predecessors in giving very little to great lords in Lothian and, during the fourteenth century, families like the Haliburtons and Douglases of Lothian made greater gains in the province than James's heirs. As a result, the Douglas earls were forced to build their lordship on a network of personal connections. Moreover, the wealth and influence of the Haliburtons of Dirleton, Hepburns of Hailes and Setons, which made them so important to the Douglases, also placed limitations on the earls' ability to control these families. They had their own ambitions and sought Douglas lordship largely to secure them. For example, between 1400 and 1402 the fourth earl provided local barons with political and military backing in their efforts to extract themselves from the earldom of March. Such goals hardly indicated a desire to become lasting adherents of another magnate, and Douglas's allies showed a limited acceptance of his lordship. His links of land with men of the province were often with lesser nobles or junior kinsmen of major families and, while John Seton became the earl's close councillor, others like Walter Haliburton retained other connections. He married Albany's daughter and, against Douglas's will, negotiated the Dunbars' return from exile, but he also supported James Douglas at Long Hermiston in 1406 and his son married

Earl Archibald's sister, the widowed duchess of Rothesay. Such a lord was a useful ally, but free from any fixed dependence on the Black Douglases.[32]

The absence of any dominant local magnate made such independence feasible. The Dunbars and, from 1388, the Red Douglases were alternatives to the Black Douglas earls, and even the Sinclair earls of Orkney and barons of Roslin possessed land and connections which could disrupt the Douglas earls' influence in Lothian. Most importantly, as David II had shown in the 1360s, such circumstances made it relatively easy for royal lordship to win support in a province where the dominance of magnates lacked deep roots. The periods of greatest Douglas influence in Lothian came when rivals were weakest. In the 1350s Earl William and his ally, Earl Patrick Dunbar, were the only sources of effective lordship for local barons, while after 1400 the eclipse of royal, Dunbar and Red Douglas interests allowed the fourth earl of Douglas a personal hegemony in Lothian. Significantly, these were also the periods during which the Douglases were masters of Edinburgh castle and, linked to this, wielded most influence with the burgesses of Edinburgh. Edinburgh provided manpower and money for its effective patrons, and Earl William witnessed burgess charters and was named in their prayers, while the fourth earl was associated with the provost and community of Edinburgh in the request to have St. Giles, then the parish church, erected into a college of priests. Since the English invasion of 1400, Douglas had been the burgh's principal guardian, residing with his household in the town, and the earl may have regarded his appropriation of Edinburgh's customs revenue as payment for these and other tasks.[33]

However, Douglas's frequent presence in Edinburgh may also have reflected his need to maintain regular, personal contact with the men of Lothian. The same motive probably influenced Earl William's construction of Tantallon. Even so the earls had no impact on local political society comparable to their lordship in the marches. Only the fall of the Dunbars in 1400 had major significance, and in Lothian this largely worked to the advantage of families like the Hepburns and Haliburtons, who were already keen to assert their independence. In Berwickshire, by contrast, the brief period of Black Douglas dominance had a long-term local significance. By making Alexander Hume his deputy in the running of the lands of Coldingham priory and by giving Hume's brother, David, the lands of Wedderburn and others, Douglas did much to create the shape of local politics. For the next century, Berwickshire would be dominated by the many-branched Hume kindred, feuding amongst themselves over the lands of Coldingham. Though the Humes possessed the resources to rise rapidly in local society, it was the promotion of Alexander by his personal lord, Douglas, which provided the spark for this success story. As with the Kerrs and Scotts further west, this border kindred advanced to prominence in the context of Douglas lordship. In Lothian itself, the earl's principal creation had huge implications for the house of Douglas. By giving Abercorn castle to his brother, James, the fourth earl rewarded his loyalty between 1402 and 1409. Unwittingly, the earl also ensured that the fate of his family would be bound even more closely with the management of the barons of Lothian. James Douglas, as head of a new cadet branch of the Black Douglases, built his career as one of these barons.[34]

James Douglas of Abercorn was also granted the barony of Strathaven and lands in Stonehouse barony in Lanarkshire from the inheritance of his mother, Countess Joanna Murray. This formed part of a wider provision for her younger son which also included a number of her north-eastern estates, but the alienation of these Clydesdale lands also indicated the limited importance of the region to the fourth earl of Douglas. He still retained the baronies of Douglas, Bothwell, Drumsergart and Crawfordjohn and other lands but, despite this territorial base, the earl made no obvious efforts to build up a major following in the area. Links with local baronial families, like the Somervilles, Hamiltons and Flemings, existed but seem less frequent than similar ties with Lothian lords. The Black Douglases do not seem to have sought support from any following associated with the Murrays as lords of Bothwell, and the old connections of the earls in upper Clydesdale never recovered after 1388. Earl William's local allies, the Lindsays of Crawford and the Sandilands, did not establish similar close ties with the Black Douglases. Instead, the Lindsays concentrated on their growing interests north of Forth while the Sandilands were left to deal with unsympathetic lords. Evidence from 1434 suggests the Douglas earls retained a hostile attitude to tenants who had presumed to aspire to their earldom, and before 1424 there can have been little escape from the hostility of the Black Douglas earl for a minor landowner when even magnates like the Dunbars were denied their full rights.[35]

This survey of the regional interests of the earls of Douglas indicates the ways in which the presentation and exercise of the family's power was tailored to the expectations and resources in different communities. The captains of Galloway responded to lordship which incorporated the traditional rights of the province's rulers. The kindreds of the middle march regarded Douglas leadership as the alternative to English allegiance, entering or deserting the family's following with regard to physical survival. In Lothian, while considerations of protection were still important, major lords saw the Douglases as one of a number of magnate patrons whose power was valuable as a means to their own ends. For lesser men of the province, however, service to the Douglases was a source of employment and hope-fully of land and status. The landed power of the earls was not the only factor in determining their pull as lords, but the estates which the Douglases had accumulated during the century up to 1424 were exploited by them to cement their personal lordship. Rents in cash and food were the basis for the maintenance of an impressive and potent household and for the construction of a network of residences which enhanced the family's image. Moreover, possession of lordships like Galloway, Annandale and the Forest gave the Douglases the leadership of local communities, networks of kinship and friendship which, despite the turmoil of the fourteenth century, retained a degree of significance. Finally, land was a reservoir of patronage. The grants of land made by the first and fourth Douglas earls in particular showed magnates deliberately forging links of lordship based on landhold-ing. The recipients of such grants were often already their servants or allies, but land was held of the earl 'for service done and to be done'. Though such services were not specified, the lord expected support in politics and, often, in war from his new tenants.

The scale of Douglas patronage and indirect support altered the balance of political society in many parts of the south. The rise of families like the Humes, Kerrs, Scotts and Maitlands to positions of local importance can be traced back to their promotion in Douglas service. Grants of land and, more importantly, of offices to these families were exploited by men who, in the absence of the earl in other lands or kingdoms, became accustomed to the exercise of local justice and government. Though such service was indispensable to the earls' running of such a wide accumulation of lands and interests, the rise of lesser families contained long-term problems for the dynasty. Their power was established and maintained through a net-work of relationships which was essentially personal in character.

The greatest of these was the alliance between Archibald the Grim and James Douglas of Dalkeith. The rise of the two kinsmen was a virtual partnership. Where Archibald gained lands and influence, James also gained a stake. By the 1400s, the Douglases of Dalkeith were among the earls' principal vassals in Galloway, Annandale and Eskdale, as well as one of the family's principal allies in Lothian. These lands were, in part, the fruits of an alliance which had provided vital support for the Black Douglases in the south-west, but, at root, it rested on the close relationship between two men. The failure of the heirs of James and Archibald to maintain the same friendship would create problems, not just for the lesser family but also for the magnate house. Across the south, Douglas lordship depended on similar, if less wide-ranging, connections. The crisis of 1388 showed the potential for the fragmentation in Douglas lordship should such bonds be broken.

If Archibald the Grim's success in reforming the earls' following in the marches was a tribute to his reputation and ability, it was also the product of the continued need, from Galloway to the Tweed, for protection and leadership in war.

War was a powerful bond between Douglas magnates and their men. The gradual cessation of major conflict in the marches between 1389 and 1420 threatened to weaken this bond. While it brought greater wealth and security from attack to the Douglas earl, peace placed new strains on the maintenance of his following. It was partly the search for new fields for his military lordship that drew Archibald, fourth earl of Douglas, to France.

NOTES

1. *The Bruce*, Book 5, lines 270–79, 292–300.
2. *R.M.S.*, i, appendix 2, no. 192; *Clement VII, Letters*, 67–68; British Library, Add. MSS, no. 6443, 19v–20r.
3. Fraser, *Douglas*, iii, no. 300.
4. Hume of Godscroft, *History of the House and Race of Douglas*, 239–40; *Melrose Liber*, ii, no. 540; *R.M.S.*, ii, no. 143.
5. *Caerlaverock*, ii, nos 21, 25; *H.M.C.*, vii, 728; *R.M.S.*, ii, no. 255; S.R.O., GD 10/19; *Melrose Liber*, ii, nos 512, 545; Fraser, *Douglas*, iii, nos 360, 373, 387; A. Agnew, *The Hereditary Sheriffs of Galloway* (Edinburgh, 1893), 239.
6. *Scotichronicon*, ed. Watt, vii, 96–97, 140–45, 154–57; *H.M.C.*, Hamilton, nos 21, 130; vii, 728; *Morton Reg.*, ii, nos., 43, 71; Fraser, *Douglas*, nos 18, 323; *Melrose Liber*, ii, nos 490, 519; *Calendar of Papal Registers, Petitions*, ed. W. H. Bliss (London, 1896) i, 200; Watt, *Graduates*, 82–84; S.R.O., GD 12/16; 32/3/2; Fraser, *Carlaverock*, ii, nos 21, 25; Fraser, *Douglas*, iii, nos 346, 353, 354, 356, 367; *Morton Reg.*, ii, nos 203–205; Fraser, *Scotts of Buccleuch*, ii, no. 22; N.L.S., MSS, 80.4.15, no. 4; *R.M.S.*, ii, nos 112b, 255, 364; *Wigtownshire Charters*, no. 133; N.R.A.S. Mansfield Muniments, no. 6; *C.S.S.R.*, i, 220–21; ii, 55; *Rot. Scot.*, ii, 230, 233, 235.
7. *E.R.*, iv, 277, 300; *Cal. Docs. Scot.*, iv, no. 782; *C.S.S.R.*, i, 102, 192–93; Fraser, *Douglas*, iii, no. 380. Livingston, like other Douglas familiars, broke customs probably with the earl's permission (*E.R.*, iv, 301).
8. *A.P.S.*, i, 573; ii, 35. For an examination of Scottish noble retinues, see J. M. Wormald, *Lords and Men*, 42–46, 91–98. For English comparisons, see M. Cherry, 'The Courtenay Earls of Devon: the formation and disintegration of a late medieval aristocratic affinity', *Southern History*, 1 (1979), 71–97; S. Walker, *The Lancastrian Affinity*; A. Goodman, *John of Gaunt*, 312–21; K. B. Macfarlane, *The Nobility of Later Medieval England* (Oxford, 1973), 110–11; C. Rawcliffe, *The Staffords, Earls of Stafford and Dukes of Buckingham* (Cambridge, 1978); C. Given-Wilson, *The English Nobility in the Late Middle Ages* (London, 1987), 87–103.
9. *Copiale*, 238; *Cal. Docs. Scot.*, iv, no. 787; Fraser, *Douglas*, nos 57, 298, 336, 362, 373, 382, 387, 401; *R.M.S.*, ii, nos 70, 119; S.R.O., GD 10/16; GD 157/75, 260; RH 6/296; *Wigtownshire Charters*, no. 131; *H.M.C.*, vi, 710; vii, 728; Milne Hume, nos 1, 2; *A.P.S.*, ii, 35. Christian Ramsay herself received lands from Douglas to be held by her and any son 'begotten by mutual intercourse' (Fraser, *Douglas*, iii, no. 60).
10. *R.M.S.*, ii, no. 254; *Scotichronicon*, vii, 403; Fraser, *Douglas*, iii, no. 342; S.R.O. GD 120/76; GD 350/1/952.
11. *Cal. Docs. Scot.*, iii, no. 746; *R.M.S.*, i, nos 696, 697; ii, nos 12, 19, 70, 143; Fraser, *Douglas*, iii, nos 360, 392, 393, 398, 401, 406; *Melrose Liber*, ii, nos 491, 541–43; S.R.O. GD 134/3; 157/75.

12. *R.M.S.*, ii, no. 143; *E.R.*, iv, 19, 42, 78, 252, 321; Fraser, *Douglas*, iii, nos 342, 356; British Library, Harleian MSS, no. 6443, 19r. –20v.; N.R.A.S. no. 832, Lauderdale Muniments, 44/36.

13. *R.M.S.*, ii, nos 143, 242; Fraser, *Douglas*, iii, nos 298, 390, 406.

14. *R.M.S.*, i, appendix 1, no. 38, appendix 2, no. 232; ii, nos 58, 59; *E.R.*, vi, 223–25.

15. *Melrose Liber*, i, no. 13; *Glasgow Reg.*, i, no. 30; *E.R.*, vi, 223–29; vii, 524–31; *R.M.S.*, ii, nos 58, 59. Deloraine in the ward of Ettrick was held by the lord in the 1450s, probably as a result of the forfeiture of the De Lorraine family in the previous century (*E.R.*, vi, 224; *R.M.S.*, i, no. 463). Melrose had a right to timber from the Forest and to pasture in lands from Gala water to the Leader granted by David I. Glasgow's tenants had rights of pasture along lower Ettrick water.

16. *Origines Parochiales*, i, 241, 246; *E.R.*, i, 568; ii, 38; vi, 225; S.R.O., GD 157/75; R. H. 6/155; *Rot. Scot*, ii, 181; *H.M.C.*, Milne Hume, no. 1; Fraser, *Douglas*, iii, nos 367, 406; *Cal. Docs. Scot.*, v, no. 809; *R.M.S.*, i, no. 22, appendix 2, no. 136; *S.H.S. Miscellany*, v, nos 13, 16.

17. *R.R.S.*, vi, no. 191; G. F. S. Elliot, *The Border Elliots* (Edinburgh, 1897), appendix no. XI; S.R.O. GD 157/75; Fraser, *Scotts of Buccleuch*, ii, no. 25; *R.M.S.*, i, no. 922, appendix 2, nos 917, 1034, 1904; Fraser, *Douglas*, iii, nos 393, 406.

18. *R.M.S.*, i, nos 208, 365, 692, 780, appendix 2, no. 1275, 1723, 1966; *Cal. Docs. Scot.*, iii, no. 1635; iv, no. 18; Fraser, *Douglas*, iii, nos 25, 365; *A.P.S.*, vii, 142; *Scotichronicon*, ed. Watt, vii, 439; viii, 53; *H.M.C.*, Roxburghe, 23.

19. *A.P.S.*, vii, 139, 142, 159; Fraser, *Douglas*, iii, no. 334; *H.M.C.*, Hamilton, nos 125, 126; *Yester Writs*, no. 56.

20. *Morton Reg.*, ii, nos 114–17; *Cal. Docs. Scot.*, iii, nos 1521, 1634; iv, nos 1, 12, 24, 62, 89; *S.H.S. Miscellany*, v, nos 13, 16; *R.R.S.*, vi, no. 130; S.R.O., R. H. 6/155.

21. Boardman, *Early Stewart Kings*, 121; Grant, 'The Otterburn War', 47; *Rot. Scot.*, ii, 60; *A.P.S.*, i, 553–54; *H.M.C.* Drumlanrig, no. 2; Fraser, *Douglas*, iii, no. 335; *North Berwick Chart.*, xxxvii; S.R.O., GD 150/78; *R.M.S.*, i, appendix 2, no. 1938; ii, no. 255; *Cal. Docs. Scot.*, iv, nos 321, 348, 426. Ochiltree had been granted to the Colvilles in the 1340s (*R.M.S.*, i, appendix 2, no. 1089).

22. *Cal. Docs. Scot.*, iv, nos 430, 523; Fraser, *Scott*, ii, no. 22; *H.M.C.*, Drumlanrig, i, no. 4; *R.M.S.*, i, appendix 1, no. 156; S.R.O., GD 157/75. The grant of a large estate in Sprouston to the Charterhouse of Perth in 1434 was clearly not made with defensive considerations in mind (Fraser, *Douglas*, iii, nos 396–97).

23. *R.R.S.*, vi, no. 451; *R.M.S.*, i, no. 507; *Wigtown Charter Chest*, no. 7; Agnew, *Sheriffs of Galloway*, i, 280–93; British Library, Harleian MSS, no. 6439; *Clement VII, Letters*, 67; *Mort. Reg.*, i, lix–lxi; ii, no. 83; N.R.A.S., 832, Lauderdale Muniments, no. 77; *H.M.C.*, Hamilton, no. 127.

24. *C.S.S.R.*, i, 320; *R.M.S.*, ii, nos 12, 70, 133, 255; Fraser, *Douglas*, iii, nos 367, 371; Agnew, *Sheriffs of Galloway*, i, 238.

25. Reid, 'Edward de Balliol'; K. Stringer, 'Periphery and Core in Thirteenth-Century Scotland: Alan son of Roland, Lord of Galloway and Constable of Scotland', in Grant and Stringer, *Medieval Scotland*, 82–115; *R.R.S.*, vi, no. 451; *R.M.S.*, i, no. 507; *A.P.S.*, i, 551; H. L. MacQueen, 'The Laws of Galloway, A preliminary survey', in Oram and Stell, *Galloway, Land and Lordship*, 131–43; *E.R.*, vi, 194, 565; Fraser, *Douglas*, iii, no. 371.

26. *H.M.C.*, 10, appendix 6, 77–78; *Scotichronicon*, ed. Watt, vii, 412–15; vii, 48–49; Boardman, *Early Stewart Kings*, 286; Fraser, *Douglas*, iii, nos 360, 368; *R.M.S.*, ii, nos 12, 133, 255.

27. *E.R.*, vi, cxiv, 556–57; *Mort. Reg.*, i, lxxv–lxxvi; *Foedera*, viii, 58–60; Fraser, *Douglas*, iii, no. 58. Kirkandrews on the border with England was granted to William Stewart of Durrisdeer in 1431 (Fraser, *Douglas*, iii, no. 68).

28. *H.M.C.*, Drumlanrig, i, nos 68, 69, 70, 71, 75, 76, 77, 79, 110; *R.R.S.*, vi, nos 165, 262; *S.H.S. Miscellany*, v, no. 17; N.R.A.S., no. 776, Mansfield Muniments, 1835/3–6. John Johnstone, Herbert Corry, John Carlyle and John Carruthers were among Archibald the Grim's sureties for keeping the 1398 truce. The lands of Douglas of Nithsdale,

Douglas of Dalkeith and Douglas of Drumlanrig in the region must also have increased Black Douglas influence in Annandale (*Foedera*, viii, 58–60).

29. *H.M.C.*, Drumlanrig, i, no. 110; N.L.S., Adv. MS. 34.1.10; N.R.A.S., no. 776, Mansfield Muniments, 1835/4; *R.M.S.*, ii, no. 85; Fraser, *Douglas*, iii, no. 382. Douglas's grant of land in Galloway to an Annandale tenant, Gilbert Grierson, may have been influenced by his continued links with the Dunbars (S.R.O., RH6/228, 260; *Lag Charters*, no. 3).
30. Fraser, *Caerlaverock*, ii, nos 21, 22; *R.M.S.*, ii, no. 242.
31. *R.M.S.*, ii, nos 119, 254, 255, 256; *H.M.C.*, vii, 728.
32. *Scotichronicon*, ed. Watt, viii, 72–75; *Wyntoun*, ed. Laing, iii, 94; *E.R.*, iv, 2, 19, 43, 52, 278, 300.
33. *St. Giles Chart.*, 7–8, 12–14; *C.S.S.R.*, i, 77. The first provost of St. Giles was Edward Lauder, a clerk who was chancellor of the fifth earl in 1426. The fourth earl also used his influence at the Papal court on behalf of Holyrood abbey on several occasions (*C.S.S.R.*, i, 131, 216, 230).
34. Fraser, *Douglas*, iii, no. 298; *H.M.C.*, Milne-Hume, nos 1, 2; *R.M.S.*, ii, no. 38.
35. *R.M.S.*, ii, nos 39, 40, 43, 49; Fraser, *Frasers of Philorth*, nos 17, 19; Fraser, *Douglas*, iii, no. 344, 374; N.L.S., Adv. MSS, 80.4.15, no. 4; S.R.O., GD 119/164. Lesser families like the Dalziels and Symingtons do seem to have transferred their loyalties to the Black Douglases after 1388 (Fraser, *Douglas*, iii, no. 385; *R.M.S.*, ii, no. 13).

CHAPTER NINE

'Eldest Son of the Pope': The Douglases and the Church

PROTECTOR AND DEFENDER

The lords and earls of Douglas made much of their fame as defenders 'of the liberty of the kingdom'.[1] This reputation was largely formed and transmitted by Scottish churchmen. Poets and chroniclers like Barbour, Bower, Wyntoun and Holland were clerics. Their view of liberty related, not just to the survival of the Scottish kingdom, but to the defence of the rights and freedoms of the Scottish church, at liberty from English domination.[2] While the perceptions and priorities of these clerical historians differed from the goals of the Good Sir James, the knight of Liddesdale and Earl William, and the experience of the communities of the south, Douglas warleaders were conscious of their duty to the church. As physical protectors and generous benefactors of many religious institutions across the south, they sought to show their worth as secular lords and a desire to cleanse their deeply besmirched souls. They also sought to harness the material resources of the church to extend their own resources, and the influence of Douglas lords with the religious communities of the south was a vital element in the accumulation of regional power by the family in the years of war after 1296.

By that date the Douglases already had a range of ties with the Scottish church. Like many other secular lords they were lay patrons of the principal church in their lordship. The church of Douglas was dedicated to the Irish Saint, Bride or Bridget of Kildare, and, like Frankish lords across the British Isles, the Douglases developed an affection for their Celtic patroness. Barbour described Sir James Douglas swearing by St. Bride, and in 1320 'on the ninth day of the blessed Bridget' in the park of Douglas, James endowed masses for his soul on St. Bride's day (1 February) to be said at Newbattle abbey. However, it was with the reformed structures of the church rather than its older traditions in Scotland that the Douglases principally identified themselves. Contacts were strongest with the local bishopric of Glasgow, and in providing one bishop and several canons for the church of Moray in the thirteenth century, the family was associated with the extension of ecclesiastical reform and royal authority to the northern province, where their kinsmen, the Murrays, were the crown's chief secular lieutenants and vassals. During the same period the family was also involved with several of the new religious houses which formed the centres of the reformed church in the south. The Tironensian abbey of Kelso founded a daughter house at Lesmahagow, whose lands neighboured Douglasdale, and one of its earliest priors was a younger son of the lord of Douglas. Kelso itself held lands and rights within Douglasdale, probably as gifts of the lords, as part of an ecclesiastical estate which included appropriated churches, rights and lands across Lothian, Clydesdale and the Borders. A second abbey also had direct early contacts with the lords of Douglas. The Cistercian house at Melrose held lands up the river Ayr which bordered Douglasdale to the west but, unlike Kelso, found such proximity a mixed blessing. Among the numerous complaints brought against Sir James Douglas's father, William, by his neighbours was one from the Melrose monks, whose officials and tenants had been harassed by the lord of Douglas in the 1290s. The rich lands of the abbeys and their limited defences would become even more tempting to their secular neighbours in the coming years of war.[3]

However, in the time of general peace in southern Scotland before 1296, the abbeys of Tweeddale, Teviotdale and Galloway were centres of economic and cultural prosperity for the region. They also symbolised continuing cross-border contacts with England. Both Melrose and Dundrennan abbey in Galloway had been founded by Cistercian monks from Rievaulx, in Yorkshire, and Melrose went on to found a daughter house at Holm Cultram, south of the Solway in Cumberland. Other communities of monks from England and northern France founded the abbeys at Holyrood near Edinburgh, and Dryburgh, Jedburgh and Kelso in the vales of Teviot and Tweed. Similarly, Scottish reverence for St. Cuthbert's shrine at Durham was reflected in Scotland in the maintenance of a cell of the cathedral priory at Coldingham staffed by English monks. As with the Scottish nobility, the Scottish church's ties to the wider Christian world were largely, though never exclusively, channelled through England in the thirteenth century, and the religious structures of Galloway and the south-east retained elements of the Anglian and Cumbrian heritage which straddled the more recent boundary between English and Scottish realms.[4]

For the church in these regions, and especially the great abbeys of the Borders and Galloway, the transformation of southern Scotland into a military frontier was a disaster from which they never fully recovered. Their ties with the church of northern England, their reliance on vulnerable lands and revenues on both sides of the border and their own existence as important centres of local society presented these religious communities with problems which surpassed those of their secular neighbours. The greatest danger was of the physical destruction of the abbey buildings at the hands of armed bands. In the 1300s, the Melrose monks complained that their dwellings had been 'burned and destroyed' by unnamed assailants while under English protection, and in 1322 the abbey suffered further and greater damage from the retreating English army of Edward II. The house was 'despoiled and looted', its prior and a number of other monks being killed in the attack. Nearby Dryburgh suffered even worse. According to Bower, the house was 'entirely destroyed by fire and reduced to ashes'. Sixty-two years later and after a strenuous rebuilding programme, the two abbeys were burnt by the invading army of Richard II. Though Richard later contributed to the abbeys' repair, the work was not completed until the next century and the permanent vulnerability of the Border abbeys was obvious. Bower's clerical horror at the 'diabolical' violation of holy places could not disguise the fact that, as centres of wealth and as easily destroyed symbols of Scottish identity, these buildings made attractive targets for English invaders.[5]

The southern abbeys were not only harmed by direct attacks. Years of war also devalued lands and reduced revenues. In the 1300s Melrose was denied its income from Eskdale by local men who drove off or imprisoned the abbey's servants, and during major Anglo-Scottish warfare the monks were clearly powerless to enforce their rights. Later in the century complaints also focused on the damage done to ecclesiastical incomes in the marches by depopulation. Though linked to war, this also reflected a wider economic downturn caused by plague and famine, which must have drawn inhabitants of the south to safer pastures. As a result monastic incomes suffered. When Sweetheart abbey in Galloway was 'accidentally destroyed by fire' in 1381, its rents proved insufficient to fund reconstruction because of depopulation in its estates. As late as 1420 the Kelso monks complained that 'on account of the unhappy outbreaks of war between the two kingdoms, it is diminished more than half in its faculties and edifices'. Even if exaggerated for effect, such complaints reflect the disappearance of the security and prosperity of the thirteenth century.[6]

War brought other problems. Like their secular neighbours, the religious communities of the south were faced with the issue of allegiance. The monks had little choice of loyalty. The abbeys in the central Borders lay open to occupation. Kelso and Jedburgh stood hard against major English garrison points and Melrose and Dryburgh were situated in open country by the Tweed. Melrose, in particular, possessed accommodation and storage which made it a useful military base for the English from the 1330s, and in 1341 and 1356 Edward III launched major campaigns from the abbey precincts. Local Scots could treat houses in the same fashion. The nunnery of Lincluden in the west march was used as a strongpoint by its neighbours, 'very evil men', against 'the enemies of Scotland'. Like other landowners in the south, the monks entered the protection of the English king in 1296 and the early 1300s, came under Scottish control after 1314 and returned to English allegiance in the 1330s and after Neville's Cross. Resistance to changes in the military balance meant expulsion. The abbot of Kelso remained in Scottish allegiance in the late 1290s but was forced to flee his abbey, which was left destitute as a result and, after James Douglas captured Roxburgh castle in 1314, the abbot and eleven canons of Jedburgh took refuge in England rather than submit. When they sought to return, they found a new abbot in residence who shut the gates in their faces. The flight of these canons was not unique. In 1316 two monks of Dryburgh were turned out of the abbey 'for being English'. Houses which possessed well-established cross-border connections clearly had mixed populations which reacted differently to new tests of allegiance.

In Galloway, the war also raised questions of loyalty. In 1347, following his triumphant return to the province after Neville's Cross, Edward Balliol granted Buittle church to Sweetheart abbey, an act which displayed his favour towards the Galwegian church. The grant was witnessed by the heads of that church, the bishop and archdeacon of Whithorn and the abbots of Dundrennan, Glenluce and Tongland. Probably, like MacDowalls and McLellans who were also present, this attendance on Edward, lord of Galloway, combined expediency with residual loyalty to Balliol. The recognition of Scottish lordship by these religious communities also came at a price. Melrose, Kelso and Jedburgh all lost estates in England following their abbots' return to Scottish allegiance after 1314 and, although, like James Douglas's lands in Northumberland, these were recovered in the peace of 1328, the restoration was cut short by renewed warfare. Jedburgh forfeited its rights to a number of northern English churches and Melrose was only restored to her estates in Northumberland when the abbot submitted to English lordship after Neville's Cross.[7]

The loss of lands and appropriated churches, the declining value of their remaining estates and the dangers of direct attack all threatened the monastic communities of the south during the fourteenth century. In insecure times they were increasingly dependent on the protection provided to them by secular lords. The traditional source of this protection was the Scottish crown, which had been closely associated with the monastic orders since the twelfth century. Despite the demands of war, the Bruce monarchs remained keen to show their ability to guarantee the security and prosperity of the southern abbeys. Failure to do so would represent a decline in the will and power of the crown to fulfill its duty to protect the church. Both Robert I and David II were active in upholding and enlarging the rights of many of the religious houses in the Borders, but their concern focused on Melrose. This special attachment probably owed much to the long-standing links between the monks and the Bruces' ancestors, the earls of Carrick. Robert, as well as confirming the monks in their existing rights, assigned incomes from the royal fermes and customs to them and, most generously, gave Melrose £2,000 sterling to be collected from the feudal casualties of Roxburghshire for 'the reconstruction of the fabric of abbey church'. In 1369 David II repeated this grant, which suggests that Melrose had suffered new damage or that Robert's scheme had not been completed. Following his return from England in 1357, David had repeatedly sought to guarantee the wealth and security of the monks. He allowed them to hold their lands in Scotland while remaining on terms with the English because their abbey 'is situated on the marches' and took the abbey into his own protection while allowing them to hold their estates in regality, free from the interference of the crown. This concern, which was greater than that shown to any of Melrose's neighbours, marked the special affection of the Bruces for the abbey, confirmed by the choice of the monastery as the last resting place of King Robert's heart.[8]

However, in the years of local war in the fourteenth century, when royal authority in the marches was exercised only intermittently, the protection and support of the crown was not enough to safeguard the religious communities of the region. Both monks and kings could see the need for more immediate protectors. Just as royal leadership in war was delegated, by choice and necessity, so the role of guardians of the local church passed in the marches to local magnates and, in particular, to the house of Douglas. Links between the secular nobility and monastic houses were not new. The de Moreville lords of Lauderdale had founded Dryburgh abbey, the Avenal lords of Eskdale had given a large part of their lordship to Melrose and the native lords of Galloway had founded and provided for at least six religious houses in their province. As heirs of all these lords, the Douglases inherited a range of links with the southern church. As with the family's secular lordship, the impact of war and absence of regular royal authority after 1330 added to Douglas influence. By the early 1420s the extent of this was remarkable. The fourth earl of Douglas was patron of the Galwegian abbeys, interceded with the Papacy on behalf of Dryburgh and many lesser churches, and was named as 'special protector and defender' of the favourite house of the Bruce dynasty, Melrose, said to lie within his 'lordship (*dominio*)'. Although these links were, in part, the product of Earl Archibald's exceptional prestige within the church from 1418, they also reflected the growth of his family's influence over the previous century.[9]

The basis of this influence, as the title from Melrose implied, was the ability of Douglas lords to protect and defend the monks, their buildings and estates. The only example of a formal grant of power by a religious house came in 1414, when the prior of Coldingham made Earl Archibald, already keeper of the priory's lands and rents, their principal bailie and governor with 'full power and auctorite'. In the years from 1400, Coldingham needed protection. As a recently restored community of English monks in the Scottish marches, the house was regarded by many Scots as a 'serpent in the bosom of the kingdom'. With their traditional patrons, the Dunbars, in exile, Coldingham turned to their supplanter, Douglas, for protection. The fee of £100 Scots which he received from the monks was worth paying to retain the service of the only man able to manage and defend the estates of the priory in war. Coldingham's choice of protector had a precedent. Over eighty years before, in an earlier crisis of war and allegiance in the south-east, the Coldingham monks had sought a Douglas defender. In return for his 'protection and counsel' James lord of Douglas was given the lands of Swinton. The first Black Douglas, like his grandson, was paid to shield the English priory.[10]

Elsewhere in the marches, James showed himself capable of providing the protection which Coldingham required. In early 1316 he was in a force which defeated a raid on Melrose by the English garrison of Berwick, and six years later Douglas routed the English as they plundered the abbey. Such demonstrations of local military power must have impressed the surviving monks, and this leadership and the powers given to James in the region by King Robert made Douglas the obvious choice to oversee the raising of the royal revenues assigned by the king for the abbey's reconstruction. As in much else, James provided a model for his successors. In 1343 the monks of Kelso rewarded the keeper of nearby Roxburgh, William Douglas of Liddesdale, for his 'counsel and aid' with a grant of land in Eskdale, and in the 1360s David II chose Archibald the Grim to take his father's place to administer the royal grant to Melrose, perpetuating Black Douglas connections with the abbey.[11]

Dependence by religious communities on their powerful secular neighbours was not without danger. After the death of their protector, the Coldingham monks found it impossible to recover Swinton from his heirs. For Melrose the establishment of the Douglases as their local protectors held other problems. Robert I's patronage had made the family landlords of the abbey's estates in Eskdale, and neighbours of their Tweeddale lands which lay between the Douglas lordships of the Forest and Lauderdale.

Even during Robert's lifetime there was friction over the abbey's rights in Eskdale, and by the time of David II's return from England in 1357 the monks may have been in serious difficulties in the middle march. During the early 1350s, when Earl William Douglas was extending his lordship, and with it the 'Scottis fay', in the region, Melrose remained in English peace. Although the monks claimed that they were 'driven by force and compelled by neccessity' into this choice, the local Scots, led by William Douglas, may well have denied them the revenues of estates beyond English control. David II's subsequent letters, which allowed Melrose to retain its lands despite adhering to England and erected the abbey's Tweeddale lands into a regality, were probably a response to local harassment. The king issued his letters of protection twice over the winter of 1357–58 and twice ordered the justiciar of Lothian to respect the regality in 1359–60. The southern justiciar was Earl William and, against this background, his generosity to the abbey in these years had a different meaning. His grants of the church of Cavers and lands in the same barony may have been acts of atonement for past actions, and a further grant for masses to be said for the soul of the lord of Liddesdale, whose tomb was in the abbey church, was a gesture of penitence designed to win the favour of God, the monks, King David and the dead knight's old partisans. The earl's patronage did not entirely heal old wounds. In 1369 the king again issued letters of protection for the abbey's lands and for its men, the latter perhaps coming under the earl's powers of leadership over men acknowledged in 1354. Two months later, it was William's cousin and rival, Archibald, whom David chose to administer the collection of funds for the abbey.[12]

Whatever difficulties the earl may have posed the monks of Melrose in the 1350s and 1360s, when he died in 1384 William chose to be buried in the abbey and, four years later, his son's corpse was also laid in Melrose. Their tombs would retain a significance for William's Red Douglas descendants, and the choice of Melrose, rather than Douglas where William died, suggests that earlier disputes were over. After 1371 both earl and monks saw the advantages of co-operation. Melrose was in Scottish allegiance but the death of David II had removed their main patron. As in the next century, Melrose lay within the '*dominio*' of the Douglas earl, and in renewed local warfare, which again led to Melrose's destruction, the monks were dependent on the family's protection. The Douglas earls recognised this duty. Earl James's immediate response to the sack of Melrose in 1385 was to threaten her English daughter, Holm Cultram abbey, with the same fate, demanding a blackmail of £200 to restrain his troops. His raid appears as a deliberate act of revenge. For Earls James and William the prestige to be gained from guarding the favourite abbey of the Bruces, the resting place of King Robert's heart, was valuable in the promotion of the Douglas family. The earl, like his uncle, the Good Sir James, was again the guardian of the casket which contained the bloody heart. Though Melrose retained its connections with the Scottish crown and in 1401 was named by David duke of Rothesay as a place secure from Black Douglas dominance, for most of the half-century from 1371 it was the power of the Douglas family which shielded and coerced the monks.[13]

This protection, at Melrose and elsewhere, involved more than just defence in war. Coldingham priory's grant of the office of 'soverayn bailye and governour of all our lordship and landes' empowered Douglas to hold courts, raise rents, levy fines, establish tenants, and punish 'trespasours' on their estates. While this was to be done to the profit of the monks, the management of the principal estate in north Berwickshire was to Douglas's financial and political advantage, especially in his rivalry with the Dunbars. However, in return, the earl's employment guaranteed the monks a powerful advocate in any disputes over their lands. In the same way, James Douglas signalled an early relationship with the monks of Melrose when he witnessed grants to the abbey in 1319, one from Thomas Randolph, the other from Douglas's own 'man', Laurence Abernethy. His attendance at an assize to determine Melrose's claims to lands in Mauchline in 1327 was certainly linked to his, by then, formal connections with the abbey and the probable use of his influence on the monks' behalf. Nearly a century later, James's grandson and great-grandson were providing the same role. They settled, in the monks' favour, a local dispute with their neighbour, John Haig of Bemersyde, and the fourth earl sponsored the abbey's request that the Pope automatically excommunicate any who carried off the Melrose monks' goods. This act specifically sought to harness the temporal sword of Douglas to the ecclesiastical sword of the Pope, and the earl was almost certainly 'the secular arm' to whom the monks turned in their dispute with the bishop of St. Andrews over lands in Wedale (the valley of Gala water). The protection of Melrose abbey, a role close to the heart of the Bruce kings, had passed fully to the earls of Douglas.[14]

UPROOTER OF SCHISMATICS

Beyond the official and unofficial protection and support which the Douglas family gave to the religious communities of the south, in the fourteenth century the family also emerged as major patrons of the church in their lands. The motive for their generosity was, partly, personal and family piety. Countess Joanna Murray in her 'pure widowhood' granted Glasgow cathedral revenues for masses to be said for 'our late lord', her husband of nearly forty years, Archibald the Grim, seven months after his death. In 1426, Margaret duchess of Touraine made a similar bequest to the Greyfriars of Dumfries in memory of her 'dread lord' and her younger son, James, killed together in France eighteen months before. Even Earls William and Archibald Douglas, endowing prayers for dead mentors and enemies, still recalled their own families in their bequests.

However, Douglas patronage of the church had wider concerns than simple piety. Like the crown, from the 1350s the Douglases sought to repair the ravages of war on the southern church. In 1379 Earl William wrote to the Pope about 'leper hospitals and other pious foundations' within his domains which, 'on account of the continual warfare in those parts have now become so ruinous that they are uninhabitable'. The earl wished to restore these foundations 'in a safer place'. Although the identity of these foundations is unclear, the earl similarly expressed a desire to repair the poverty caused by war in a grant to Sweetheart abbey two years later, as did his cousin, Archibald, in bequests to two foundations. These Douglas lords felt reconstruction to be their inherited duty.

The importance of these duties, at work in the family's dealings with Melrose, was clearest in the patronage of the church in Galloway. The province's lords were traditional patrons of the local church and, as lord of Buittle and lord of Galloway, William and Archibald sought to play the role still claimed by Edward Balliol in the late 1340s. Earl William deliberately repeated Balliol's grant of rights to the church of Buittle to Sweetheart in 1381. His attempt, in the late 1360s, to give lands in Buittle barony to the abbey were blocked by David II, who was perhaps keen to divert influence over the local church from William to the new lord of Galloway, Archibald the Grim. Archibald certainly appreciated the value of these ecclesiastical connections. In 1372 he refounded and provided for the hospital at Holywood north of Dumfries, built by Edward Bruce but left incomplete by its patron's death in Ireland and the wars which followed. By fulfilling the wishes of his predeccesor as lord of Galloway Archibald was living up to his inherited duties. His son and grandson followed in the same tradition, demonstrating their generosity across the province in gifts to the abbeys of Glenluce and Sweetheart and the cathedral priory at Whithorn.[15]

These connections also encouraged attempts by Douglas lords to increase the status and standards of the local church through new foundations. In 1378 Earl William sought permission to found a chapel dedicated to the family's patron, St Bridget of Ireland, and to transfer lands from other, now desolate, houses to the new church. Though war and the earl's death seem to have forestalled this foundation, his impulse was shared by his cousin Archibald the Grim. Like many of his contemporaries, Archibald sought to limit the costs of his piety by building on existing structures, turning churches into colleges of priests. In 1398 Bothwell was founded as such a collegiate church, Archibald undertaking to endow it with £20 worth of land, displaying further affection for his earliest lordship through his patronage. It was Douglas's second foundation. His first was Lincluden near Dumfries, where a Benedictine nunnery had been founded by the lords of Galloway in the twelfth century. In 1389, as patron of the house, Archibald petitioned Rome for its refoundation as a college of priests. To justify this action he painted a picture of moral and physical decay in the nunnery during which the beautiful buildings had fallen into ruin, the local neighbours, 'very evil men', used the house as a fortress and the nuns, reduced to four in number, employed the revenues to dress their daughters, 'born in incest', in rich clothes. The petition had the desired effect. The nuns were expelled, a provost and eight secular priests were put in their place and the new priory was united with the hospital at nearby Holywood. While showing a concern for the church, for which Archibald received praise, the treatment of Lincluden shows a robustness in dealing with the local church comparable with his handling of his lay vassals in Galloway. Archibald's son, the fourth earl of Douglas, showed a similar link between secular and ecclesiastical attitudes in his attempt to found a priory of Carthusian monks. His decision, like many of his actions in Scotland, displayed a consciousness of European trends and contacts. His 'singular devotion' was for an order whose ascetic way of life had attracted Duke Philippe of Burgundy and the English kings, Richard II and, most importantly for Douglas, Henry V. Although the earl never fulfilled his plan, his goal was achieved by James I, whose foundation of a charterhouse at Perth was generously patronised by the earl's son and successor, out of filial piety as well as the desire to please his king.[16]

With traditional responsibilities, Douglas lords also inherited traditional powers over the church in their lands. In many of their new lordships the Douglases acquired rights as lay patrons to appoint clerics to benefices on their estates. In the baronies of Cavers, Smailholm, Sprouston, Bedrule and Hutton in the marches, and in the lordships of Bothwell, Cortachy and Cambuslang further north, Douglas lords possessed this right of presentation. Similarly, the church of Yarrow or St. Mary in the Forest by St. Mary's Loch was in their gift as lords of Ettrick Forest, while as lords of Galloway the family's rights of patronage extended to at least four parishes in the province. Although some of these rights were granted away to religious houses, Cavers church to Melrose and Buittle in Galloway to Sweetheart for example, Douglas lords could and did continue to influence appointments in these parishes and other appointments made by religious houses within their orbit. The erection of Lincluden and Bothwell into colleges may have been linked to the issue of patronage. These Douglas foundations, better endowed than ordinary parishes, attracted able and ambitious priests but remained fully within the gift of the founder and his heirs.[17]

Rights of presentation were worth protecting. They provided Douglas lords with a reservoir of patronage with which to draw clerics into their service and promote them within the hierarchy of the church. By contrast with other magnate houses, however, patronage was not used by the Douglases to advance their own family in the church. From Hugh the Dull until the 1440s, there were no clerics in the immediate Douglas family. Even Hugh was forced after Halidon to take a secular path as titular lord of the Douglas estates, and from the 1330s even bastards, like William Douglas of Nithsdale and William Douglas of Drumlanrig, sought their fortunes, not in ecclesiastical politics, but in border warfare. However, for those who did seek advancement in the church, the Douglases were valuable patrons who rewarded their clerical councillors and allies with the benefices in their gift. In 1339, for example, Douglas of Liddesdale, usurping the rights of his absent kinsmen, presented William Bullock, the multi-talented adherent of Edward Balliol, to the rectory of Douglas, securing Bullock's support for both the Bruce party and Douglas. Later, after Bullock's violent death, the rector was Richard Fogo, Earl William's principal clerical councillor.[18]

This connection between service and clerical patronage was even clearer from the 1390s, after which point a number of clerical careers were built solely in Douglas employment. Alexander Cairns, for example, who was councillor and later chancellor to the fourth Douglas earl, was rector of the Forest and then provost of Lincluden. He retained this latter benefice with a canonry in the church of Dunbar and the rectories of Wigtown and Kirkinner in Galloway. These offices were in the gift of the earl or, in the case of Wigtown, within his orbit, and Cairns was not unique. The fourth earl's principal clerical servants, Cairns, Gilbert Cavan, who preceded Cairns as rector of Kirkinner, Metthew Geddes, rector of the Forest and Cortachy, whose career was cut short by leprosy, and John Merton, provost of Bothwell, all received their principal clerical offices through Black Douglas patronage.[19] This ecclesiastical affinity was formed alongside the Douglas's secular following and was affected in the same way by political upheaval like the succession dispute of 1388. While some of Earl James Douglas's clerical servants, like John Merton and Bishop Matthew Glendinning, forged new links with Archibald the Grim, others chose to remain with old patrons. William Broun, who had entered Douglas service with

Margaret Stewart countess of Angus when she became Earl William's mistress, remained with her after 1388, acting as her councillor and tutor to her young grandson, William earl of Angus, in the 1400s. Matthew Glendinning's close links to Earl Archibald from 1388, which reflected the continuation of his service to Earls William and James, were based, in part, on the acceptance of Black Douglas lordship by his kinsmen in Eskdale. In the early 1420s a similar link between ties of secular and ecclesiastical lordship lay behind the fourth earl's patronage of George and John Borthwick. George, the rector of Douglas, and John, rector of the Forest, were kinsmen of the earl's close councillor, William lord of Borthwick, and their promotion represented Douglas's treatment of a favoured family within his following.[20]

The value of attracting the service of churchmen has already been discussed. It was on such men that the Douglases relied for skilled, literate officials to staff their household and council, providing the 'sagacity and eloquence' for which William Bullock was famed. In the 1420s the vicar of Selkirk, William Middlemast, was performing his family's traditional office as master of the ward of Yarrow in the Forest, emphasising the links between local church and local lordship in the marches. Churchmen were also well-equipped to act as emissaries for their patrons and Alexander Cairns, Matthew Glendinning and Gilbert Cavan all acted in this role. Most of the chief clerical servants of the Douglases were university graduates, up to 1413 trained predominantly at Paris and maintaining connections in the wider Christian world. Both education and contacts could be exploited in diplomacy. In 1401 John Merton debated rival authorities for the political structure of Britain with the bishop of Bangor, a fellow Oxford graduate. His arguments proved the cloak behind which the earl of Douglas prepared to break off negotiations and launch war against England. Even in war clerics were not excluded. William Bullock had defended Cupar castle for Edward Balliol in the 1330s and, on switching sides, provided the brains behind William Douglas's capture of Edinburgh castle in 1341. Household clerks often bore arms with their lords. In 1327 the chaplain of James Douglas was killed in the attack on the English camp in Weardale, and at Otterburn in 1388 Earl James's chaplain was cut down defending his lord. At least four Scottish clerics accompanied the armies which went to France after 1419. One of them, Hugh Kennedy, temporarily abandoned the church to be a mercenary captain, while John Kirkmichael, according to one source Douglas's secretary, bore arms at Baugé.[21]

Powers of ecclesiastical patronage could have a significance beyond drawing individual churchmen into aristocratic service. When Archibald, fourth earl of Douglas, received the lordship of Dunbar on the forfeiture of the earl of March in 1401, he also became patron of the collegiate church of Dunbar. By the time March was restored to his lands and rights in 1409, Douglas had packed the college with his own adherents, Alexander Cairns, George Borthwick and the dean or 'archpriest' of the college, Robert Young, a distant kinsman of Douglas. For the next twenty-five years there would be friction between these men and the frustrated Dunbar family which, though it centred on the college, involved other benefices in Lothian. This friction could erupt into violence and, as late as 1433, the earl of March 'laid violent hands' on Young within the church, his action revealing hostility towards clerics who drew their income from March's patronage but were no friends to the founder of their house.[22]

The strength of these ties of ecclesiastical lordship should not be exaggerated. The university-trained clerical elite who received major Douglas benefices were an ambitious, professional group. While the Douglases could be valuable patrons in a career in the Scottish church, greater rewards were available from other, greater patrons. The papal curia and the royal court remained the normal means to obtain a bishopric or other high ecclesiastical office which formed the summit of clerical ambitions, while for a small group of Scots, including ex-Douglas servants like John Gray, the fourth earl's physician, and John Kirkmichael, a future bishop of Orléans, the king of France's favour proved worth seeking. During the last decades of the fourteenth and first decades of the fifteenth centuries, however, both crown and papacy exerted only limited authority within the Scottish church. The politics of the early Stewart period did not make for any certain royal dominance in episcopal elections, and the papacy was, from 1378, weakened by the Great Schism which divided the church's authority between two Popes. In these circumstances, the regional influence of the Douglases could bear fruit in elections of bishops which were undertaken, in the first instance, by cathedral chapters. In the diocese of Glasgow and of Whithorn in Galloway, many cathedral canons had benefited from Douglas patronage, and their choice of bishop may have reflected their patron's wishes. Such influence, and the goodwill of the lieutenant, John earl of Carrick, lay behind the election of Matthew Glendinning to the see of Glasgow in 1387 and that of the local man, Eliseus Adougan, provost of Lincluden, to the bishopric of Whithorn in 1408. Yet such local decisions could be overturned. In 1415 the Whithorn chapter chose Gilbert Cavan, tutor to Douglas's son, to succeed Bishop Eliseus, but Pope Benedict instead provided another local man, Thomas Butil, who had been at the curia, to the see. Though Butil had benefited from the patronage of the earl's father, his appointment showed the limits to Douglas influence in the church.[23]

Limits could also be revealed from below. Like secular lordship, influence depended on active maintenance of the patron's rights and could be weakened by the absence or indecision of the presenter. For example, while the fourth earl was in English captivity in the 1400s, a three-cornered dispute erupted over the church of Bothwell, with two Douglas servants claiming their master's backing in the deposition of a third, John Merton, as provost. The dispute was only resolved with the earl's return and his restoration of Merton to his benefice. The conflict may have arisen from the lobbying of Douglas by clerics ambitious for promotion. Something of this kind occurred in 1427 when two clerics in the service of the earl's widow, John McCulloch and Gilbert Park, both sought the provostship of Lincluden. Gilbert agreed to step aside in return for the rectory of Kir-kandrews but then secretly broke his agreement 'and remained near the duchess' for the purpose of persuading her to grant him Lincluden. How-ever, 'by persevering with the duchess' John obtained his promised reward. Clerical patronage, and secular lordship too for that matter, was at least as much about seeking favour from the great as winning and maintaining alliances with lesser men.[24]

The summit of Douglas influence in the church was reached in the decade from 1413. It was no coincidence that this corresponded with the highpoint of the family's power in Scotland and its influence in western Europe. In secular and ecclesiastical politics in both Scotland and beyond it, the reputation of the house of Douglas was raised by the part which Archibald fourth earl of Douglas played in the ending of the Great Schism in the western church. By 1418 this schism had lasted forty years. The unity of western Christendom had been broken and the church divided into two hostile camps supporting rival popes in Rome and Avignon. Initially the Scots sided with their French allies, and against their English enemies, in giving their allegiance to the Avignon pope, but by 1418 the Scottish kingdom was virtually alone in continuing to support this party, now headed by Benedict XIII. England, France and the Empire all recognised Benedict's rival, Martin V, as the true head of Christendom. Martin had been elected pope in late 1417 by the general council of the church assembled at Constance. From its first meeting in 1414 , the council had dominated the politics of the western church, but the Scots largely stayed away. Scotland's stubborn adherence to Benedict and lukewarm response to the council was particularly associated with the will of the aged governor, Robert, duke of Albany. Although the duke's policy had brought benefits to both church and kingdom, including the foundation of Scotland's first university at St. Andrews, the governor's stance was increasingly criticised, even from within his own circle of ecclesiastical councillors.[25]

By 1418 Archibald earl of Douglas was established as the secular leader of opposition to Albany's policy within Scotland. The role had fallen to Douglas, not just because of his extensive contacts beyond the kingdom. Most importantly, the earl had spent the early months of 1413 as the guest of the duke of Burgundy in Paris. While there Douglas had dealings with the University of Paris, the most famous and influential north of the Alps and the *alma mater* of many of the earl's servants. One of these, John Gray, may well have accompanied the earl there before leaving his service to become dean of the faculty of medicine during 1413. Paris was not just an educational magnet for Scots. The university was also the centre of the effort to end the schism which would lead to the calling of the council of Constance in 1414. Probably while in Paris, Douglas agreed to support these plans and, during 1414, he received letters from the university, perhaps from the hand of John Gray, who was sent to Scotland as an envoy of the king of France and the university at this time. Similar approaches were made to Albany, March and others, but the university clearly pinned its hopes of support on Earl Archibald. Addressed as 'magnificent prince', Douglas was told that 'we place great trust in your gracious promises' to give help 'in the cause of God'. The earl was invited to work for 'the return of the peace and unity of the holy church in Scotland ... in which your lordship is and will be the leader and director'.[26]

Such an approach must have appealed to Douglas's desire to win prestige and influence in western Europe, an ambition which increasingly determined his activities from 1412. A role in the ending of the schism would guarantee him fame beyond Scotland. It might also bring political advantage within the kingdom. As all papal candidates except Benedict XIII submitted to the council in 1415, Scotland's loyalty to him looked increasingly untenable. Though Douglas continued to adhere formally to Benedict, the earl was watching events at Constance carefully. A growing number of Scottish churchmen were taking the same course while, from 1415, others were attending the council in defiance of the governor. These actions probably enjoyed the support of their king. From England, which had already recognised the authority of the council, King James was in full contact with the proceedings at Constance and increasingly keen to make his influence felt in Scotland. Such influence would have been at the expense of Albany, who had left the king in captivity while securing the release of his own son, Murdoch, in 1415. James saw the schism as a means to weaken Albany and heighten his own status, motives which Douglas probably appreciated on his own behalf.[27]

Douglas certainly gave his support to the first open defiance of the governor on the issue within Scotland. In 1417 the majority of the cathedral chapter of St. Andrews overrode the supporters of Benedict amongst them and elected as their prior James Haldenstone. Haldenstone sought confirmation of his appointment from the newly elected Pope Martin in Constance. Douglas supported this action. He had probably recognised Martin soon after the latter's election and in April 1418 was using his influence with the new Pope to obtain recognition of the arrangements by which Prior Haldenstone compensated a disappointed rival for his office. Douglas was to work closely with Haldenstone during 1418 in efforts to induce the rest of the church to recognise Martin. He may even have given Haldenstone physical protection from the hostility of the governor, who continued to give strong support to Benedict. In this, Albany was increasingly isolated, and in August he recognised the strength of opposition and called a general council to discuss the issue. Douglas and his allies lobbied hard before the council met. The University of St. Andrews declared the intention to 'induce the governor to declare the obedience of the Scottish church for Martin V', while Douglas wrote to the prelates of the kingdom declaring the support of 'King James, us and many others' for Martin. When the estates met at Perth in early October, the debate was carried out by churchmen but, as Walter Bower made clear, it was Albany who chose Robert Harding, an English friar, to speak for Benedict. The successful refutation of Harding by John Fogo, a monk of Melrose, and the faculty of St. Andrews and the decision to recognise Pope Martin was a defeat for Duke Robert and his family.[28]

Though many churchmen, led by Bishop Henry Wardlaw of St. Andrews and Bishop William Lauder of Glasgow, took credit for the recognition of Martin, the greatest secular beneficiary was the earl of Douglas. In 1419 Haldenstone brought back letters from the papal curia which left no doubts about Martin's gratitude. Douglas, described as 'high before God', was praised for 'the consummation of the union of the church and the utter uprooting of ill-favoured schismatics' which he achieved without 'sparing labours, dangers or costs'. A year later Douglas was described as 'the devoted and eldest son of the pope in the realm of Scotland', praise which meant more than empty words. From 1418 the earl enjoyed heightened prestige in his dealings with the Scottish church, a status which may have rivalled or even surpassed that of the Albany governors. Between 1419 and 1424 Douglas was frequently sought by churchmen and institutions to sponsor their supplications to the Pope, and his appearances as protector of Melrose and Holyrood may have owed something to new-found prestige as well as established influence. Douglas's links with the curia certainly drew into his orbit others hoping to benefit from their support of Pope Martin. In 1419 the earl took John Fogo, Martin's chief advocate at the general council, as his confessor and similarly retained close ties with the chapter, masters and scholars of St. Andrews. William Croyser, one of the initial masters of the university, had visited England on Douglas's behalf in 1413 and may have represented the earl at Constance in 1417–18. He remained at Martin's curia for most of the next six years, acting in 1424 as Douglas's envoy to the pope at the height of the earl's ambitions. William Fowlis, a member of the chapter which elected Haldenstone, was even more closely connected to the earl, rising by 1424 to become his chancellor, while John Cameron, a student of the university during the debates of 1418, also saw his brightest prospects in seeking Douglas's patronage. By 1422 he had become secretary to the earl's son and heir, Wigtown, the start of a rapid rise in the church. If Douglas's allies prospered, some of his enemies suffered. In 1420 the Dunbars came under renewed pressure which led to the arrest of a papal official seeking to enforce unwelcome judgements. The unfortunate messenger had his letters taken from him and was beaten by the countess, while her son, Columba, then archdeacon of Lothian, said that his holiness the pope had little holiness about him. The words and actions were duly reported to Pope Martin in letters from his 'devoted' servant, Earl Archibald.[29]

These advantages in Scotland were matched by those in neighbouring realms. In England, King James had looked on his Douglas brother-in-law as an ally in overturning Albany's ecclesiastical policy. Both men may have recognised the benefits of maintaining this connection. Douglas's claim to have re-united the Scottish church with the rest of western Christendom brought him prestige beyond the papal curia. The fame which he achieved for his championing of Pope Martin added to the growing reputation of his house in Europe as princes of international status. Inspired by this success, in the six years from 1418 Archibald earl of Douglas increasingly saw the future of his dynasty in terms of international power and status. The uprooter of schismatics aimed to become the founder of a European princely dynasty.

NOTES

1. *Scotichronicon*, ed. Watt, vii, 138–39.
2. A. Grant, 'Aspects of National Consciousness in Late Medieval Scotland', in C. Bjørn, A. Grant and K. Stringer (eds), *Nations, Nationalism and Patriotism in the European Past* (Copenhagen, 1994), 68–95.
3. *Newbattle Reg.*, no. 123; *The Bruce*, XIX, line 302; Väthjunker, thesis, 134; *Moray Reg.*, 61, 81; *Rot. Scot.*, i, 2.
4. Duncan, *Scotland, The Making of the Kingdom*, 147–51, 410–20, 423–26.
5. *Scotichronicon*, ed. Watt, vii, 10–13, 406–409; *Cal. Docs. Scot.*, ii, no. 1982; iv, no. 397; *C.S.S.R.*, i, 309. For a general discussion, see A. Goodman, 'Religion and Warfare in the Anglo-Scottish Marches', in R. Bartlett and A. Mackay, *Medieval Frontier Societies* (Oxford, 1989), 245–66.

6. *Cal. Docs. Scot.*, ii, no. 1981; *Clement VII, Letters*, 67–68; *C.S.S.R.*, i, 176–78.
7. *Scotichronicon*, ed. Watt, vii, 138–39; *Scalachronica*, 299; *Chron. Knighton*, ii, 85; *Clement VII, Letters*, 145; *Cal. Docs. Scot.*, iii, nos 893, 894, 962, 1561, page 322; v, nos 509, 1087, 1329; *Calendar of Papal Letters*, ii, 245; iii, 396. The pattern of losses was not one-sided. Melrose's daughter-house at Holm Cultram lost its lands at Kirkgunzeon in Galloway to a local lord, John Herries, and its rights to the church to Scottish clerics (*R.M.S.*, i, no. 282; *C.S.S.R.*, i, 22).
8. *R.M.S.*, i, no. 19; appendix 1, nos 12–16, 19; appendix 2, nos 83–86, 1039; *R.R.S.*, vi, nos 151, 164, 168, 194, 219, 227, 237, 254, 450; *The Bruce*, XX, lines 608–11.
9. *C.S.S.R.*, i, 8, 21, 69, 77, 106, 131, 197, 216; Fraser, *Carlaverock*, ii, no. 21; *Melrose Liber*, ii, no. 512; *R.M.S.*, ii, no. 12.
10. *Cold. Corr.*, nos xxi, xcviii; Fraser, *Douglas*, iii, no. 298; *Scotichronicon*, ed. Watt, vi, 68–71.
11. *Cal. Docs. Scot.*, iii, no. 470; *The Bruce*, XV, lines 320–68; *R.M.S.*, i, appendix 1, no. 12; Fraser, *Douglas*, iii, no. 17; *R.R.S.*, vi, no. 450.
12. *Cold. Corr.*, no. xxi; *R.R.S.*, v, 151; vi, nos 151, 164, 168, 194, 219, 227, 237, 254, 442, 450; *Melrose Liber*, ii, nos 461, 462; Fraser, *Douglas*, iii, no. 23; *E.R.*, ii, 77, 82.
13. *Scotichronicon*, ed. Watt, vii, 402–403, 406–407; *Cal. Docs. Scot.*, iv, no. 343; *P.P.C.*, i, 127.
14. *Cold. Corr.*, no. xcviii; *Melrose Liber*, ii, nos 408, 421, 422, 540, 541, 542; *C.S.S.R.*, i, 23, 69.
15. *Glasgow Reg.*, i, no. 321; W. M. Bryce (ed.), *The Scottish Greyfriars*, 2 vols (Edinburgh and London, 1909), ii, 101–102, no. 1; *Clement VII, Letters*, 34, 67–68; *Calendar of Papal Letters*, ii, 245; *Calendar of Papal Petitions*, i, 538; *R.R.S.*, vi, no. 469; *R.M.S.*, i, no. 483; *Wigtownshire Charters*, no. 37; S.R.O. GD72/2; Fraser, *Carlaverock*, ii, no. 21.
16. *Calendar of Papal Petitions*, i, 537; *Clement VII, Letters*, 145; *Benedict XIII, Letters*, 83; C.S.S.R., i, 68; iv, no. 591; Fraser, *Douglas*, iii, nos 396–97; W. N. M. Beckett, 'The Perth Charterhouse before 1500', in *Analecta Cartusiana*, 127 (1988), 1–74.
17. *Clement VII, Letters*, 68, 69, 145, 191; *Benedict XIII, Letters*, 83; *C.S.S.R.*, i, 278; ii, 23; iv, no. 591; I. B. Cowan, *The Parishes of Medieval Scotland*, Scottish Record Society (Edinburgh, 1967), 25, 36, 117; *Melrose Liber*, ii, nos 463, 491, pages 478–80.
18. *Morton Reg.*, ii, no. 41; *Melrose Liber*, ii, nos 489, 490; Fraser, *Douglas*, iii, nos 23, 323. For the career of Hugh Douglas in the church see Watt, *Graduates*, 152.
19. Fraser, *Carlaverock*, no. 21; *C.S.S.R.*, ii, 94, 220–21; *Calendar of Papal Letters*, vii, 269; Watt, *Graduates*, ii, 82–84, 93–94; Cowan, *Parishes*, 24, 117; Fraser, *Douglas*, iii, no. 342.
20. *C.S.S.R.*, i, 94–95, 160, 278; Fraser, *Douglas*, iii, no. 51; *Benedict XIII, Letters*, 187–88. Merton had received the rectory of Cambuslang from either Earl William or Earl James before 1387. He went on to be the second provost of Archibald the Grim's college at Bothwell by 1400 and a councillor of the third and fourth earls. Glendinning similarly counselled Douglas earls before and after 1388 (Watt, *Graduates*, 65–66, 391–93; Fraser, *Douglas*, iii, no. 342).
21. *Rot. Scot.*, ii, 4, 190, 223; Stones, *Anglo-Scottish Relations*, 173–82; *Scotichronicon*, ed. Watt, vii, 140–47, 154–57; viii, 120–21; *Chron. Knighton*, i, 445; Froissart, *Chroniques*, iv, 30; Watt, *Graduates*, 311–12.
22. *Benedict XIII, Letters*, 247, 384–86; *C.S.S.R.*, i, 36, 140, 224–25; iv, 30, 46.
23. Watt, *Graduates*, 70–71, 93–94, 233–36, 311–12; Dowden, *The Bishops of Scotland* (Glasgow, 1912).
24. *Benedict XIII, Letters*, 198–99; *Calendar of Papal Petitions*, i, 594–95; *C.S.S.R.*, ii, 170.
25. J. H. Smith, *The Great Schism* (London, 1970), 151–54; R. Nicholson, *Scotland: The Later Middle Ages*, 243–46; R. Swanson, 'The University of St. Andrews and the Great Schism, 1410–1419', in *Journal of Ecclesiastical History*, 26 (1975), 223–45. Finlay of Albany, a close adherent of the governor's family, led an embassy from the council in late 1416 to persuade the duke to recognise its authority (Watt, *Graduates*, 5).
26. *Copiale*, 238–40, 243–48; Watt, *Graduates*, 235–36; *Rot. Scot.*, ii, 209.
27. Watt, *Graduates*, 72, 132, 396; Swanson, 'St. Andrews and the Great Schism', 228, 237;

Calendar of Papal Letters, vii, 6; J. M. Anderson, 'The Beginnings of St. Andrews University, 1410–18', in *S.H.R.*, 8 (1911), 225–48 and 333–60, 349–51; Nicholson, *Scotland*, 245.

28. R. K. Hannay, 'A Chapter Election at St. Andrews in 1417', in *S.H.R.*, 13 (1916–17), 327; *C.S.S.R.*, i, 8; *Copiale*, 5, 18–19; *Scotichronicon*, ed. Watt, viii, 87–93; Swanson, 'St. Andrews and the Great Schism', 238–42. Haldenstone certainly felt Albany's displeasure in 1418 and has been identified as the author of Douglas's letter to the prelates in the same year (*Copiale*, 15, 20, 387, 395).

29. Watt, *Graduates*, 132–33, 333; Swanson, 'St. Andrews and the Great Schism', 231, 237; *Copiale*, 27–29; *C.S.S.R.*, i, 8, 21, 69, 77, 92–94, 102, 106, 131, 133, 136, 140, 216, 224–25, 230; Fraser, *Douglas*, iii, no. 57, 380; *R.M.S.*, ii, no. 13.

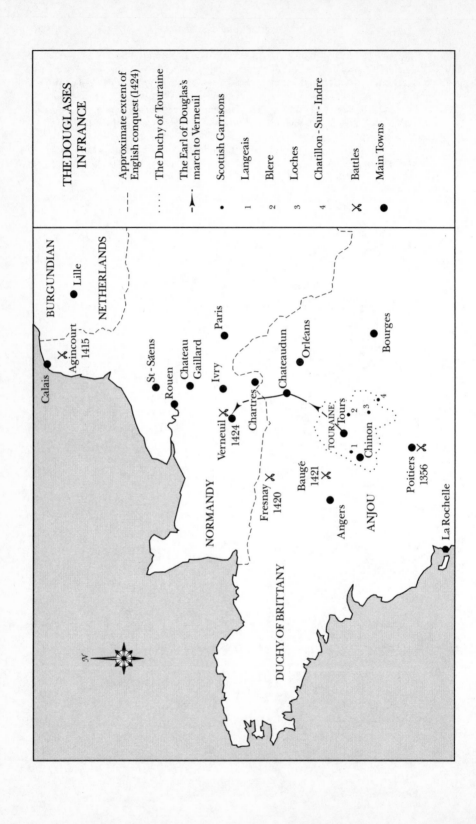

THE DOUGLASES
IN FRANCE

--- Approximate extent of
    English conquest (1424)

..... The Duchy of Touraine

→ The Earl of Douglas's
    march to Verneuil

• Scottish Garrisons

1 Langeais

2 Blere

3 Loches

4 Chatillon - Sur - Indre

✕ Battles

● Main Towns

BURGUNDIAN
NETHERLANDS

• Lille

Calais

✕ Agincourt
   1415

St - Saens
Rouen

Chateau
Gaillard

Paris

Ivry

Bourges

Orléans

Verneuil ✕
       1424

Chartres

Chateaudun

TOURAINE

Tours
      2
1   3
Chinon   4

Fresnay
1420 ✕

Baugé
1421 ✕

NORMANDY

Angers

ANJOU

Poitiers ✕
        1356

La Rochelle

DUCHY OF BRITTANY

N

CHAPTER TEN

'Known Through all Christendom':
The Douglases and Europe

'SONS OF THE SAXONS'

The *Buke of the Howlat*'s claim that the arms of the house of Douglases were 'knawin throw all Christendom' reflected justifiable pride in the international fame of the family. This European reputation was greater than that of any other Scottish noble house and rivalled that of the Stewart kings and governors. Like the family's Scottish fame, international recognition came primarily from the military successes of Douglas lords. Their normal victims, the English, learned early on to recognise the name of Douglas. Barbour remarked on the dread of Sir James Douglas which earned him and his line the name of Black Douglases and made him the bogeyman for generations of northern English children. Contemporary English writers ascribed James's ability to wreak havoc to the help of the diabolic spirits with which he communicated. The success of Douglas was spread to the continent, in less supernatural terms, by the Hainaulter, Jean le Bel. Le Bel had a taste of the Black Douglas at first hand. He had served with Edward III in the campaign of 1327, in which the English were outmanoeuvred by a Scots army under Randolph and Douglas, 'the greatest barons of the whole kingdom of Scots'. Le Bel's account of Douglas's night attack on the English camp described James's skill to a continental audience deeply interested in martial prowess.[1]

When Jean Froissart, who incorporated much of Le Bel's work into his chronicle, wrote of James's successors, he wrote about a family which was well-known and well-connected in Europe. Froissart himself knew the Douglases, not as opponents in war, but as their guest. In the 1360s he visited Scotland and resided in the household of Earl William Douglas at Dalkeith castle. Froissart clearly left with a favourable impression of the earl and his kinsmen. While much of his description of the Scots and Scotland was unflattering, coloured by the unhappy experience of the French knights sent to the kingdom in 1385, Earls William, James and Archibald the Grim earned favourable descriptions. All three were described as models of knightly prowess, and the battle of Otterburn formed a chivalric set-piece to delight Froissart's aristocratic patrons. Froissart regarded these Douglas lords less as defenders of Scottish liberties in a clash of national communities than as part of a common aristocratic world which encompassed the whole of western Christendom. Froissart could reside at Dalkeith in an aristocratic household, just as he had visited the court of the count of Foix in the Pyrenees, the Black Prince at Bordeaux, the English royal court or the various princely households of his native Low Countries. It was to the credit and greater prestige of the Douglases to be placed in this company and, like much of the rest of the English- and French-speaking nobility of Scotland they did not regard the severing of ties of land and blood with the Plant-agenets as the severing of ties with western European culture and society. As we shall see, if anything, the breaking of these bonds with England encouraged families like the Douglases to seek European fame and fortune more actively and identify themselves as Scots with greater stridency.[2]

The pride felt by the Douglases in the last exploit of James Douglas, his expedition to Spain, was part of this process. Its value was not solely as a link between the Douglas dynasty and King Robert. By departing for the war against the Moors of Granada, James and his followers were living up to the ideals of international chivalry, serving their lord in a Christian cause. The journey to southern Spain was intended only as the first stop on a pilgrimage to Jerusalem and was also undertaken to enhance Scottish prestige in Christendom. Jean le Bel reported that *en route* Douglas enter-tained the locals 'as if it was the king of Scotland'. For the first time a Douglas appeared on the continent as a great Scottish lord. Though the Spanish were possibly less impressed after James's death in his first skirmish, for his kinsmen and compatriots, dying in an attempt to save a fellow knight and in a war against 'Goddis fayis' was itself worthy of praise.[3] The glamour asssociated with James's crusading expedition encouraged other Scots to repeat his venture during the following century. For the lords of Douglas crusading had a continuing importance, though their involvement corre-sponded with periods of Anglo-Scottish truce and of wider crusading popularity, especially in the 1360s and 1390s. In 1356 William lord of Douglas led his retinue on a pilgrimage, which may also have had Spain as its goal, but which got no further than Poitiers. Eight years later, one of those involved, William Ramsay, departed Scotland a second time, on this occasion bound for the war against the pagan Prussians in the Baltic region, a favourite stamping ground of late-medieval crusaders. Ramsay's motivation in 1363 came more from his new master, King David, who, like his father, crusaded by proxy. His crusading associates included other former Douglas adherents, but not Archibald the Grim. As both men consciously recalled the bond between their families, Archibald's lack of interest in following his father's example is striking, though, given James's fate and his own dogged pursuit of power in the south-west, not surprising. Douglas interest in the crusade revived with the truce of 1389. In 1390 two Douglas bastards made plans to crusade. George, earl of Angus, probably never set out to join the duke of Bourbon's expedition to Tunis, but Archibald the Grim's son, William Douglas of Nithsdale, shared the character and crusading zeal of his grandfather. However, his experience of crusading was equally anti-climactic. He departed for Prussia with a party which included his companion in Anglo-Scottish warfare, Robert Stewart of Durisdeer, and two nephews of James Douglas of Dalkeith, but at Königsberg in the summer of 1391 Douglas of Nithsdale was caught up in a quarrel with a group of English crusaders. 'He defended himself manfully when he fell with one leg into a hole, and was killed there together with one of his household.'[4]

Even at war in the cause of Christendom, a lord like William Douglas could not forget the conflict with England, in which he had spent much of his life. It was this conflict and the alliance with England's enemy, France, which provided the framework for the activities and ambitions of the house of Douglas in a European context. However, even in their dealings with England there were still elements of the Douglases' identification with a wider aristocratic class. They were certainly not constrained in their outlook by propaganda claims that the family's power was a product of unremitting hostility to their southern neighbours. The nobility of England were not simply the enemy. They were also members of an international elite with common values and interests. In 1381 the Peasants' Revolt caught John of Gaunt, duke of Lancaster, at Berwick in his role as lieutenant of the marches. Gaunt, the English king's uncle, was a particular target of the rebels. In these circumstances the Scots, although they toyed with the idea of exploiting English disunity, also offered Gaunt security and hospitality in Scotland as a nobleman beset by the uprising of lesser men. It was to James 'master Douglas', the future earl, that Gaunt gave the richest of a number of gifts bestowed in gratitude to his Scottish hosts on his safe return to England. The treatment of the fourth earl of Douglas in captivity owed something to the same ethos. His initial captors, the Percies, released the earl to fight alongside them in rebellion, valuing his military prowess above national enmities. Even after his second capture, Douglas was allowed his own 'hôtel' in London, ran up debts, was served by a large household and allowed frequent access to his lands on the promise of his return, a promise he ultimately broke. William Scolalle, the servant of one of Douglas's London creditors, had the unenviable task of following the earl to Scotland to seek repayment of his debts. Although the English hoped to win the earl's goodwill by treating him well, it was also honourable behaviour towards a captive enemy lord.[5] Earl William Douglas, Archibald the Grim and the fourth earl all visited England, not just as diplomats or prisoners, but on private business. In periods of truce the land route through England was probably the most reliable means of reaching the continent, and the shrine of St. Thomas at Canterbury was a favourite centre of pilgrimage for Scots. While the Douglases maintained no links with English religious houses, like those fostered by the Dunbars, both Earl William and his countess made pilgrimages to Canterbury during the truce of the 1360s. This truce also illustrated continued economic links across the border. From the 1360s Earl William and Lord Archibald Douglas, in common with many Scottish magnates, sought to export wool safely and received licences to buy grain, cattle, wine, even vegetables, from England, sending men of their households south for this purpose. In searching for foodstuffs and luxuries which symbolised their membership of an international class, Scottish nobles were also restoring trading contacts disrupted by decades of major war. Even Douglas lords whose careers were built on war with England could not afford to regard their neighbour with one-dimensional hostility. The dictates of survival, rather than common values, forced Douglas of Liddesdale to seek English help to rebuild his power in the marches. His action forced both enemies, like the lord of Douglas, and friends, like Archibald the Grim, to see English resources as a factor in the political balance of southern Scotland to be used or opposed according to personal needs. Though Liddesdale, and later exiles like Drummond, Dunbar and the last earl of Douglas, failed to return with English backing, none of them forfeited his right to Scottish support by seeking foreign help.[6]

Anglo-Scottish conflict had a place in the wider world of aristocratic politics and society. The medieval nobility identified themselves as a military caste. War was their principal purpose and, ideally, was fought according to set rules. Froissart described the warfare of the 1380s in these terms, favourably comparing the ferocity of the Scots and English in battle, and their chivalrous treatment of prisoners, with the behaviour of Saracens and, worse, of Germans. Such ideals were not purely the result of chroniclers' interpretations. In 1341, Henry earl of Derby challenged Douglas of Liddesdale to a joust at Roxburgh, impressed by his deeds in war. Though Douglas proved less successful in tournament, returning home 'without honour', Alexander Ramsay did better in a second joust at Berwick. These tournaments were not games, but 'jousts in war'. William Towers and two English knights were killed in what was an organised fight between enemies. As in Anglo-French warfare, the existence of rules did not limit the scale or intensity of the violence. Nor did such rules apply uniformly. Froissart admitted that Earl William 'preferred death in battle' to captivity, knowing that the English 'would never have agreed to ransom him', and the fate of William Stewart of Teviotdale, executed by the Percies after Humbleton on the grounds that he had been born in English allegiance, showed that, into the fifteenth century, the English viewed certain Scots as rebels excluded from chivalric treatment.[7] While it suited the nobility to see war as a pursuit worthy of their status, victory was achieved, not in tournaments, but by inflicting maximum damage on the enemy. For the Douglases, defeat of the 'sonnis of the Saxonis' was a matter of duty, pride and survival. A disaster like Humbleton, a century after the struggle had begun, still threatened the family's existence in the marches. Such a threat made chivalric rules a luxury and bred in the Douglases an attitude to war symbolised by James Douglas's mutilation of captured English archers. Although this practice was copied by the French, the Scots were famed abroad for their special hatred of the English. Stewart of Durisdeer encouraged Earl Archibald in his support of the Percies' rebellion in 1403 with the thought that, whoever won, Englishmen would be killed. Stewart's special hostility went back to the fight at Königsberg, in which Douglas of Nithsdale was killed. This clash was sparked by English and Scottish adherence to rival Popes in the Great Schism, but amongst crusaders of different papal allegiance from across Europe, it was the knights from Britain who started a mêlée which probably owed more to national antagonisms than ecclesiastical loyalties.[8]

Nithsdale's death may also have been the result of a more personal conflict. His enemy was Lord Clifford, and the hostility between the two may have included the inherited feud arising from the Englishman's claim to Douglasdale, granted by Edward I to his ancestor. This dispute was not unique. From the 1350s the Dacres claimed Liddesdale as heirs of Mary, daughter of Sir William Douglas, adding an extra element to the Scottish feud over the lordship. Most famous was the rivalry between the Percies and the Douglases over Jedworth forest. This was seen as the basis of much conflict in the marches, especially in the 1370s, when the English and Scottish lieutenants presented local war in terms of this private dispute. Later border writers took this further, characterising Anglo-Scottish war as the competition of the houses of Percy and Douglas in a common cross-border world.[9] In the fourteenth century, the experience of England north of the Tyne certainly mirrored that of the Scottish marches. Both regions experienced a crisis of allegiance and internal structure in the years of war which followed 1296. During the thirteenth century, when the king of Scots was lord of Tynedale and many families, including the Dunbars, Umfravilles and, on a lesser scale, the Douglases held lands in both Northumberland and Scotland, the English borders had developed strong ties with Scotland. The loyalty of the local community was, therefore, severely tested, both by the initial outbreak of war, and when subjected to sustained Scottish pressure after 1314. Like Teviotdale and the Merse in the 1330s and 1340s, military pressure forced local men to enter the protection of their attacker. This crisis of allegiance proved disastrous for many established families. For lords like the Umfravilles, who chose the English side, and the Dunbars, who, eventually, chose the Scottish, the loss of part of their lands and the continued ambivalence of their position led to a decline in their local standing. On both sides of the border power passed to men committed to Anglo-Scottish conflict. In England this group was dominated by the house of Percy.[10]

From the 1300s to the 1400s the rise of the Percies mirrored that of the Douglases. Both families originated, not in their kingdom's marches, but in neighbouring, more secure, regions. As the Douglases continued to draw support from Clydesdale and Lothian, so the Percies included many York-shiremen in their retinue. The power of the Percies to the north of their Yorkshire estates was a product of the war. From 1296 onwards the family acted as military lieutenants of the English crown in Scotland and the north, receiving lands in Galloway and Angus for their services. However, more significant gains were made in Northumberland. The purchase of Alnwick in 1310, the acquisition of Warkworth and other estates in return for royal service and the grant of the forfeited Dunbar barony of Beanley in 1335 were the basis of a landed stake in the county, which was increased further in the 1370s at the expense of the Umfravilles. When the Percies received Jedburgh castle and forest in 1342, it was due to their ability to defend these lands, much as Bruce expected Douglas to hold on to the border lordships granted to him. The creation of Henry Percy as earl of Northumberland in 1377 acknowledged his leading role in the northern marches. Like the Douglases in Scotland, the Percies were crucial to the English war effort in the north. The earl's retainers provided many con-stables for English-held castles and constituted a significant proportion of the English armies of the 1380s. This military role was recognised by the family's prominence as march wardens, which reached its height in the early 1390s when Earl Henry and his son 'Hotspur' had a monopoly of the office.[11]

In both English and Scottish marches, the crisis of war and allegiance had produced new families of military magnates. These families gradually established their leadership of local border communities, themselves increas-ingly structured around the needs of warfare. Douglases and Percies supplanted older forms of lordship, represented on both sides of the border by the Dunbars. The hostility of both new families towards the Dunbars after 1400 stemmed from fears that Earl George Dunbar was seeking to regain lost cross-border influence at the expense of both Douglases and Percies. More usually the Douglases and Percies were hereditary enemies in the marches, a conflict which seemed to reach its height in early 1403 when Henry IV granted the earldom of Douglas to Henry Percy. Conflicting claims gave their conflict the appearance of competition for lands and rights between the two families. The impression of private war was enhanced by the resistance of both Douglases and Percies to what was considered interference with their place in the marches. Richard II and Henry IV, and Carrick and Rothesay all found that attempts to control and contain the role of these marcher magnates aroused hostility sufficient to threaten the survival of a royal prince.[12]

However, parallels of attitude and ambition should not obscure crucial differences between the houses of Douglas and Percy and the societies they dominated. To see the families as involved in an exclusive private competition for disputed lands ignores their purpose in Anglo-Scottish conflict. The power of the Douglases and Percies did not come from their membership of a common marcher society but from their success in binding frontier regions to the national communities they represented. While questions of allegiance affected other families, Douglas and Percy lords became identified as the upholders of Scottish or English claims, drawing their support from beyond the disputed zone. The truces made by men of the two families as wardens represented the regulation of war in the marches, not doubts about loyalties. Likewise disputes over specific lands represented, not a private feud, but the physical manifestation of wider loyalties to Scottish and English crowns. That Douglas and Percy were the only lords restored to forfeited lands after the peace of 1328 signified that the two houses were already seen, by their royal masters, as fixed in their allegiance.[13]

Secondly, although the experiences of the English and Scottish marches in the fourteenth century were similar, they were not identical. Despite intense raiding in the 1370s and 1380s, only in the decade from 1314 was the allegiance of the northern English under pressure. By comparison, between 1296 and 1314 and from 1332 into the 1380s the loyalty of southern Scottish communities to magnates and kings was the subject of warfare. While the successes of the Douglases in returning and holding these communities to Scottish allegiance brought new men to their following, the Percies remained within a more structured, if still highly militarised, local society, in which, despite their influence, war and local politics were directed by the crown. As elsewhere in the Plantagenet dominions, these functions were performed by paid, contracted officers. Men like Anthony Lucy, John Copeland and the exiled Scot, John Stirling, made their careers in the marches as agents of the crown. Though the murder of Copeland shows the unpopularity of such men with English borderers, they were the product of a system which had no counterpart in Scotland. The Percies functioned within this system. Their desire to monopolise the march wardenships reflected competition for the increased salary and powers of the office from the 1370s. By contrast, the Scottish wardenship conferred no power, but relied on existing structures of border lordship. The English marches formed only one of the Plantagenets' borderlands and, like Gascony and Ireland, the crown ran the north through a combination of local magnates and paid officials with special powers. In Scotland, while the marches with England formed a region of special sensitivity, they were not the military frontier of a centralised realm. Instead Scotland south of Forth was one of several militarised, magnate-dominated regions which made up fourteenth-century Scotland. Unlike England, where, even in the far north, politics revolved around the search for royal commissions and rewards and the fear of these going to rivals, the competition for power in Scotland was played out in regional society. Though in 1400 the earls of Douglas and Northumberland seemed to exercise matching dominance in the marches of their kingdoms, in Scotland this power rested on the leadership of local communities established over a century, in England it depended on the exceptional circumstances of the 1390s. Rather than confirming them as leaders in a common border world, the rise of the Percies and Douglases and their place in their respective realms reflected the responses of the, very different, English and Scottish political societies to the problems posed by a long-standing war on their frontier.[14]

'TRÈS CHERS COUSINS ET ALLIEZ': THE FRENCH CONNECTION

If contact and conflict with England forged the character of Douglas lordship in Scotland, France held a place as the family's second homeland. In the fourteenth century, France was the nursery of Douglas power. According to *The Bruce*, the young, disinherited James Douglas spent three years as an exile in Paris. Though this story cannot be proved, at the very least Barbour built his tale on the well-known friendship of the Douglases for France. This special relationship dated back to the 1330s when, in company with King David, the young heirs of Douglas came to France for safety. William, the future first earl, and Archibald, his cousin, probably spent the 1330s in David's household at Château Gaillard by the Seine. The two principal architects of Douglas power grew to manhood in this

Franco-Scottish environment and, when William returned to Scotland after thirteen years abroad, he was as much a French as a Scottish nobleman. He held the lands of St-Säens in Normandy, was referred to with favour in the Norman *Chroniques des Quatre Premiers Valois* and, in 1356, once secure in Scotland, William returned to France, abandoning his pilgrimage to serve his lord, King John II of France, against the English. At Poitiers John knighted Douglas and took his advice to dismount the bulk of his forces, a decision which led to the king's capture in defeat, but which showed William's standing with the French. For William and Archibald, the first generation of Scottish noblemen with no experience of the aristocratic connections established within the Plantagenet dominions, this lifelong alliance with France had great significance. It gave them and their compatriots from the distant, isolated rim of Europe a direct link to the heartlands of Western Christendom. The visit of Froissart to Dalkeith took place against the background of a magnate seeking to maintain these ties and create the life of a French lord on his Scottish estates.[15]

These cultural links were maintained by William's successors. In the early fifteenth century Earl Archibald was probably one of the employers of John Morrow, the French master mason who rebuilt the south transept of Melrose abbey in the French Flamboyant style. Morrow left an inscription at Melrose which stated that he had also worked in other places including Glasgow, Nithsdale and Galloway. Given Douglas's links with Melrose, the earl may well have employed Morrow at Lincluden in Nithsdale, Threave in Galloway and, perhaps also, Bothwell castle. All were under reconstruction after 1400 and the styles of all three share comparable elements. At Bothwell in particular the new hall, lit by large, arched windows, suggests a conscious modelling of the surroundings of the earl's household. If the source of inspiration for this was France, rather than the English palaces – Westminster, Windsor and others, with which Earl Archibald was more familiar – then the employment of Morrow shows Douglas's continued identification with the culture and style of the French nobility.[16]

The identification of the house of Douglas with France was much more than an alliance of convenience based on shared hostility towards England. However, it was in the context of the rivalry and conflict between Valois and Plantagenet throughout western Europe that Douglas links to France functioned. From the 1330s to the 1420s, the house of Douglas, more than any other Scottish dynasty, saw the war with England in a European context, playing a consistent role in the maintenance of Franco-Scottish military and political alliance. A landmark in this process was provided by the visit of William Douglas of Liddesdale to France in 1338–39. With the permission of David II and, presumably, of Philip VI of France, Douglas recruited a force of French men-at-arms and crossbowmen and five armed barges commanded by the 'dangerous pirate', Hugh de Hautpol. Backed by these soldiers, Douglas played a leading role in the capture of Cupar and Perth in 1339, and the presence of French troops in the British Isles was seen as especially ominous by English chroniclers. Archibald the Grim may well have served in this Franco-Scottish army and his connections with France, established in the 1330s and 1350s, led to his role in diplomacy between the two realms. In 1371 Archibald was chosen to head the embassy to renew the French alliance after the change of king and dynasty in Scotland. As Archibald had represented David II in France, his presence implied continuity in Scottish relations with their ally. Douglas left his mark on the negotiations. He returned to Scotland with not just a defensive alliance, but a French offer of troops and money to open a second front against England. In 1369 Charles V of France had renewed open war with England and in 1371 he clearly encouraged Archibald to continue the attacks in the marches which he and George Dunbar had been making since 1368. Though the active military alliance was rejected by Robert II, who followed a more cautious policy, for the next seventeen years Douglas magnates would show an awareness of the connection between the war in the marches and the, usually favourable, course of the conflict in France and Spain.[17]

This awareness fuelled the aggression of Earl William Douglas in particular. While his fellow march wardens and the crown tailored their attitudes to England with an eye to the war in France and Spain, it was William who actively sought allies in war on the continent. In 1378, the son of one of William's servants, John Mercer, led a combined Scottish, French and Spanish fleet in an attack on English shipping, and this connection may have led to a military alliance between Douglas and a Castilian fleet based in Normandy the following year. Like Douglas of Liddesdale, William was using contacts in France to obtain the promise of active military help. These continental agreements may have been as important in the earl's increasingly bellicose attitude as the indirect support of the lieutenant in the marches, Carrick. Earl William's massive raid of 1380, when he devastated the west march as far as Penrith with a force numbered by Wyntoun at 20,000 men, marked an increase in the scale of border war, an increase possibly designed to impress Douglas's watching allies.[18]

French volunteers may have taken part in the 1380 raid. In 1355 Earl William had led a Franco-Scottish force in an attack on Norham and in the 1380s his family's fame attracted new recruits. French knights served in the war retinue of Earl James in his raids on England in 1384 and 1388. It was from these French knights, as well as those who took part in the

1385 expedition, and from Scottish servants of Douglas who had travelled to the continent after 1388, that Froissart drew his information. While the source of his information clearly affected Froissart's perspective in favour of Douglas, it also confirmed the status and connections of the earl in Europe. Froissart implied that Earl James used the French in his company to suggest that the French government send men and money to Scotland, and Douglas was described as one of only two Scottish magnates who welcomed the arrival of Admiral de Vienne and his army in 1385. Froissart's chronicle was not the only proof of the significance of the earls in European war and politics. The £7,500 *tournois* paid to Earl James by the French confirmed the key role he played in the Franco-Scottish war effort, while discussions about renewed military co-operation between the king of Castile and the house of Douglas, probably in 1386, showed Earl James to be a prince whose alliance was valued by the kings of western Christendom. The Douglases were seen by England's enemies on the continent as having the power and independence to orchestrate major warfare in the north. These perceptions were not damaged by Otterburn. Earl James's death gave him a heroic reputation and enhanced his family's status as the scourge of the English, while his successor, Archibald the Grim, was already well known on the continent as a soldier and diplomat.[19] When Archibald's son, the fourth earl of Douglas, launched a full-scale invasion of England in 1402, the Douglas name attracted some thirty French knights to his support, despite the truce between England and France. After Humbleton the French sought to negotiate the release, not just of their own knights, but of Douglas, because of his 'attachment to France and fidelity to their king'. Despite his continued captivity, Earl Archibald continued to impress the French, and in 1404 Charles VI's council considered Douglas as likely to succeed to the Scottish throne. Although this was a misconception of Douglas's ambitions and resources, the French verdict paralleled the English perception that Douglas ruled southern Scotland. As in the 1380s, from the perspective of Scotland's neighbours, Douglas was the most prominent Scotsman of the day, a leader who could launch war against England and a prince with ambitions in the politics of Europe. Neither the duke of Albany nor, less still, Kings Robert III and James attracted the same attention from foreign observers, and it was a marriage alliance with the house of Douglas, not of Stewart, that the French considered to be most prestigeous in 1404.[20]

Since Poitiers, Douglas magnates had built their reputations in Europe on their military power in the marches. However, a significant number of Scots had a more direct experience of Anglo-French conflict, which they regarded, not as an adjunct of Scottish resistance to the Plantagenets but as a source of employment and profit. With the end of major Anglo-Scottish warfare in the late 1350s, these men looked to the continent. In 1358 a large company of Scots passed to France and some at least found their way into the Free Companies, which terrorised the south in the 1360s, before joining the 'Great Company' led by the constable, Bertrand du Guesclin, to fight in the civil war in Castile. This force may have included John Stewart of Darnley, whose grandson and namesake in the 1400s showed a developed attachment to France. For many Scots, though, English service held greater promise of rewards. Two of Earl William Douglas's connections, his brother-in-law, Thomas earl of Mar, and his ally, Thomas earl of Angus, agreed to provide troops for Edward III in France, and in 1359 Edward sent two of his Scottish partisans to recruit their compatriots into English military service. As the career of John Swinton showed, the primary attraction of service to the English king was the money to be made in the French war, which Swinton later used to fund his role in war against the English. Even with the cessation of major warfare in 1389, Scottish magnates still found employment on the continent. In 1402 David earl of Crawford became the military retainer of Louis duke of Orléans, the brother of Charles VI of France, and, six years later, Alexander earl of Mar was employed by John duke of Burgundy in his war with the men of Liège. Both lords were paid well for their service and led companies of men-at-arms and archers which, like the retinue of the Douglas earls, were built around their private followings. The deeds of Mar in particular made an impression on French and Scots alike and confirmed Scotland as a source of 'excellent soldiers'. For the hugely ambitious Earl Archibald Douglas, this reputation would combine with the fame which his own family enjoyed in France as war leaders and allies, to draw him in search of the rewards to be won in continental warfare. From 1413 France would become the focus of Douglas ambitions.[21]

DUKE OF TOURAINE, LIEUTENANT-GENERAL OF FRANCE

In the autumn of 1412, Earl Archibald 'crossed over for the first time to France along with an honourable band'. This expedition was no pilgrimage or pleasure trip but, despite some casual piracy *en route*, concerned international politics. Douglas landed in Flanders and, not without local difficulties, made straight for Paris, at that time under the control of John the Fearless, duke of Burgundy, who dominated the regency for the mad king, Charles VI. In late 1412, Burgundy's position was under threat. His enemies, led by Charles duke of Orléans, had negotiated English support, and in August Thomas duke of Clarence, second son of Henry IV, had landed in Normandy with an army of 4,000 men. John, recalling Mar's support in 1408, may have sought Scottish allies as a counterweight. At least one of his principal captains, Jacques sieur de Heilly, had fought under Douglas in 1402, and the earl's arrival in Paris may have been the result of Burgundy's invitation. During the spring of 1413 Douglas remained with Duke John and, although the English army returned home, in early April the two magnates negotiated a 'treaty of alliance and confederation'. By the terms of this Douglas promised to bring 4,000 men-at-arms and archers to serve John in Flanders or Artois in return for payment of the costs of passage and the wages of this force. For the first time, a French prince sought to hire a large Scottish contingent for continental warfare and looked to the house of Douglas to provide this force. As events would show, the promise of the earl to produce an army to match the English was no empty boast. Instead it reflected Douglas's power as a lord capable of raising armies from his private following and the first indication that this power, established in the Anglo-Scottish marches, had significance in continental warfare.[22]

Douglas's alliance with Burgundy was quickly rendered void, largely as a result of the accession, in May 1413, of a new English king, the young and aggressive Henry V. Unlike his father, Henry favoured the Burgundians and immediately opened negotiations with Duke John. With the threat of English support to his enemies removed, Burgundy had no need of Douglas's aid. After a year away from Scotland, the earl returned home with gifts from the French court but little else except a lesson in European power politics. When Douglas visited the English court on the way back to Scotland, there may already have been signs that Henry V intended to exploit French dissent to the full by renewing claims to the lost estates of the Plantagenets. In 1415 he backed these claims with force. The victory at Agincourt and, even more, the English conquest of Normandy in 1417–19 were achieved against a French realm still divided between Burgundy and his enemies, now led by the Dauphin Charles.[23] The Scots sought a role in this renewed Anglo-French conflict. In 1415 Archibald burned Penrith in a raid which coincided with Henry V's first French campaign. When Henry launched his second expedition two years later, Douglas and Albany co-operated in an even more ambitious attack. Roxburgh was besieged by Douglas while Albany assaulted Berwick. However, the English proved to be a much more effective force than in the 1380s and, despite the absence of the king, their military organisation functioned well. In 1415 Dumfries was burned in revenge for Penrith while in 1417 the English border captains, the march wardens and the regent, Henry's brother, John duke of Bedford, all responded quickly to the attack. Despite Douglas's confidence that he could take Roxburgh, both it and Berwick were easily relieved and the

English followed up by raiding Douglas lands in the middle march. Though Archibald himself led forays in 1419 and 1420, the 'foul raid' of 1417 showed that the successes of the 1380s could not be extended and Douglas's efforts were token attempts to display leadership without running real risks or causing major damage.[24]

Instead the opportunity for Scottish, and in particular Douglas, involvement in the war came directly from France. In 1419 the situation of the Dauphin Charles was desperate. Normandy was lost to the English and Paris to the Burgundians, and the assassination of Duke John of Burgundy in the Dauphin's presence created a firm alliance between Charles's enemies. In these deteriorating circumstances, Charles appealed to the Franco-Scottish alliance. Although the request for military assistance was delivered to Albany and discussed by the estates, Charles's intention was to win specific magnates to his service as Burgundy had earlier attempted with Douglas. The leadership of the Scots force reflected this. Albany's younger son, John earl of Buchan, was made commander with Douglas's heir, Archibald, given equal rank by his creation as earl of Wigtown, as his 'comrade and partner'. As his son and son-in-law, both were acceptable to Douglas, but, more importantly, the two earls could draw on the military lordship of their fathers to recruit men for continental warfare. From 1419 to 1424 Charles directed his appeals to Albany, Douglas and a third lord with a European reputation and military following, Alexander earl of Mar. Charles asked these men 'for help against the English, our ancient and common enemy', 'by bringing the greatest possible number of people which you are able to furnish', including 'kinsmen, vassals and subjects'. The dauphin wanted, not the small companies of Scots already arriving in France in 1418, but a ready-made army. The French knights at Humbleton had seen the ability of Albany and Douglas to produce such an army, and Burgundy had been keen to recruit this following for warfare in France. In 1419 Charles simply followed suit on a larger scale.[25]

Charles got his ready-made army. During 1419 some 6,000 Scots entered his service, the bulk of them arriving with Buchan and Wigtown in October. These numbers were maintained, despite losses, by further recruiting in Scotland, and in 1424 the Scottish army consisted of 6,500 men. As many as 15,000 Scots served in France between 1419 and 1424, the leaders, and probably the men, remaining in arms for years, producing a professional force which in numbers and length of service was comparable with English field forces in France. Though some Scots served in garrisons or small companies, unlike other foreigners in Charles's employ, the bulk of these men formed the 'army of Scotland', a separate force led by the earls with, beneath them, a constable, the veteran John Stewart of Darnley, and a marshal, Thomas Kirkpatrick. These men led an army which was composed of separate retinues provided and led by minor nobles. Bower described the two earls choosing 'captains ... with bands of fighting men; each was in charge of his own troops with his little banners and blazons'. The service of many of these 'captains' was, as at Humbleton, based on their connections with Douglas and Albany, reflecting the domination of Scottish political society by these families. The 'kinsmen, vassals and subjects' of Douglas were certainly present in the army. William Douglas of Drumlanrig, who led a Scots force to defeat at Fresnay in 1420, fitted into all three categories. John Swinton was the son of the earl's close ally of 1401–2, William Crawford, Douglas's long-standing deputy in Edinburgh and Thomas Kirkpatrick, the marshal, from a family whose connections to the Black Douglases went back to the 1360s. The service of John Turnbull, Thomas Colville, Robert Pringle, William Glendinning and John, son of William Stewart of Teviotdale, represented the involvement of the border affinity of the Douglas earls, while John Haliburton, John Cockburn, Maxwell of Calderwood and Thomas Seton came from Lothian and Lanarkshire families with landed and political connections to Earl Archibald. Most of these families had similarly provided men for the Douglas-led armies at Otterburn and Humbleton and were natural followers of Wigtown. The inevitable presence of two kinsmen of Douglas of Dalkeith confirms this. His nephews, Henry and William Douglas, looked to Wigtown to attest arrangements between them. Though such bonds of lordship did not account for the whole army, the connections of Douglas and Albany must have provided a solid core to the force.[26]

The army was made up of the followings of these lesser nobles. Drumlanrig brought a company of 450 men, Thomas Seton one of 160 soldiers, William Bell, an Annandale vassal of Douglas, had 110, while many captains brought retinues of just 30 men. These contingents, from all over Scotland, reflected the militarised character of Scottish lordship. As he had promised Burgundy, a magnate like Douglas could raise men to fight, not just on the borders, but in France. While the continued attraction of such service was linked to the rewards of continental warfare, the initial force was largely built from the structures of Scottish regional lordship. The equipment of the army was also a product of the European experience of magnates like Douglas. Unlike the spear-armed hosts which fought at Neville's Cross and Humbleton, the army in France contained a large number of archers. The companies detailed in 1419 contained two archers to each man-at-arms, and in 1424 Douglas led 2,500 men-at-arms and 4,000 archers. At Humbleton Douglas's army had been destroyed at a distance by English archery, but in 1408 Burgundy had specifically praised the bowmen in Mar's small company. Both French desire to recruit archers able to compete with the English and Douglas's own experience of the longbow's effectiveness led Earl Archibald and Robert of Albany to anticipate James I's legislation promoting archery, itself inspired by the king's experience in France with Henry V. The 'army of Scotland' in France was led, raised and equipped by the great magnate families which dominated the structures of government and warfare in the kingdom.[27]

From late 1419 to 1421 this Scottish army was the mainstay of the Dauphin's defence of the lower Loire valley. It saw its first fighting in the absence of Buchan and Wigtown, when Douglas of Drumlanrig led a large Scottish force to defeat at Fresnay in Maine in early 1420. However, the following year the Scots seized the chance to redeem themselves. In response to a raid by Thomas duke of Clarence into Anjou, Buchan and Wigtown led their army and some French contingents to block his retreat at the village of Vieil Baugé east of Angers. Clarence, whose army of 4,000 had dispersed plundering the countryside, came upon the Scots and French camped round the village. Rejecting calls for caution, the duke cried 'Let us go against them, they are ours', and attacked with the men-at-arms around him. In the confused fight which followed in the village streets and a swampy river bed, it was the English who suffered from Scottish archery. Once Clarence's initial charge had been halted, Buchan gathered his own company and, according to Bower, it was the earl who struck down Clarence, who had already been wounded by John Swinton.[28]

Though the Scots earls failed to exploit their victory, allowing much of the English army to escape, Baugé and the death of Henry V's lieutenant and heir was the first victory for the Dauphin and secured the military reputation of the Scottish army. When Buchan and Wigtown wrote to Charles on the night after the battle, they were already thinking of the rewards owing to the Scots for the victory. They asked the Dauphin to appoint Thomas Seton, 'your servant and our kinsman', as seneschal of Berri, a position he claimed in right of his wife, Isabeau Goyon, widow of the viscount of Thouars. Seton's marriage had already given him the lands of Langeais near Tours, and he was not the only Scot seeking lands in France. Buchan's lordship of Châtillon-sur-Indre acted as the Scots' base in Touraine, while Wigtown received Dun-le-roi in Berri. For lesser captains and their men the rewards were also considerable. In 1423 the Scots received £30,000 *tournois* as their wages for two months, and Charles was forced to seek loans and dismiss other troops to maintain the army. The Scots also took direct payment. They repeatedly plundered the district round Tours, even taking money from Charles's officials, inspiring fear and hatred in their French hosts. Baugé delivered further rewards to the victors. The captors of English nobles gained instant wealth, one squire, Laurence Vernor, receiving money and lands for delivering the earl of Somerset to Charles. The leaders also profited. Wigtown was made count of Longueville, a lordship occupied by the English, while Buchan was made constable of France. Unlike Longueville, Buchan's promotion was no empty honour but made him the chief military officer in France. His new importance was apparent in the campaigning of 1421–2 and in his presence at a council of French princes held at Blois in August 1421. Although there was French discontent at the prominence of the Scots, they had shown themselves to be the most active and aggressive element in Charles's army and the Dauphin rewarded them accordingly.[29]

The news of Baugé spread quickly beyond France, exciting Pope Martin to exclaim that the Scots were 'an antidote to the English'. In Scotland Archibald earl of Douglas reacted at once to the news. Although he had ignored the Dauphin's appeals to join the army, Douglas's ambitions were focused on its fortunes. While Buchan, who had limited prospects in Scotland, sought a career in France and brought his wife, Elizabeth Douglas, out to join him, Wigtown acted as his father's deputy. Douglas, who had long been tempted by service in France, was timing his move. Baugé encouraged him to act. However, Douglas offered his support, not to the Dauphin, but to Henry V. In May he was at the English court agreeing to become Henry's retainer for life, serving him in war with 200 men-at-arms and 200 archers for a pension of £200 sterling. King James supported this arrangement and in return Douglas obtained permission for James to spend three months in Scotland. It is hard to take these plans at face value. James's release never took place and Douglas would hardly have wished to meet his own adherents in battle. Instead Douglas was playing a longer game. He showed himself to be a potential English ally and, in contrast to the governor, Murdoch of Albany, concerned to secure James's release. His dealings also increased the Dauphin's anxiety to secure the service of a magnate whose connections permeated the Scots army in France. James went to France, not Scotland, in 1421, but Douglas maintained contact with him and with the English through his secretary, William Fowlis, while his son and son-in-law provided a channel to the Dauphin.[30]

The deaths of Henry V and of Charles VI of France in late 1422 opened up new opportunities for Douglas. Increased English fears and French hopes during 1423 focused on efforts to recruit a 'new army of Scotland' to the service of the Dauphin, now Charles VII. Both sides sent embassies to Scotland armed with offers far in excess of previous rewards. Central to this diplomatic bidding war was the earl of Douglas. Charles probably sent Buchan specifically to win Douglas's service, empowering him to offer a royal duchy and supreme military command. The English, for their part, now offered to release King James permanently for ransom and for a truce which ended Scots support for France. Douglas wanted to secure both offers and probably led pressure on Murdoch to negotiate James's liberty. He left Buchan, whose future in France depended on the continued strength of his army, to find a way of making Murdoch agree truce terms which would allow Douglas to serve in France. In early October, Buchan induced Murdoch's son and heir, Walter, to champion the French alliance, and while Walter threatened war, Murdoch, with Buchan in attendance, instructed his embassy to negotiate a truce which would safeguard Douglas's border lordships but excluded the Scots in France from its terms. By the end of the month Douglas formally agreed 'to pass into France ... with our son-in-law, John earl of Buchan ... leading many lords in my company ... in order to support monsignor the king against his ancient enemies of England'.[31]

The lords in Douglas's company were already assembling in late 1423 as the earl progressed around his southern Scottish lordships. With him were his sons, Wigtown, back from France and now being prepared as his father's lieutenant in Scotland, and James, who would accompany the army to France. Also in attendance were John Swinton, also back from France, Alexander Hume, Douglas's 'lufit squier', and William Seton, eldest son of one of the earl's closest councillors, all of them captains in Douglas's army. In February 1424 these men brought their companies to embark on the French and Spanish ships assembled by Charles VII in the Clyde. On 7 March Douglas landed at La Rochelle and, gathering his 'new army' and the remnant of the old, he led the Scots to Charles's court at Bourges. On 24 April the king reviewed 6,500 Scottish troops, and five days later Douglas received his promised reward. Charles issued letters granting Douglas the duchy of Touraine, including 'our castle, town and city of Tours' and 'our castle and town of Loches'. Only Chinon and a few other estates were excluded from the grant, and the officials and vassals of Touraine were ordered to obey their new duke. Even before this, Douglas had been named by Charles as his 'lieutenant-general in the waging of his war through all the kingdom of France', an office which implied viceregal powers in war against the English. The king of the richest realm of the west had handed vast lands and powers to a Scottish earl.

No other foreign noble and no Frenchman unconnected to the royal house was given ducal rank by a king of late-medieval France. Touraine had been held in the previous century by the sons of kings, including Charles's brother.[32] The grant to Douglas appalled many French, especially the inhabitants of Tours and the surrounding province, whose experience of the Scots since 1419 had not been pleasant. Despite the appeals of the citizens of Tours to his council, Charles handed them over to their new duke's care. His decision rested on military necessity. The grant of Touraine rehearsed the achievements of Buchan and Wigtown in Charles's cause, and a duchy may have seemed a reasonable price for an army capable of halting the English advance. Moreover, Touraine was already under virtual Scottish occupation. Though Thomas Seton had been killed in 1423, his garrison remained in Langeais, while Buchan's men held Châtillon and Scottish troops under Maxwell of Pollok occupied Bleré. As in the Scottish marches, there was a sense in which Touraine, defended from the English by the men of the Scottish army, was being granted as a reward for these military efforts. Douglas military lordship was again bound up with the family's acquisition of lands and wealth.[33]

The citizens of Tours were forced to accept that this Scottish occupation would now extend to within their walls, and prepared to win the favour of their new duke. In early May, when Douglas entered his duchy and took up residence in Loches, Tours sent a deputation of clergy and notables to invite him to the city. At three in the afternoon of 7 May 1424 Douglas received the city's keys and entered by the *Porte de la Riche* escorted by French nobles and townsmen. Processing through streets hung with banners, Douglas was led to the cathedral, where the archbishop of Tours and the dean of the cathedral chapter installed him as a canon. The new duke took up residence in the city and stationed a garrison in the castle under Adam Dalgleish. Over the next few months, Douglas lived up to the fears of his new vassals. Processions were all very well but, as throughout his career, Douglas wanted cash and credit. During May, June and July Douglas ran up debts of £6000 *tournois* in Tours, exceeding the annual returns of the city in three months, as well as drawing on the regular income of his duchy. In addition the duke billeted his men on the hapless citizens. For the Scottish soldiers, the opportunity to live off the rich valley of the Loire was probably worth the risks of war. For Archibald Douglas the wealth and stature of a European prince was certainly worth those risks. In terms of his dynasty's status, not just in Scotland but in western Christendom, the early summer of 1424 marked the high point of Douglas power. Both England and France were waiting for Douglas to enter the war.[34]

The new duke of Touraine seemed on the point of turning the Douglases into Franco-Scottish princes. His family arrangements prior to his departure in 1424 suggest that Archibald planned a long future for himself and his younger son, James, in France. With his extensive French lands and an army of Scottish adherents, Douglas had the means to play a major role in political and military conflict on the continent and, as the diplomacy of

1423 had shown, he was aware of his value to both England and France. His rise to the rank of duke and his role in securing James I's release were the product of exceptional circumstances. As the war in northern France approached a military balance, Charles's search for troops and English attempts to forestall his efforts gave Douglas huge leverage on both parties. However, Douglas's entry into Tours also symbolised relationships in existence since the 1330s, when the family first gained fame on the continent. From the 1370s the earls of Douglas were recognised as lords whose military power had a value in war and politics and, during the next half-century, the kings of England, France and Castile, the duke of Burgundy and the earl of Northumberland had all sought Douglas allies. But France had a special relationship with the house of Douglas, and the homage of Archibald to King Charles renewed a bond of service established when Earl William received knighthood from Charles's great-grandfather, King John, at Poitiers.

Despite his appearance as a battle-scarred giant, Douglas had gained his duchy in negotiation, not combat. He had promised to fight for the French king, and in early August 1424 he prepared to fulfil his promise. On 4 August he led the Scots army north from Tours leaving behind French subjects whose relief at his departure was tempered by the garrisons he left behind and the £1800 *tournois* he extracted on leaving, to pay his troops. At Châteaudun the Scots joined a French army and a force of Italian mercenaries and the allies marched to the relief of the small Norman town of Ivry. However, on 13 August Ivry fell to an English army led by John duke of Bedford, regent of France, causing the French and Scots to seek easier targets, and within two days they captured the nearby town of Verneuil-sur-Avre by a ruse. The fall of Verneuil brought Bedford in pursuit, and on 16 August the allies decided to offer battle outside the town.

French accounts stress that Douglas forced this decision on the French leaders, the duke of Alençon and the count of Aûmale, when they argued for withdrawal. He may have wanted to exercise his power as lieutenant-general and to confirm his value to Charles VII by defeating Bedford's smaller army. Bower recorded that Douglas ordered the Scots to take no prisoners and exchanged letters with Bedford, whom he knew as march warden, which communicated this order. The battle which followed on 17 August was the hardest fought of the war. Jean de Waurin, who took part on the English side, identified the Scots archers as particularly 'murderous' and stated that the 'greatest brunt' was borne by Thomas earl of Salisbury, who faced the Scots and came near to defeat at the hands of the 'earl of Douglas and his troop'. As at Shrewsbury, Douglas was impressive on the battlefield, but, like Shrewsbury, the battle was decided elsewhere. While the Scots fought on, the rest of the army, led by the Italian mercenaries, broke in rout. Surrounded by English troops shouting 'Clarence, Clarence' in memory of Baugé, the Scots were refused quarter. Out of over 6,000, only 40 Scots escaped the field. There was no ransom for Douglas in his third defeat. He was killed along with his son, James, and son-in-law, Buchan, and all the leaders of the 'new army':

> The bodies of the noble duke of Touraine, the constable of France and James de Douglas were brought to the cathedral ... of St Gatien in the city of Tours ... where they were buried side by side in the choir: dearly loved and delightful they were in life, in death they were not divided.

Bower saw the burial of Douglas and his kin as an appropriate mark of their comradeship. It was appropriate too that Douglas's resting place was in France, the summit of his ambitions. The scale of these ambitions, which led him to dominate the south of Scotland more fully than his predeccesors and culminated in the French adventure, marked Archibald as the most exceptional expression of aristocratic power in late-medieval Scotland. The irony of his career was that his greatest successes were achieved in diplomacy and politics, not on the battlefield. The 'giant warrior' lost all his battles but still established the reputation of a warlord, which proved his greatest political weapon.[35]

Although they treated their dead lord with respect, the citizens of Tours were less friendly to his live followers whom they blockaded and expelled from the castle. If Verneuil did not prove to be a major turning point in the war or end Scottish participation in French armies, it did spell the end of the Scots' occupation of Touraine and hopes of a Douglas principality on the Loire. Charles VII expressed his sorrow at Douglas's death and returned his goods to Scotland, but ignored the claims of his lieutenant's heirs to Touraine. Wigtown, now fifth earl of Douglas, used the title, but the duchy had been granted by Charles to his sister-in-law only two months after Verneuil, while both the office of constable and Wigtown's own lands of Dun-le-Roi passed to Charles's new hope in war, the Breton, Arthur count of Richemont. The Scots earls had won rewards for their leadership in war. With the end of that leadership, their rewards were taken from them. It was a lesson with ominous significance for the Douglases' role in Scotland.

The loss of their lands in France did not mean the end of the family's reputation on the continent. The career of the fifth earl and the fate of his son were followed by French and Burgundian chroniclers, and in the 1440s the Douglases were still the flower of Scottish chivalry to be challenged by Burgundian knights to a tournament. In 1450 the kings of England and France, the duke of Burgundy and the Pope all knew the arms of Douglas, as Holland boasted, and received the earl of Douglas with respect as he progressed on pilgrimage to Rome. The massive retinue in his train must have reminded these princes of the power of the Black Douglases, which, only thirty years before, had seemed capable of determining the fate of western Europe.[36]

NOTES

1. *The Bruce*, XV, lines 537–38; *Chron. Le Bel*, i, 52–70.

2. *Chron. Froissart*, ed. Johnes, iii, 223–25; iv, 12, 30. See also P. Contamine, 'Froissart and Scotland', in G. Simpson (ed.), *Scotland and the Low Countries, 1124–1994* (East Linton, 1996), 43–58.

3. A. Macquarrie, *Scotland and the Crusades 1095–1560* (Edinburgh, 1985), 74–79; Väthjunker, thesis, 135–45; *Chron. Le Bel*, i, 83–85; *Foedera*, ii, 770.

4. A. Macquarrie, *Scotland and the Crusades*, 79–88; S. Boardman, *Early Stewart Kings*, 45–47; *Scalachronica*, 125; *Rot. Scot.*, i, 877; Francisque-Michel, *Les Ecossais en France. Les Français en Ecosse*, 2 vols (London, 1862), i, 91; *Westminster Chron.*, 474–77; S.R.O., AD1/27. One of those at Poitiers with the lord of Douglas, Patrick Dunbar, continued with his pilgrimage, dying on the journey in 1357 (*Rot. Scot.*, i, 707–709; *Chron. Fordun*, i, 377).

5. A. Goodman, *John of Gaunt*, 80–84; S. Boardman, *Early Stewart Kings*, 117; *Cal. Docs. Scot.*, iv, nos 707, 729, 736, 752, 762, 782, 788; *Rot. Scot.*, ii, 177, 180. Men from the Douglas household took part in a joust with English knights organised by the earl of Mar in 1406 (*Wyntoun*, vi, 420–21; N.L.S., Adv. MSS, 80.4.15).

6. *Rot. Scot.*, i, 752, 819, 839, 841, 877, 915, 931, 960, 963, 968, 969; ii, 2, 3, 7, 41.

7. *Chron. Knighton*, ii, 23; *Wyntoun*, vi, 102–103; Froissart, *Chronicles*, ed. and trans. G. Brereton, (London, 1968), 139 ; *Scotichronicon*, ed. Watt, viii, 46–49.

8. *Chron. Knighton*, i, 460; *Scotichronicon*, ed. Watt, viii, 58–59; *Westminster Chron.*, 477–77.

9. *Rot. Scot.*, i, 832, 955; ii, 3; A. Goodman, 'Introduction', in Tuck and Goodman, *War and Border Societies*, 1–10.

10. J. A. Tuck, 'Richard II and the Border Magnates', in *Northern History*, 3 (1968), 27–52, 27–34; J. A. Tuck, 'Northumbrian Society in the Fourteenth Century', in *ibid.*, 6 (1971), 22–39; J. A. Tuck, 'The Emergence of a Northern Nobility, 1250–1400', in *ibid.*, 23 (1986), 1–17; R. Robson, *The Rise and Fall of the English Highland Clans* (Edinburgh, 1989), 1–66.

11. Tuck, 'The Emergence of a Northern Nobility', 5–16; Tuck, 'Richard II and the Border Magnates', 35–52; J. A. Tuck, 'The Percies and the Community of Northumberland in the Later Fourteenth Century', in Tuck and Goodman, *War and Border Societies*, 178–95.

12. *Foedera*, viii, 313; Tuck, 'The Emergence of a Northern Nobility', 9; McNiven, 'The Scottish Policy of the Percies'; Tuck, 'Richard II and the Border Magnates'; A. Goodman, *John of Gaunt*, 81–82, 89–91.

13. Grant, 'The Otterburn War', in Tuck and Goodman, *War and Border Societies*, 31–34; Stones, *Anglo-Scottish Relations*, 171–72; *Foedera*, ii, 804.

14. Tuck, 'The Emergence of a Northern Nobility', 13–17; Tuck, 'Northumbrian Society', 30–38; Storey, 'The Wardens of the Marches', 594–95, 600; J. M. W. Bean, 'Henry IV and the Percies'. The differing relationship between crown and border magnates is also illustrated by the early 1380s. While Carrick, the Scottish lieutenant, bought support with exceptional patronage, his English counterpart, Gaunt, used his position to undermine Percy's local lordship (Tuck, 'The Percies and the Community of Northumberland', 187–90).

15. *The Bruce*, Book I, lines 313–28; *Scalachronica*, 125; Francisque-Michel, *Les Ecossais en*

France, i, 62–64; *Chroniques des Quatre Premiers Valois*, ed. S. Luce, 284; *E.R.*, i, 507; G. du Fresne de Beaucourt, *Histoire de Charles VII*, 6 vols (Paris, 1881–91), i, 332.

16. J. S. Richardson, M. Wood and C. J. Tabraham, *Melrose Abbey* (Edinburgh, 1981); *R.C.A.H.M.S.*, Roxburghshire, ii (Edinburgh, 1956), no. 567; *ibid.*, Galloway, ii (1914), 242–53; I. Campbell, 'A Romanesque Revival and the Early Renaissance in Scotland, c. 1380–1513', in *Journal of the Society of Architectural Historians*, 54 (1995), 302–25.

17. *Scotichronicon*, ed. Goodall, ii, 330–31; *Wyntoun*, vi, 126–27; *Chron. Lanercost*, 318; *A.P.S.*, i, 559; Archives Nationales, J677, nos 12, 13; Boardman, *Early Stewart Kings*, 109–10; P. Contamine, 'Froissart and Scotland', 46–47. French troops had also campaigned in Scotland in 1355 (*Scotichronicon*, ed. Goodall, ii, 351–54; J. Campbell, 'England, Scotland and the Hundred Years' War', 199–200).

18. A. Goodman, 'A Letter from an Earl of Douglas to a King of Castile', in *S.H.R.*, 64 (1985), 68–78, 70–71; *Chroniques des Quatre Premiers Valois*, 284; Wyntoun, *Chronicles*, ed. Laing, iii, 14–15.

19. *Chron. Froissart*, ed. Johnes, ii, 20–22, 35; iv, 30; S. Boardman, *Early Stewart Kings*, 124, 136; Archives Nationales, J677, no. 15; Goodman, 'Letter', 74–77.

20. MacDonald, thesis, 147; *H.M.C.*, 15, appendix 10, 77–78; Francisque-Michel, *Les Ecossais en France*, ii, 104–105; *Henry IV Letters*, i, 205.

21. *Rot. Scot.*, i, 824, 826, 836, 840; Goodman, 'Letter', 69; Francisque-Michel, *Les Ecossais en France*, i, 101–102; Swinton, 'John of Swinton'; Archives Nationales, K57, no. 9/12; C. J. Ford, 'Piracy or Policy: the Crisis in the Channel, 1400–1403', in *Transactions of the Royal Historical Society*, 29 (1979), 63–78, 71–72; R. Vaughan, *John the Fearless* (London, 1966), 55, 58, 260; *Wyntoun*, vi, 420–22.

22. *Scotichronicon*, ed. Watt, viii, 81–83; *Copiale*, 238–40, 400; *Cal. Docs. Scot.*, iv, no. 834; L. de Laborde, *Les Ducs de Bourgogne. Etudes sur les lettres etc*, 3 vols (Paris, 1849–52), i, 96–97; Francisque-Michel, *Les Ecossais en France*, i, 113–14; R. Vaughan, *John the Fearless*, 94–102, 146, 260. Douglas's retinue of 50 men included Henry earl of Orkney, William Hay and, probably, one of the earl's sons. In the agreement with Burgundy, the duke also promised to lead 300 men to Scotland if required to do so.

23. *Rot. Scot.*, ii, 207; *Calendar of Signet letters of Henry IV and Henry V* (London, 1978), no. 762; R. Vaughan, *John the Fearless*, 193–223; C. T. Allmand, *Henry V* (London, 1992), 66–101.

24. *Scotichronicon*, ed. Watt, viii, 82–87; Walsingham, *Historia Anglicana*, ii, 325–26; J. H. Wylie and W. T. Waugh, *The Reign of Henry the Fifth*, 3 vols (Cambridge, 1914–29), iii, 88–90.

25. *Scotichronicon*, ed. Watt, viii, 112–15; Beaucourt, *Charles VII*, i, 332–33; Francisque-Michel, *Les Ecossais*, i, 114–115; B. Chevalier, 'Les Ecossais dans l'armée de Charles VII jusqu'à la bataille de Verneuil', in *Jeanne d'Arc: une époque, un rayonnement* (Paris, 1982), 85–94.

26. Beaucourt, *Charles VII*, i, 332; ii, 59–63; *Scotichronicon*, ed. Watt, viii, 112–15, 118–19; W. Forbes Leith, *The Scots Men at Arms and Life Guards in France*, 2 vols (Edinburgh, 1882), i, 153–54; *Letters and Papers illustrative of the Wars of the English in France during the Reign of Henry the Sixth, King of England*, ed. J. Stevenson, Rolls series, 2 vols (London, 1861–64), i, 15; ii, 385; Fraser, *Douglas*, iii, nos 62, 63; W. Fraser (ed.), *The Maxwells of Pollok* (Edinburgh, 1863), no. 28; Chevalier, 'Les Ecossais ...', 87–88. For a full account of the war in France between 1419 and 1424, see R. A. Newhall, *The English Conquest of Normandy, 1416–1424* (Yale, 1924). The fullest examination of the Scots in French service occurs in B. G. H. Ditcham, 'The Employment of Foreign Mercenary Troops in the French Royal Armies, 1415–1470' (unpublished Ph.D thesis, University of Edinburgh, 1978), 14–52.

27. Bibilothèque Nationale, Fond Français, no. 7858, fo. 346v, 361r; no. 32510, fos 360r, 362v; Ditcham, thesis, 14–16; *Scotichronicon*, ed. Watt, viii, 46–47; *Letters and Papers illustrative of the Wars of the English in France*, i, 15; M. Brown, *James I*, 115.

28. Wylie and Waugh, *Henry the Fifth*, iii, 293–310; *Scotichronicon*, ed. Watt, viii, 118–21; Walsingham, *Historia Anglicana*, ii, 339.

29. Beaucourt, i, 220–21, 230; Chevalier, 'Les Ecossais ...', 90–91; Bibliothèque Nationale,

Fonds Français, no. 32511, fo. 140r; Fonds Clairambault, 41, no. 144; Ditcham, thesis, 23, 27, 30, 32, 33, 34–37; *Scotichronicon*, ed. Watt, viii, 112–15.

30. *Scotichronicon*, ed. Watt, viii, 120–21; Ditcham, thesis, 34; *Foedera*, x, 99, 123–25, 174, 230; *Cal. Docs. Scot.*, iv, no. 905; *Rot. Scot.*, ii, 230, 233, 235, 238. The hostages for James's release included five earls, four bishops and Walter Stewart, Murdoch's son, who was openly hostile to the king. Walter's consent would have been unlikely, especially as he was in open dispute with his father in 1421–22.

31. Beaucourt, *Charles VII*, ii, 336–40; *Foedera*, x, 294–95, 299–300, 301–308; *Rot. Scot.*, ii, 233; R. A. Griffiths, *The Reign of Henry VI* (London, 1981), 154–56; M. H. Brown, 'Crown-Magnate Relations in the Personal Rule of James I of Scotland, 1424–1437' (unpublished Ph.D thesis, University of St. Andrews, 1991), 12–59; M. Brown, *James I*, 24–31; Archives Nationales, J677, no. 20; J680, no. 71.

32. *R.M.S.*, ii, nos 13, 59, 85, 143, 256; Fraser, *Douglas*, iii, 380; *Melrose Liber*, ii, 507; Archives Nationales, J680, no. 71; X1a no. 8604, fo. 65r–66v; Bibliothèque Nationale, Fonds Latins, no. 10187, fos 1r, 2r, 2v; *Scotichronicon*, ed. Watt, viii, 124–25; *Liber Pluscardensis*, Book 10, Chapter 28. Even before reviewing the Scots, Charles had issued orders about Douglas's rewards (Archives Nationales, J680, no. 70; Bibliothèque Nationale, Fonds Latins, no. 10187, fo. 2v).

33. Ditcham, thesis, 34–37, 46; Chevalier, 'Les Ecossais', 91; Archives Nationales, X1a, no. 8604, fos 65r–66v. Charles may have seen the promotion of the Scots as a further element in his policy of maintaining his authority by playing off rival factions in his following. See M. G. A. Vale, *Charles VII* (London, 1974), 31–44.

34. Ditcham, thesis, 47; Forbes Leith, *Scots Men at Arms*, i, 26–27; Chevalier, 'Les Ecossais …', 91.

35. *Scotichronicon*, ed. Watt, viii, 124–27; Jean de Waurin, *Chroniques et Anchiennes Istoires de la Grant Bretagne*, 5 vols (London, 1864–91), v, 67–82; Thomas Basin, *Histoire de Charles VII*, 2 vols (Paris, 1933), i, 93–101; Jean Chartier, *Chronique de Charles VII, Roi de France*, ed. Vallet de Viriville (Paris, 1858), i, 42; Chevalier, 'Les Ecossais …', 91; M. A. Simpson, 'The Campaign of Verneuil', in *English Historical Review*, 49 (1934), 93–100; A. H. Burne, *The Agincourt War* (London, 1956), 196–215.

36. Bibliothèque Nationale, Fonds Latins no. 10187, fos 5v, 6v; *Letters and Papers Illustrative of the Wars of the English in France*, i, 15–16; B. G. H. Ditcham, ''Mutton Guzzlers and Wine Bags': Foreign Soldiers and Native Reactions in Fifteenth Century France', in C. T. Allmand, *Power, Culture and Religion in France* (Woodbridge, 1989), 1–13, 3; R. Vaughan, *Philip the Good* (London, 1970), 112; C. A. McGladdery, *James II*, 41–43, 164. The expulsion of the Scots from Tours, Langeais and Châtillon took until December 1424 and was achieved by a combination of bribery and force (Chevalier, 'Les Ecossais …', 92–93; Ditcham, thesis, 51). Douglas had been granted Touraine as an apanage, a royal principality held in the male line of descent, failing which it reverted to the crown. The efforts of the Douglases to recover the duchy after 1440 were, therefore, doomed (Archives Nationales, X1a 8604, fo. 65v; Bibliothèque Nationale, Fonds Latins no. 10187, 4, 13).

CHAPTER ELEVEN

The Rise of James the Gross

'OUR DEAREST NEPHEW': JAMES I AND THE EARL OF DOUGLAS

Less than six weeks after the fleet of hired ships had carried Douglas, Buchan and the 'new army of Scotland' out of the Clyde, James king of Scots crossed into his realm on the road to Melrose. The departure of the great lord and the return of the king were closely bound together. Douglas's French ambitions had led him to take a major role in winning James's release from captivity since 1421. Though both goals would prove disastrous for the house of Douglas, the earl clearly saw no threat in the return of his brother-in-law. Douglas had no reason to fear a Stewart king. He may rightly have deduced that James had greater grievances against his Albany Stewart cousins, themselves weakened by internal conflict. In the earlier rivalry between the descendants of Robert II in the 1390s and 1400s, the Black Douglases had gained much by holding the political balance and could reasonably hope to do so again. Earl Archibald had shown himself to be a friend in need to King James and seems to have left his family, headed by Countess Margaret and the earl of Wigtown, James's sister and nephew, with instructions to work with their royal kinsman. James's friendship offered protection from rivals in the south, new rewards and peace in the marches. Douglas, who saw his future as duke of Touraine and King Charles's lieutenant, regarded James as a means of securing his interests in Scotland.[1]

During the spring and summer of 1424, as he prepared for his last campaign, the expectations of the new duke of Touraine seemed justified. While making clear his political independence, the newly returned king looked on the Black Douglases as close allies. He turned for counsel and help to the family and advisers of the departed Douglas earl, and made the castle of Edinburgh, still held by its Douglas keepers, his principal residence. By contrast, King James showed his early suspicions of his Albany Stewart cousins, and the arrest of Walter Stewart, heir to Murdoch of Albany, though supported by the duke, seemed to promise a return to the internecine struggles of the 1380s, 1390s and 1400s.[2] However, the return of King James meant more than just the renewal of Stewart family rivalries.

TABLE 5. THE DOUGLASES OF ABERCORN

Archibald 'the Grim'
3rd Earl of Douglas

James 'the Gross' = Beatrice Sinclair
Lord of Balvenie and Abercorn   sister of William 3rd Earl of Orkney
1st Earl of Avondale   (d. pre-1463)
7th Earl of Douglas (d. 1443)

William, 8th Earl of Douglas
2nd Earl of Avondale
(k. 1452)

(1) = Margaret of Galloway = (2)
daughter of 5th Earl of Douglas

James, 9th Earl of Douglas
3rd Earl of Avondale
(d. 1491)

Archibald, Earl of
Moray (k. 1455)
= Elizabeth Dunbar
Countess of Moray

Hugh, Earl of
Ormond
(exec. 1455)

Hugh, dean of Brechin

John, Lord of
Balvenie
(exec. 1463)

Margaret =
Henry Douglas of Borg,
brother of James (III)
Douglas of Dalkeith

Beatrice =
William Hay of Errol
1st Earl of Errol

Janet =
Robert Fleming of Biggar
1st Lord Fleming

Elizabeth = John Wallace
of Craigie

Henry
(d. 1450 ?)

TABLE 6. THE RED DOUGLASES

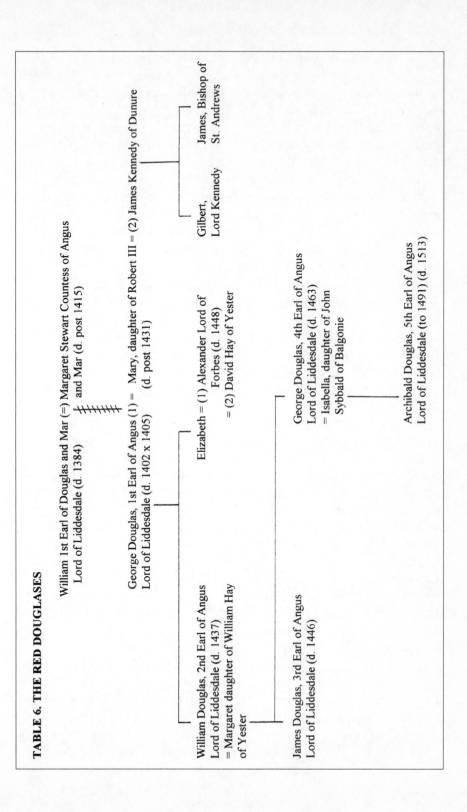

William 1st Earl of Douglas and Mar (=) Margaret Stewart Countess of Angus
Lord of Liddesdale (d. 1384)   and Mar (d. post 1415)

George Douglas, 1st Earl of Angus (1) = Mary, daughter of Robert III = (2) James Kennedy of Dunure
Lord of Liddesdale (d. 1402 x 1405)   (d. post 1431)

Gilbert,   James, Bishop of
Lord Kennedy   St. Andrews

Elizabeth = (1) Alexander Lord of
     Forbes (d. 1448)
     = (2) David Hay of Yester

William Douglas, 2nd Earl of Angus
Lord of Liddesdale (d. 1437)
= Margaret daughter of William Hay
of Yester

James Douglas, 3rd Earl of Angus
Lord of Liddesdale (d. 1446)

George Douglas, 4th Earl of Angus
Lord of Liddesdale (d. 1463)
= Isabella, daughter of John
Sybbald of Balgonie

Archibald Douglas, 5th Earl of Angus
Lord of Liddesdale (to 1491) (d. 1513)

It also marked the return of kingship as the major political force in the kingdom, and in the years to come, the Douglases would find themselves victims of their own propaganda. Pride in the service of Black Douglas to King Robert Bruce, which was used to justify the exceptional powers and status of the Douglas earls, was also an acknowledgement of the place of kingship as the heart of Scotland's political identity. As the heir of Bruce, King James could claim Robert I's leadership of the realm and the service of the heirs of Douglas, and unlike his father and grandfather, James possessed an acute sense of the rights and respect due to the crown from even its greatest subjects. From the day of his return the king showed an awareness of his public powers as guardian of Scotland's peace, powers which could be used to present the extension of royal authority as the protection of the community from the 'thieving, dishonest conduct and plundering' of great men. Despite the king's shortcomings, his message held attractions for contemporaries who saw effective kingship as the return to a natural order and as a genuine source of order.[3] For rulers like James I and his son, determined to stamp their authority on their realm without regard for claims of special rights and status, any dependence on great men like the earls of Douglas represented a check on that authority. Such an alliance would never be more than a matter of neccessity, born of short-term weakness, to be redrawn or cast aside when circumstances made it possible. Within months of his return James I would seize an opportunity to change the character of his dealings with the Black Douglases for good.

For Archibald, earl of Wigtown, leader of the Black Douglas connection in Scotland, the news of Verneuil and the death of his father, brother and brother-in-law was a disaster. Though Verneuil was fought far from the marches and brought no threat of an English attack on Scotland in its wake, the consequences for the house of Douglas were as great as those of Halidon Hill, Neville's Cross, Otterburn or Humbleton. Equally, while there was no struggle for the control of Douglas estates like those which followed the capture of Douglas of Liddesdale in 1346 and the death of Earl James in 1388, the killing of the fourth earl did major damage to the fortunes of the family. His death removed a figure of influence in the south and across western Europe, and severed the connections and bonds of lordship which the earl had built up since 1400. His offices and powers as keeper of Edinburgh castle, as protector of Melrose and Holyrood abbeys, as bailie of Coldingham and as justiciar south of Forth were lost to his family at his death. Of equal or greater significance was the destruction of the earl's army. The dead of Verneuil numbered thousands of Scots, many of them Douglas adherents and their followers. The loss of men like John Swinton, Alexander Hume, William Seton, Robert Pringle, Walter

Bickerton and Adam Dalgleish, added to the earlier deaths of William Douglas of Drumlanrig, John Stewart of Teviotdale and Thomas Colville, deprived the Black Douglases of many locally significant supporters. If many of those killed under the earl's banner at Verneuil were younger sons who, like their leader's second son, James Douglas, had gone to France in the hope of lands and wealth, their deaths still weakened the political and military resources of the new Black Douglas earl. Moreover, in contrast to the aftermath of previous Douglas defeats, in 1424 there was a king ready and able to exploit the weakness of the family to his advantage and to prevent Wigtown, now the fifth earl of Douglas, from recovering the influence and independence of his father. The quiescence of the earl in the face of this royal interference was a mark of the damage done to the Black Douglas affinity in the pursuit of his father's continental ambitions.[4]

Verneuil gave James I the opportunity to alter the political balance of the kingdom. In the seven years after Verneuil, the king, while never provoking a major clash with Douglas, sought to bind the new earl into a royal-dominated polity and restrict his power and independence. Appropriately enough, the first act in this process was played at Melrose, where the heart of Bruce and the bodies of the first two Douglas earls rested as symbols of the connections of both King James and Earl Archibald with the abbey. In early October the two men met there. Though the bonds between them were not broken by news of Verneuil, the battle changed the character of their relationship and freed James to pursue an increasingly aggressive policy towards the Albany Stewarts, themselves weakened by Verneuil and the deaths of Buchan and many of his followers. Within weeks of the Melrose meeting, a number of Duke Murdoch's kin and allies were imprisoned. Tensions between the king and Albany escalated during the winter, and after an acrimonious parliament at Perth in March 1425 the duke was arrested and his castles seized, provoking open rebellion in Lennox and Argyll. During the months from October to the trial of Murdoch and his family the following May, the new earl of Douglas gave the king his support. In April 1425, James placed the captive duke in the hands of Douglas's servant, Herbert Maxwell. The king clearly regarded the earl, his lordship and adherents as allies. At the trial of Murdoch, Douglas again gave full backing to his uncle, sitting on the assize which condemned the duke and his sons to death.

The support of the earl was crucial to the king's victory. In 1402 Douglas's father had similarly co-operated in the removal of James's brother, Rothesay, by Murdoch's father. Had the new earl of Douglas switched support to the house of Albany, the future of James's fragile regime must have become suddenly uncertain. In 1402 Black Douglas support had come at the high price of Albany support for the fourth earl's dominance south of Forth. In 1425, the fifth earl received nothing concrete from James. In return the king had eliminated his principal Stewart rival, ending the conflict within the royal house which had been waged sporadically since the 1380s, doubling the territorial resources of the crown, and, most importantly, establishing his personal primacy in the kingdom as a ruler to be feared and a lord worth following.[5]

Earl Archibald, uncertain of his own strength after Verneuil and perhaps cowed by James's physical presence, had helped ensure his uncle's success. This, in turn, ended the king's remaining dependence on the Douglas earl. Instead of rewarding him, James's treatment of the magnate was designed to undermine his regional power. In 1426 the king confirmed the earl's mother, Margaret, duchess of Touraine, as lady of Galloway for her lifetime. His charter extended and formalised Margaret's custody of Galloway, which she had received from her husband, the fourth earl, before his departure to France. The king's intention was to weaken the resources of the earl, and his action had the long-term effect of exacerbating new rivalries within the house of Douglas in the coming decades. Although it was not mentioned in the royal charter to Duchess Margaret, the earl must have hoped to recover the lordship on his mother's death, and this expectation, dependent on royal generosity, was designed to encourage Archibald's desire for his uncle's favour. Over the next three years, the king's plan worked and the earl served James as a councillor, courtier and military retainer.[6]

In these years, Douglas was not denied a place as one of the king's greatest subjects and closest kinsmen, but, unlike the 1380s and 1390s when these family ties were first established, James I identified kinship with the royal house as the basis and not a reflection of aristocratic power. In 1428 the king presented the magnates of royal blood as a special group, with the earls of Atholl, Angus and Douglas swearing the oath to observe the French alliance ahead of the bishops. For Douglas such status, though it gave him a place as the king's 'dearest nephew', confirmed his loss of his father's influence. The king forged a marriage alliance with King Charles VII of France and obtained the promise of the county of Saintonge in return for promised military aid to the French against the English. Only four years earlier, the Black Douglases could negotiate such a deal themselves. The occasion marked James's domination of foreign policy, and Archibald, denied the title of duke of Touraine which he claimed in his own documents, was made aware of the reduction of Douglas standing with France. In Scottish terms too, the grouping of Douglas with the earls of Angus and Atholl, men whose power had grown with the return of the king, marked Archibald's decline.[7] Douglas felt this loss of power, but his efforts to revive it failed. In late 1429 he sought to influence the king's policy towards the marches and towards the politics of the Kennedy family in Carrick. His lack of success led him to take direct action. In 1430 he opened direct negotiations with the English warden of the west march, while the following year he was associated with a possibly violent feud between John Kennedy and his uncles. The private talks led to Douglas's temporary exclusion from his march wardenships, and his involvement in Carrick led to his arrest and incarceration in the castle of Lochleven.[8] Though lasting only a matter of weeks, the earl's imprisonment was one more blow to his pride. Not since the 1360s had a Douglas magnate suffered such a humiliation and, even then, Earl William Douglas had been in full rebellion with an aristocratic coalition. The fall of Albany had deprived Douglas of any similar allies against James. The remaining magnates, like Atholl and Mar, worked with the king to their advantage in the 1420s. Yet, like David II, James made no direct attack on the Douglas earl, but sought to weaken and control him without an open breach. Douglas, without major allies and with his following weakened by Verneuil, was ill-equipped to resist. Real and lasting damage was being done in these years to the power of the Black Douglases, less by brief spells of imprisonment than by the exercise of royal lordship.

KING'S MEN: JAMES THE GROSS AND THE ROYAL AFFINITY

A significant part of this royal lordship was directed at the Douglas family. As in the previous century, difficulties experienced by the head of the kindred were treated as opportunities by others of the Douglas name. The Douglas who gained most from the king's return was James, younger son of Archibald the Grim, and another uncle of the fifth earl of Douglas. Known in his old age as 'the Gross', James was a political, as well as a physical, heavyweight. Like his father, James's long career as a junior kinsman of the Douglas earls had a huge impact on the family's fortunes and ended with his winning, by dubious means, the principal lands and titles of the senior branch of the dynasty. As with his father, the sponsorship which he received from a royal lord was vital in his rise to a position of influence beyond the Douglas following. However, while Archibald the Grim emerged from this royal sponsorship as a great border magnate, James's power continued to lie elsewhere, deriving from other connections and interests.

The importance of James Douglas for all but the last three years of his life sprang from the influence and rewards he built up as the councillor and agent of the great men of the kingdom. The origins of this role lay in the defeat and capture of his brother at Humbleton in 1402. During the next seven years, James acted as his brother's principal lieutenant, responsible for the defence of the family's interests in politics and warfare, and holding the key office of warden of the marches. Though only a squire, James provided the focus of lordship for his brother's tenants and adherents, and when the earl returned to Scotland in 1407 and 1408, his younger brother took pride of place on his council. This experience of lordship, allied to his proximity to the earl in blood, meant that James could never be regarded as just another retainer. The lessons of their father's career were clear to both Archibald and James, and once the earl was securely back at the head of his following, his younger brother was found less frequently at his side. However, there was no breach between the brothers and James continued to wield influence in his family and the kingdom. Moreover, unlike his father, James was well rewarded for his service to the Douglas earl. In 1408 the fourth earl granted James an apanage, principally from the lands of their mother in the north. Lands in Moray, Aberdeenshire and on the Black Isle and the castle of Balvenie in Banffshire were made over to James, and provided him with new status. His brother may also have hoped that northern interests would draw his ambitions away from the family's heartlands. In the event, the lord of Balvenie devoted little time to what was, by the 1400s, a difficult inheritance. Instead, from 1408 James's activities centred on his southern lands, also received from the earl, the barony of Strathavon and part of the barony of Stonehouse in Lanarkshire, and the castle of Abercorn, situated by the Forth in West Lothian.[9]

Abercorn castle became James Douglas's principal residence. From 1408 it served as the base for its new lord's systematic plundering of nearby Linlithgow. Up to 1420 the burgh accounts recorded how James Douglas had extorted goods and money by threats and force from 'diverse merchants and burgesses', dragging away those who resisted to the dungeons of Abercorn. James was more than just a robber baron. With his brother's tacit support, James was demonstrating his power in West Lothian, probably using the funds he extracted to build connections in the district. His chief partner in crime was Walter Haliburton, the second husband of James's sister, Mary, widowed duchess of Rothesay. The Haliburtons had supported James in his defeat of David Fleming at Long Hermiston in 1406, and from 1408 Walter acted in concert with his brother-in-law in despoiling Linlithgow. The connection established between the two families lasted into the 1450s. However, if James's success in 1406 had cemented this alliance, his defeat of 'the leading men of Lothian' also made enemies. It was partly to end the legacy of hostility that, in late 1423, James married Beatrice Sinclair, sister of William earl of Orkney and his own great niece. The match carried personal considerations. James was probably a widower but at the age of nearly fifty remained childless. Over the next eighteen years his young wife would produce six sons and four daughters, a family which may have stirred in James the ambition to found a new dynasty. In the short term, though, the marriage forged an alliance with Earl William Sinclair, an extremely well-connected figure in Lothian and secured the lord of Abercorn's place as, not just 'the brother of the earl', but as part of the province's political community.[10]

From 1424, membership of this community made for close contact with James I. As has been stressed, the predominance of the fourth earl in Lothian rested entirely on the personal connections he had built up since 1400. With his departure and death in 1424, the pull of Douglas lordship was greatly weakened. King James replaced the earl in Edinburgh, and his court and council became the natural political focus for local nobles who traditionally preferred royal lordship to that of a great magnate whose principal estates lay beyond the province. Like David II, James I saw the knights and barons of Lothian as the core of a new royal affinity, a body essential to the extension of the king's authority after years in exile. From the opening years of James's rule, Lothian men like Robert Lauder of Bass, the sheriff of Lothian, who replaced Douglas as justiciar, were among James's administrative workhorses. In 1429 James relied on Lothian lords for his military retinue in his war against the Lord of the Isles. Though the earl of Douglas was also present, the service of his father's old squires, Adam Hepburn, Walter Haliburton and William Borthwick was no longer to the house of Douglas. In war, as in politics, the king was now the principal source of lordship in Lothian.[11]

James Douglas of Abercorn and Balvenie fitted into this royal following. Balvenie, who in 1406 had killed James's guardian, David Fleming, and precipitated the king's flight into exile, clearly took pains to ensure good relations with the new regime. Marriage to Beatrice Sinclair linked him to a family with close ties of service to the king, and even before this, James I identified the uses of a figure like Balvenie. His experience, not just within the Douglas connection, but also, before 1424, as an occasional councillor of the Albany governors, was invaluable for a king with only limited knowledge of his realm, and from the opening months of the reign Balvenie was a royal councillor, often alongside his brother-in-law, the earl of Orkney. If his initial presence also owed something to his Black Douglas blood, by 1426 Balvenie was a king's man. While Earl Archibald experienced royal pressure, Balvenie stood by his new lord, serving on the royal council and probably acting as King James's agent when he met with his nephew at Newark in July 1426. When the earl again fell under the king's displeasure in 1430, Balvenie played a similar role. While his nephew negotiated in secret with the English, James Douglas was one of the king's most active diplomats and was named as one of the keepers of the truce agreed in 1430. The absence of the earl from both the talks and list of march wardens guaranteeing the truce may even suggest that, once again, Balvenie had replaced his senior kinsman as warden. This time, though, it was at the urging, not of the Douglas earl, but the king.[12]

These services won Balvenie royal gratitude, and although new lands were granted to him in 1426 and 1427, Balvenie's chief reward was his place among the king's principal baronial adherents. King James's reliance on men like James Douglas left a lasting impact on Scottish political society. Before 1424, middle-ranking nobles, the holders of one or more minor lordships, were largely dependent on their connections with the great magnates Albany, Douglas, Mar and a few others. The death of David Fleming in 1406 showed the dangers facing even the best connected baron in challenging a major magnate house, while the adherence of powerful Lothian lords to the earl of Douglas indicated the primacy of these great aristocratic dynasties before 1424. The return of King James and his success in disrupting the structures of regional lordship in Lothian and elsewhere changed this. James Douglas was not alone in being won to royal service. The king employed men of similar resources in vital roles. In his attack on the Albany Stewarts, the Lothian baron Robert Lauder was custodian of Walter Stewart, and a knight from the Lennox, John Colquhoun of Luss, secured Dumbarton for James in the face of the king's local enemies. The Lennox was finally subdued for James by a force led by John Montgomery and Robert Cunningham, two Ayrshire barons. This reliance on minor nobles extended into the exercise of royal justice. Robert Lauder was succeeded as southern justiciar by another baron with loose connections to the Douglases, Thomas Somerville, and Somerville, with two other lords, Robert Stewart of Lorn and Walter Haliburton, provided the aristocratic element in the assize which forfeited George Dunbar in 1435.[13]

King James identified such men as powerful enough to be effective lieutenants but not beyond his control. The services they performed, especially in the arrest and trial of magnates like Albany and March, must have had an effect on their political perceptions and ambitions. James I may well have fostered such feelings deliberately. At the baptism of his twin sons in 1430 he chose to knight the heirs of numerous lords. All but one of the Scottish youths knighted were the sons of Lothian barons and included Balvenie's son, William. The exception was the elder son of Archibald earl of Douglas. All the boys were to be 'fellow soldiers' of the princes in the future. The king may have hoped to emphasise that service to the crown, not private power, would be the basis for status in his kingdom and that, despite his lands and following, Earl Archibald was guaranteed no special honour at the royal court. Service to James I certainly gave significant influence to a number of baronial families of limited resources. The king's political successes brought local communities under direct royal lordship, excluding or limiting the influence of great magnates, and like the Douglas earls, King James's exercise of power on a wide scale depended on a group of trusted deputies able to maintain royal authority and hold royal castles in his absence. By the mid-1430s he had recruited a group of key lieutenants, sheriffs and castle keepers, in the lands from Dunbar to Stirling which formed the most secure regions of royal authority. One of these men was Adam Hepburn of Hailes. In 1434 he aided in the downfall of the earls of March, repeating the deeds of his father three decades before. Unlike 1400, however, it was the king and not the Black Douglases who instigated the seizure of Dunbar castle, and as a result, it was Hepburn who achieved the greatest local gains. Though his offices of keeper of Dunbar and steward of the earldom of March were held at the king's pleasure, in practice they gave Hepburn extensive powers in the south-east, powers which in 1400 had been assumed by the earl of Douglas. For Hepburn, royal lordship had obvious benefits.[14]

King James's lordship held even greater benefits for the two minor Lothian families who by 1437 dominated royal office-holding in the rest of Lothian and Stirlingshire. In 1424 William Crichton and Alexander Livingston were figures of purely local significance, both of whom had links to the house of Douglas. Their subsequent rise was almost entirely achieved in royal service. Crichton built his position on intimate contact with King James, first as his personal chamberlain and then as master of his household, combining these duties with the offices of sheriff of Edinburgh and keeper of Edinburgh castle. These offices gave Crichton influence amongst the burgesses of Edinburgh and with the king's household knights. This latter group included William's cousin, George Crichton, another rising star in royal service who, by the mid-1430s, was sheriff of West Lothian and lord of Blackness castle, the port for the king's new palace at Linlithgow. By comparison, the rise of the Livingstons was less spectacular. It depended on the family's custody of Stirling castle, a role which may have begun before 1424 and which they certainly held by the mid-1430s. By that time others of Alexander's kin also held royal offices. His brother, John, once steward of the old earl of Douglas, was provost of Edinburgh and a royal servant, while his cousin, Robert, was master of works at Linlithgow palace, a post of importance to the king.[15]

James Douglas of Balvenie was one of this band of trusted royal agents. His appointment as sheriff of Lanark was designed to increase the local influence of both king and baron, perhaps especially at the expense of their Douglas nephew. However, Balvenie's connection to the Crichton and Livingston families went beyond service to the king as sheriff, keeper and councillor. With great significance for the house of Douglas and the kingdom, these men formed an identifiable clique in the royal affinity before 1437. The first connection was geographical. All had interests in West Lothian. Balvenie at Abercorn, George Crichton at Blackness and the Livingstons in Linlithgow and elsewhere in the sheriffdom were close neighbours. In political terms, all had links with William Sinclair earl of Orkney, Balvenie's brother-in-law; and in 1424 Orkney, Balvenie and Alexander Livingston travelled together to meet King James at Durham on the eve of his release from custody. Like Balvenie, Livingston may have had his path into royal service smoothed by Orkney, and contact was maintained between the two barons during the next dozen years. In the same period, Balvenie also retained connections with his neighbour, George Crichton, while John Livingston's position as an Edinburgh burgess would have brought him into contact with William Crichton. Though on a small-scale initially, as the interests and ambitions of these men grew in royal service, these personal contacts would grow in significance. Unlike David II, James did not make magnates. Instead he turned a group of local landowners and officials into men ready and able to exercise power at the highest level but still dependent on access to the resources of the crown. For their old lords, like the Black Douglas earls, this transformation would have serious consequences.[16]

It was not just secular lords whom James sought to win into his service from that of the earl. Like the Douglases, the king also sought the skills of talented churchmen as councillors and agents, and in this search the crown possessed a reservoir of patronage in church benefices and offices of state which far exceeded the resources of even his greatest subjects. Nowhere were the attractions and possible gains of royal service for an ambitious cleric clearer than in the case of John Cameron. In early 1424 Cameron was the secretary of the earl of Wigtown and provost of Lincluden college, a benefice in the gift of the Douglases. However, even before Verneuil, Cameron had become the king's secretary, the first step in a meteoric rise which by 1427 saw him established as bishop of Glasgow, chancellor of the realm and the chief agent of royal ecclesiastical policy. Though it was exceptional for the scale and speed of his promotion, Cameron's career was not unique. Like him, William Fowlis had entered Douglas service at the highpoint of the family's ecclesiastical influence. Despite his rapid rise to be the fourth earl's chief clerical councillor, Fowlis was also drawn to the royal affinity, and in the 1430s, in Cameron's absence on the continent, Fowlis, as keeper of the privy seal, was the bishop's effective replacement. Such service did not sever links with the Douglases, and Fowlis in particular continued to counsel the earl. Douglas even gained clerical servants from the king. Fowlis's successor as the earl's chancellor, Edward Lauder, was a royal protégé. However, such connections no longer reflected the attractiveness of the earl's employment but rather the king's pervasive influence amongst aspiring clerics and in directing the disposal of Douglas patronage. In the 1430s this extended to the family's prized foundation at Lincluden, the provostship of which had been reserved for favoured ecclesiastical adherents. It was given in succession to John Winchester and John Methven, churchmen whose careers lay primarily in royal service.[17]

There was a similar overlap between the earl's secular advisers and the king's affinity. In the 1430s, James the Gross of Balvenie was again regularly in attendance on the head of his family and, on two occasions, William Crichton sat alongside him on the earl's council. In part, such contacts demonstrated improved relations between Earl Archibald and King James. Links with two of the king's principal servants gave Douglas new influence in royal circles. However, any such gains came at a price. Crichton and Balvenie sought and gained influence of their own with the earl, who, despite his setbacks, remained the greatest lord south of Forth. The earl's new councillors were men with their own interests in the network of Douglas lands and followers, Balvenie as Earl Archibald's closest adult kinsman, Crichton as a tenant and neighbour of the earl in the marches. In the coming years these interests would bind the fate of the Douglas earldom and its border estates to conflicts over the exercise of royal rights and powers and to the ambitions of men like Balvenie and Crichton, who had learned to combine their roles as valued servants of king and earl with the search for influence in the kingdom.[18]

MARCHMONT: JAMES I AND THE DOUGLAS HEARTLANDS

While the king turned Lothian into a royal stronghold, reducing the power of rivals like the earl of Douglas and building his own following of trusted lieutenants, he had different aims in the marches and Galloway. These were the heartlands of Black Douglas power. On the eve of James's return the family's regional dominance, the power of the 'great guardian' of the marches, seemed at its height. It was this virtual Black Douglas monopoly in border war, diplomacy and politics which the king sought to break, to prove to his nephew that the dales of the south were no longer beyond the reach of Stewart kingship. But instead of direct confrontation the king sought to extend his authority through the assertion of dormant royal rights, by the exercise of his personal lordship in the marches and, once again, through the promotion of rivals to the Black Douglas earl.

The king signalled his intentions early on. James's presence at Melrose in October 1424 showed the royal banner to the men of the middle march for the first time in forty years. Unlike Robert II, who visited the Forest in 1382, James I came south, not to confirm the dominance of the Douglas earl, but to show that there were no territorial limits to his own authority. In the east march this ambition had already been demonstrated. Here Black Douglas influence had been the creation of the fourth earl and it died with him at Verneuil. The right to act as the superior lord of his adherents in the earldom of March lapsed at his death, while the most important of these adherents, John Swinton and Alexander Hume, were slain alongside their lord. Their heirs and the remaining associates of the earl in the south-east, like David Hume of Wedderburn, had little reason to look to the new Douglas earl for leadership. Instead David Hume, like his landlord Adam Hepburn of Hailes, saw Verneuil as an opportunity to extend his own interests. David had been left by his brother, Alexander, to run the estates of Coldingham priory for the fourth earl, and on news of Verneuil, David sought to secure his future in this office. However, Coldingham's owners, the monks of Durham, saw their best hopes in another protector. They sought to replace the earl of Douglas as bailie, not by Hume, nor by their traditional patrons, the Dunbars, but by the king. In the east march King James was quickly identified as having the means to defend the monks, which since 1400 had been the task of the Douglases. The king used his powers as protector to put his own mark on local political society and confirm the exclusion of Black Douglas influence from the east.[19]

In the rest of the marches, from Lauderdale to the Rhinns of Galloway, the influence of Douglas magnates stretched back over eighty years and could not be disrupted so easily. The king lacked the physical and political resources to uproot Douglas lordship in these regions. Instead he aimed to place checks on the dominance of his Black Douglas nephew in the traditional strongholds of his house. In the west this meant Galloway, won for the family by Archibald the Grim fifty years before. In May 1426 the king granted the province to his sister, Margaret duchess of Touraine, the mother of Earl Archibald. She was to hold Galloway for life as it had been held by her husband. By formalising whatever rights Margaret had received from the fourth earl in 1424, the king effectively separated Galloway from the other principal lands of the Black Douglas family. Earl Archibald may have feared for his recovery of the lordship intact. Although the terms of the charter confirmed the powers of the Douglas lords, they also implied that, on his mother's death, the earl would have to sue for a fresh royal grant of the province. In those circumstances, the rights of the earl would be open to scrutiny. Anomalies like the usurpation of regality powers and the treatment of Wigtown and Galloway as a single lordship would be exposed by a king who was always greedy to extend his own legal rights.[20]

In the meantime, Duchess Margaret was placed in control of Galloway. From 1424 until the late 1440s she possessed the revenues and government of the lordship, which she ran from the island castle of Threave, apparently with the support of both the chief local lords and of her husband's principal officers in Galloway. She also promoted her own men. Andrew Agnew, a household servant of the duchess, may have been from a Wigtownshire family which had been expelled by Archibald the Grim. His fortunes certainly revived under Duchess Margaret. In 1426 he was made constable of Lochnaw castle and given lands in Wigtownshire by the powerful local baron, William Douglas of Leswalt. Leswalt's generosity may have been forced, and his relationship with Agnew was the pretext for Earl Archibald to exert his own influence in Galloway. To accept his exclusion from the province would have jeopardised Archibald's prospects of rebuilding his local influence should he recover Galloway from his mother, and he may have seized on his chance. In October 1426 he appeared at Threave backed by an impressive company of men from the west march to support a grant of the duchess compensating Leswalt. His presence made clear to the men of Galloway that he was their future lord, and showed his ability to intervene in the lordship. In 1432, with similar aims, he was at Wigtown, attended by the local community. The attachment of Galloway to the Black Douglases had not been severed. This had not been the king's aim. Instead James I had given the earl problems in the south-west and had limited his resources. In the coming decades, the detachment of Galloway would prove more troublesome to the house of Douglas than the king could have planned.[21]

In the middle march too, the king sought to loosen the ties of Black Douglas dominance through small-scale interference. James's presence at Melrose in 1424 began this process. His visit coincided with the election of a new abbot. The choice of John Fogo was a reminder of the co-operation between the king and the Douglases on the issue of church unity, but by 1426 Fogo, who had been confessor to the fourth earl, was confessor to the king. Like Cameron, he had entered royal service. Led by their abbot, the monks of Melrose looked to the king as their principal protector, and James I resumed the interest of his Bruce forebears in the abbey. Though the earl of Douglas settled the long-running case between the monks and Haig of Bemersyde in 1425, in 1428 and 1431 it was the king's council which arbitrated in similar boundary disputes.[22] The attitude of the Melrose monks to the king and the earl reflected local perceptions of the relative power of the two lords. The principal shift in this balance occurred in 1426. In moves which coincided with his grant of Galloway to his sister, King James interfered with Douglas's rights in the Forest of Ettrick and Selkirk. The Forest was the core of Douglas power in the marches, and had been untouched by the family's Scottish rivals for a century. For this very reason the king sought to leave his mark on the lordship. Selkirk burgh was incorporated into the structures of royal administration, having almost certainly been removed from Douglas control, and Earl Archibald was made to give his royal uncle certain rights within the Forest itself. These were probably rights to pasture rather than a reduction of Douglas's territorial powers, but they symbolised a change in the Forest which was not lost on local men. In August 1426 William and George Middlemast, the rangers of Yarrow ward, received the king's confirmation of their offices. This unprecedented act suggests the penetration of royal influence into the Forest whose inhabitants had for decades looked no further for lordship than the house of Douglas.[23]

The king was very close to the heart of Douglas power, the personal relationship between the earl and his trusted servants in the Forest. Elsewhere in the middle march the king sought inroads into the following of the Douglas earl. He depended not only on his own influence but on a network of allies. The Lothian lords in James's service had their own links with the marches, many of which had been fostered by the Douglases in their search for support. In the 1420s and 1430s the interests of Walter Haliburton, William Borthwick and William Crichton in the borders were far less clearly at Earl Archibald's disposal. Given the history of his house, however, Archibald's greatest anxieties were with another nephew of the king. William Douglas, earl of Angus, was a natural rival. He inherited the ambitions of the Red Douglases to recover the influence and lands of his ancestor, William earl of Douglas, and had his own grievances against his Black cousins who had denied him his rights in the marches since 1402. The king was an ally in pursuit of these goals while, for James, Angus was a means to open old wounds in the Douglas family. In 1426 Angus asserted his own claims in the marches, renewing his rights to Jedworth Forest and exercising effective lordship in Liddesdale. However, his activities focused on the east march, where the king handed him control of Coldingham's estates and backed him in his feud with the Dunbars. The feud culminated in the seizure of Dunbar castle by Angus and two royal lieutenants, Adam

Hepburn and William Crichton, in 1434 and in the second forfeiture of
the earls of March. The further rise of Angus, even at the expense of
March, was no comfort to the earl of Douglas. The latter had reason to
fear an alliance between the crown and the Red Douglases which had
worked against his grandfather in the 1390s and had shown its potential
against the, admittedly limited, power of the Dunbars. With Angus consciously recalling his descent from the first Douglas earls through their
connections to Melrose, the current earl of Douglas was made aware of a
challenge to his own position in the Douglas kindred.[24]

In the 1420s and 1430s the challenge of Angus and of the king took the
form, not of violent attack, but of competition for local adherents. King
James and his allies offered an alternative source of lordship and promotion
for men in the marches like Nicholas Rutherford, a junior member of an
extensive Teviotdale kindred. Like the Middlemasts, Nicholas sought royal
confirmation of his lands in 1426, an act which marked his entry into the
service of both the king, who granted him new lands in 1430, and Angus,
who employed him as bailie of Liddesdale and Jedworth in the coming
years. Angus already had a local associate in the person of his cousin,
Archibald Douglas of Cavers, bastard son of Earl James Douglas, and in
1428 cemented the relationship with a grant of lands in Liddesdale. Cavers
had never been closely associated with the Black Douglases, and as hereditary sheriff of Roxburgh may have seen advantages in the increased power
of the crown locally. His ties to Angus and the king may have led others
into similar connections. Thomas Cranstoun of that ilk was a tenant of
Cavers who, from the 1430s, had links with Angus and the Crichtons. These
links were possibly also fostered by Thomas's influential kinsman and
namesake who was, by 1436, Crichton's deputy in Edinburgh castle and
royal receiver-general of rents south of Forth, a key role in the running of
the king's growing demesne. Andrew Kerr and Walter Scott also received
grants of lands from the Crichtons, and by 1437 east march men like David
Hume and Adam Hepburn were given interests in the region. The marriage
of Simon Glendinning to a kinswoman of Hepburn marked a further link
between an adherent of the king and Angus and a central figure in the old
Black Douglas affinity.[25]

These contacts should not be exaggerated. There was nothing new in
men from the marches having their rights confirmed by the crown, and
links between Lothian barons and border kindreds were natural and of
long standing. The new connections forged by the Crichtons in the 1430s
did not affect relations between Earl Archibald and the Kerrs, Scotts and
Cranstouns, who all benefited from Douglas's patronage and employment.
Similarly in the west march, the keeper of Lochmaben castle, Michael

Ramsay, combined his office with duties in the king's household without apparent problems, and his ties to both king and earl must have aided his role as customer of exports in the western borders. In 1430 he was named as one of the assessors of the Anglo-Scottish truce in the west along with Matthew Glendinning and Simon Carruthers, two other tenants of Douglas. In the middle and eastern marches the same responsibilities were held by David Hume, John Cockburn and Alexander Murray, also Douglas vassals, though, in the case of Hume and Cockburn, men whose connections to the earl had been superseded since 1424. Royal and Black Douglas lordship were therefore far from exclusive and in the west march, the Forest and much of the middle borders the earl of Douglas remained the principal source of local protection and patronage, a mighty lord whom his men approached 'on bended knee', and who could give or delay justice in his massive regalities in the south. However, the years from 1424 to 1436 had seen the erosion of the family's dominance in the marches. The death of the fourth earl, the re-emergence of royal lordship and the rival claims of the Red Douglases represented fundamental checks on the power of the Douglas earls in the borders. The connections between the king and his allies and key adherents of the Black Douglases, like the Scotts, Kerrs, Cranstouns and Glendinnings, though initially of limited significance, would have major long-term importance. They presented these families with alternatives to Black Douglas lordship, and in future their support of the earl would be measured against other connections and interests.[26]

The re-assertion of royal leadership in the marches owed much to the absence of war from the region. Since 1415, English resources were fully diverted towards France and, after Verneuil, their concern was to prevent further Scottish support for Charles VII. Though James I sought an active role in European politics, in 1424 and 1431 he agreed to truces with England which held firm until 1435. The relative peace of the marches had an effect on the power of the earl of Douglas. It released the king from his fears about antagonising a powerful border magnate. Only in 1430, when negotiations for a new truce were briefly in difficulties, could Douglas seek to deal with England as a means of threatening the king, and the threat only precipitated further royal mistrust. For Douglas's tenants and neighbours, the absence of war reduced their dependence on the earl's protection. Douglas lordship had been built on defence and leadership of the south in war. Without major war, the pull of the earl's leadership was weakened as it had been in the 1390s.[27]

The king also exploited the peace to establish royal control of the machinery of border war and diplomacy, which centred on the march wardenships. From 1400 to 1424 control of all three marches had been in the hands of the Black Douglases, and King James refused to accept this monopoly. Even before Verneuil the east march had probably been assigned to George Dunbar earl of March, an act which recreated the distribution of the wardenships in the 1390s. By 1430 James was ready to go further than this and appointed his ally, William earl of Angus, a march warden, limiting Earl Archibald to custody of the west march alone. The appointment of Angus was designed to increase royal influence in the region, an aim also reflected in the 'item on the marches', a set of statutes issued by parliament in March 1430, during a period of Anglo-Scottish friction. The item dealt with the organisation of the marches for war, drawing on earlier royal legislation and on the ordinances of the march wardens themselves and, in particular, the house of Douglas. Though the warden retained a key role in the marches, these laws made clear that, as lawgiver and warleader, King James possessed ultimate power in the borderlands of his realm.[28] In 1436 the king was ready to go beyond legislation. The attack on the house of Dunbar had shown James the continued power of march magnates. In response to the loss of their lands, the Dunbars again sought English aid, as in 1400 turning an internal feud into a situation of Anglo-Scottish conflict. Again, in this crisis, the crown turned to a Douglas warlord. In 1435, though, it was the Red Douglas, William earl of Angus, who defeated the Dunbars' English allies at Piperdean near Cockburnspath. Angus's success confirmed his position in the marches, not as a prince like the fourth earl of Douglas, but as the king's chief lieutenant, warden of east and middle marches. However, the king wanted to make his own mark in the war with England. His chosen target in August 1436 was the castle of Roxburgh, one of two remaining English strongholds in Scotland. The campaign was no ordinary military venture but a display of royal leadership and resources in a war waged as part of a European coalition. For the first time in ninety years a Scottish king was to lead an army against the English enemy, and the renewal of kingship in the mould of the Bruce was accompanied by a burst of royal propaganda. The anticipated royal capture of Roxburgh castle, the Marchmont, was commemorated by the creation of a Marchmont herald and signet. The army raised was huge and, like the Scottish forces in France, contained many archers. It also included the king's artillery train, which was expected to win the castle taken by the Black Douglas in 1314 and which the fourth earl had failed to capture in 1419. Both Black and Red Douglases were excluded from command of what was to be a royal triumph.[29]

In the event, however, there was no royal victory. Effective English defence and 'detestable splits' in the king's host caused the siege to be abandoned, leaving the artillery to the enemy and King James to face a growing crisis. In the marches, the presence of the king and his host may have increased tensions created by the intrusion of royal lordship into the area. In early 1437 Walter Scott of Buccleuch was rewarded by the king for the capture of Gilbert Rutherford, and had reputedly killed William Rutherford of Eckford, who had earned royal hostility, perhaps in connection with the events at Roxburgh. At the same time, Gilbert and William's kinsman, Nicholas Rutherford, was paid for his services in defence of the marches. These represented very different reactions to the return of royal influence and large-scale war to the middle march from within the same Teviotdale kindred, but for the earl of Douglas the whole incident was ominous. Scott, a hereditary servant of the Douglas earls, was patronised by the king and William Crichton immediately after his attack on members of a family which had its own strong connections to the Black Douglases. Crichton's grant to Scott was witnessed by the head of the Rutherford family, James lord of Rutherford, whose presence may have denoted his submission to royal authority. Despite the failure of the Roxburgh campaign, the attack on the Rutherfords spoke of continued royal influence and continued erosion of Black Douglas power in the middle marches.[30]

LIEUTENANT-GENERAL OF SCOTLAND

The murder of King James I in the Dominican friary at Perth on the night of 20 February 1437 seemed to change everything. Though Earl Archibald was not involved, the king's death was a consequence of his interference with the power of his great subjects. Like Douglas, the instigator of the king's death, Walter earl of Atholl, felt the threat posed by James's lordship. Assassination removed the focus of this threat. The new king was James's six-year-old son. In the month following the murder, custody of the young James II was a matter of conflict between Atholl and Joan Beaufort, James I's English queen, who had the support of William earl of Angus. Like many major figures, Douglas played no active part in the conflict, which ended in late March with the capture of Atholl by Angus and the triumphant coronation of the new king. The queen's victory brought about Douglas's entry into regency politics. At the general council which met at Stirling in early May 1437, Archibald fifth earl of Douglas was chosen as lieutenant-general of the realm, to govern until the king reached the age to govern. The appointment was a defeat for Queen Joan, who had been associated with her husband's regime and who had headed the royal council since his death. Douglas cannot have relished a minority government headed by the queen and in which his main rival, Angus, wielded major influence. Many present at the council had similar anxieties. Douglas represented a conservative choice as lieutenant. He was the only figure who, like the guardians of the previous century, combined membership of the royal kindred with the regional power of a great magnate. By contrast, the queen derived her power solely from the royal establishment. Her search for authority presaged later political queens, but in the 1430s represented something new. In 1437 her ties to her dead husband's policies were unappealing, even within the old king's affinity, and at the general council, probably at Douglas's instigation, the young king was made to reverse his father's restrictions on the rights of the new lieutenant's sister, Elizabeth, in the lordship of Garioch in Aberdeenshire. The dead king's word was no longer law. The act also sealed an alliance between Douglas and his sister's third husband, William earl of Orkney, who, despite the treatment of the Garioch, had remained a close councillor of James I. In May 1437 connections with Orkney and with Douglas of Balvenie, Crichton and Bishop Cameron worked to Douglas's advantage. To these men, who dominated the royal council, Earl Archibald was a well-known lord, and despite the associations of these men with the old king's policies, Douglas, unlike the queen, was identified as a lieutenant willing to reward support and extend the powers of his closest adherents.[31]

The queen kept custody of her son and received a pension of 4,000 marks, but the new lieutenant held power to enforce the law and raise 'the kingis proffettis'. These powers were the same as those held by lieutenants from 1384 to 1406, which, allied to their resources as magnates, had been the basis for the political primacy sought by Atholl in his attack on James I. There is no reason to think that Douglas became lieutenant with any lesser ambitions. He exercised his right to call meetings of the estates which made clear that the lieutenant, assisted by the 'kyngis chosyn consal', was responsible for justice and lordship in the kingdom. This council, though dominated by the old king's men, was not static. In early 1439 John Cameron, bishop of Glasgow, was removed from the office of chancellor and replaced by William Crichton, a move which only Douglas had the power to arrange. It was Douglas too who issued orders revoking grants from the royal demesne, and who was given the responsibility of bringing to justice 'rebellys or unrewlful men' holding castles illegally or despoiling church lands, by raising 'the cuntre' against them.[32]

These demands for the lieutenant to use force against rebels may suggest anxieties about increasing local violence in several parts of the kingdom. Many of these had their roots in James I's forceful interference with the structures of regional lordship. Douglas lacked the means and inclination to adopt his royal uncle's forceful search for 'firm peace'. Instead the earl sought to manage Scottish politics in the fashion of previous lieutenants.

Rather than raising the country and laying siege to rebels, Douglas, like Robert duke of Albany, worked through the sponsorship of others and the recognition of the *status quo*. This approach was clearest in the north, a region where Douglas's interests were largely indirect. Since the death of Alexander earl of Mar in 1435, north-east Scotland had been the scene of conflicting ambitions which James I had only made worse. As lieutenant, Douglas sought, not personal dominance in the north, but to limit conflict. The full grant of Garioch to Orkney gave him a base in Aberdeenshire, while by assigning Kildrummy castle in Mar to Alexander Seton of Gordon, Douglas was supporting a rising local lord who had a recently forged marriage alliance with William Crichton. However, like Albany, Douglas tempered support for allies by seeking improved relations with their rivals. In 1438 Douglas negotiated a marriage between his elder son, William, and Janet Lindsay, daughter of David earl of Crawford. Their wedding at Dundee, recalled in later garbled accounts as an occasion of unparallelled 'triumph and pompe', sealed a bond between the lieutenant and the leading magnate of Angus, whose ambitions in the north-east had the potential to exacerbate regional rivalries.[33] It was probably also in 1438 that the lieutenant sought a similar understanding with Alexander lord of the Isles in a meeting on the Isle of Bute. Alexander was also earl of Ross and, following the death of Mar, was the greatest lord in the north with influence from the Hebrides to Moray. The lieutenant did not seek to challenge these ambitions but to channel and confirm them. By appointing Alexander justiciar north of Forth, Douglas increased his formal powers but placed them within the framework of royal government, just as fourteenth-century regimes had recognised Douglas power in the south. The lieutenant's approach marked a return to more limited arts of political management. The meetings with Crawford and Ross, though they produced no increase in the power of the crown, may have restricted regional conflict in the north in the late 1430s.[34]

Where his interests were not directly involved, this approach probably represented the extent of Douglas's ambitions. In areas like the Lennox and even March, where there was a superficial stability, the lieutenant left the old king's agents in their positions of local authority. Instead Douglas's ambitions focused on the expansion of his own lordship. After thirteen years in which the king had sought to contain the earl's power, the new lieutenant was keen to recover lost influence. He received a crucial advantage in this with the death of his chief rival in the marches, William earl of Angus, in late 1437. William was succeeded by his teenage son, James, whose custody of his border estates may have been delayed deliberately by the lieutenant. While James received sasine of the earldom of Angus in

January 1438, in November the lieutenant's uncle, Balvenie, was holding his court as justiciar within the regality of Jedworth Forest, indicating that it was in the hands of the crown, not of its new lord. It was not until July 1439 that Earl James was active in Jedworth. As in the 1400s, a young Red Douglas earl was denied his rights in the marches by the Black Douglases as a result of the removal of royal support.[35] Angus was not alone in experiencing pressure from Earl Archibald, who allied his powers as lieutenant to his search for dominance in the south. In early 1438 Douglas's cousin, Egidia, dowager countess of Orkney, complained to the council that her lordship of Nithsdale had been plundered by the lieutenant, who also held, or planned to hold, royal courts in the lordship. Such interference renewed local rivalries from before 1424. Nithsdale, though of limited value, was positioned between Annandale and Galloway, and the lieutenant's actions may be connected to later accounts of his reign of terror in Annandale. The countess's complaint makes clear that Earl Archibald was flexing his muscles in the south-west, and his efforts may have extended to his mother's position in Galloway where, after 1437, though she retained formal custody of the lordship, evidence of her administration is hard to find. Douglas was clearly seeking to claw back lost influence. His family's recovery of the southern justiciarship and of the wardenship of the middle march, offices central to the rise of the Douglases, and signs of the earl's widening influence in the marches and south-west, suggest that the lieutenancy was successfully exploited to renew Black Douglas power in the south.[36]

The primacy of Douglas in the kingdom did not mark a return to early Stewart politics. Douglas was not Albany and the rule of James I had left a deep impression on Scotland. The disappearance of the houses of Albany, Mar, Dunbar, Atholl and Lennox changed the shape of the nobility and the exercise of local power in the kingdom. The leading men in much of Scotland were now drawn from those families which, before 1424, had been adherents of greater lords. These men, Gordon and Forbes in the north-east, Hepburn and the Humes in March, Murray and Drummond in Strathearn, Boyds and Stewarts of Darnley in Ayrshire, now competed for local power. While Douglas dealt with great lords like Ross and Crawford, it was a local coalition which backed the elevation of Robert Erskine to the earldom of Mar in 1438. Despite James I's murder this diffusion of regional political power could not be reversed, nor could the increased royal authority which accompanied it. In lands, revenues and relative power, the crown had become a more potent force in Scottish politics. Although, like the early Stewart lieutenants, Douglas did not consider custody of the king as essential to his powers of regency, from 1439 physical control of James II would become increasingly important. Douglas was the last magnate whose obvious power made this unnecessary, and who was both close to the centres of royal authority but not dependent on them for his rule in the kingdom.[37]

However, as lieutenant, the earl's power was heavily influenced by the impact of his uncle's regime. His renewed authority was a product of his possession of those royal powers which James's forceful rule had restored to potency. At the same time, demonstrations of effective royal lordship had cut into Douglas's relations with secular and ecclesiastical tenants and neighbours, and James I's death did not relegate men like Walter Scott to their previous dependence on their lord. Still less did the earl's apparent primacy reduce the power and expectations of William Crichton and James the Gross of Balvenie. The interests of both men, and of their associates, pervaded Douglas's lieutenancy from its outset, and their hopes of reward in return for support were not disappointed. Crichton retained his old offices, received custody of escheated lands across the south from the lieutenant in 1437, and his appointment as chancellor in May 1439 confirmed his value to Douglas. Balvenie prospered equally. He was made justiciar south of Forth and given the rank of earl. As earl of Avondale, Balvenie was given the rank to claim to be his nephew's deputy. The influence of both men also directed the lieutenant's policies. Douglas's involvement in the north reflected the interests of these councillors whose lands and allies were probably protected by his actions. The new earl of Avondale and the new chancellor may have been key supporters and principal beneficiaries of Douglas's lieutenancy, but their earlier careers suggest that their service to the lieutenant was based on expediency, not affection. A decade earlier both men had backed King James's harassment of Earl Archibald, and his reliance on them from 1437 was an uncertain alliance. Through his pursuit of royal rights, his interference in local politics and his recruitment of a group of able adherents from the earl's affinity, the old king had opened cracks in the Black Douglas following in council and locality which were not closed by the earl's two years as lieutenant. Instead, by forging an alliance with Crichton and Avondale, Douglas bound his dynasty to their pursuit of continued power during the minority. When Earl Archibald died suddenly of a fever on 26 June 1439 at Restalrig near Edinburgh, he left his young sons exposed to the ambitions of their powerful great-uncle, James the Gross.[38]

NOTES

1. *R.M.S.*, ii, nos 12, 143. For further discussion of James's return, see M. Brown, *James I*, 24–33.
2. *Scotichronicon*, ed. Watt, viii, 240–41; M. Brown, *James I*, 40–52.

3. *Scotichronicon*, ed. Watt, viii, 322–23; M. Brown, *James I*, 1–8, 201–207.
4. Ditcham, thesis, 49–50; Leith, *Scots Men at Arms*, i, 153–54; *Letters and Papers Illustrative of the Wars of the English in France*, i, 15–16; Waurin, *Chroniques*, v, 67–82.
5. *R.M.S.*, ii, no. 11; *Scotichronicon*, ed. Watt, viii, 242–47; M. Brown, *James I*, 52–73. The king jointly granted Douglas and his wife, Countess Euphemia Graham, the lordship of Bothwell in April 1425 (*R.M.S.*, ii, no. 19).
6. *R.M.S.*, ii, nos 47, 127; M. Brown, *James I*, 76–79, 102, 125.
7. Archives Nationales, J678, nos 21–25; Beaucourt, *Charles VII*, ii, 397; L. A. Barbé, *Margaret of Scotland and the Dauphin Louis* (London, 1917), 14–19; M. Brown, *James I*, 87–88.
8. *Scotichronicon*, ed. Watt, viii, 264–65; M. Brown, *James I*, 125–36, 139–40, 145–46.
9. *Morton Reg.*, ii, 203–205; Fraser, *Douglas*, iii, nos 351, 353, 354, 357; Fraser, *Frasers of Philorth*, ii, nos 18, 19; *H.M.C.*, Drumlanrig, i, no. 110; *R.M.S.*, ii, nos 38, 40, 43, 49.
10. *E.R.*, iv, 2, 54, 193, 216, 244, 270, 296, 301, 365; *Wyntoun*, ed. Laing, iii, 95; *Scotichronicon*, ed. Watt, viii, 60–61; *R.M.S.*, ii, nos 39, 40; Fraser, *Douglas*, i, 443–44. Between 1408 and 1420 Balvenie and Haliburton effectively divided the proceeds of the Linlithgow customs between them. Balvenie's first wife may have been a sister of Murdac, duke of Albany, who called James 'my brother' in the early 1420s (*A.P.S.*, i, 589; W. Fraser (ed.), *History of the Carnegies Earls of Southesk, and of their kindred*, 2 vols (Edinburgh, 1867), ii, 510; W. Fraser (ed.), *The Elphinstone Family Book*, 2 vols (Edinburgh, 1897), ii, 226–28).
11. *R.M.S.*, ii, nos 13, 46, 127.
12. *R.M.S.*, ii, nos 3, 15, 16, 17, 38, 39, 40, 49; Fraser, *Carnegies*, ii, 510; Fraser, *Elphinstone*, ii, 226–28; Fraser, *Scotts of Buccleuch*, ii, no. 25; *Foedera*, x, 487; M. Brown, *James I*, 130–31.
13. *E.R.*, iv, 380, 386, 390, 414; *Scotichronicon*, ed. Watt, viii, 246–47; M. Brown, *James I*, 56, 65; *Laing Charters*, no. 113; *A.P.S.*, ii, 23; *H.M.C.*, Milne Hume, no. 631.
14. *E.R.*, iv, 620; *H.M.C.* Milne Hume, no. 601; *Scotichronicon*, ed. Watt, viii, 262–63; 290–93; M. Brown, *James I*, 150–56, 161.
15. *E.R.*, iv, 277, 300, 395, 554, 555, 607, 609, 610, 658, 659, 660; v, 22; Fraser, *Scotts of Buccleuch*, ii, no. 33; *H.M.C.*, 14th report, appendix 3, no. 10; *R.M.S.*, ii, no. 206. Crichton witnessed a charter of the fourth earl of Douglas in 1423 while Livingston was bailie of Herbertshire in Stirlingshire, held from Douglas by the earls of Orkney. Alexander Livingston was present in Stirling castle when it was held by the Albany Stewarts and when Duke Murdac was tried. No other keeper of the castle is named before Livingston (*R.M.S.*, ii, no. 13; *H.M.C.*, vii, *Scotichronicon*, ed. Watt, viii, 244–45).
16. *Cal. Docs. Scot.*, iv, no. 943. Alexander's son, James, and kinsman, Patrick, witnessed charters alongside Balvenie in 1426 and 1432 while Balvenie witnessed a transaction involving George Crichton in the 1420s and Crichton returned the favour in 1439 and 1440 (*R.M.S.*, ii, nos 246, 370).
17. Fraser, *Douglas*, iii, nos 57, 380; *R.M.S.*, ii, nos 4, 13, 14, 22, 25, 32, 33, 34, 54, 60, 89, 118; *Glas. Reg.*, ii, 616; E. W. M. Balfour-Melville, *James I King of Scots*, 139–40; *C.S.S.R.*, ii, 92, 130–31; *E.R.*, iv, 630; *H.M.C.*, 11th report, 6th appendix, no. 19.
18. Fraser, *Douglas*, iii, nos 68, 392, 393, 396; *H.M.C.*, 11th report, sixth appendix, no. 19.
19. *Cold. Corr.*, nos cix, cx, cxiv; *H.M.C.*, Milne Hume, no. 3; *A.P.S.*, ii, 25; S.R.O. GD 12/20, 22, 24. For a fuller discussion of politics in south-east Scotland between 1424 and 1437, see M. Brown, *James I*, 52–54, 149–56, 160–63.
20. *R.M.S.*, ii, no. 47.
21. *R.M.S.*, ii, nos 86, 87, 183, 184, 185; A. Agnew, *The Hereditary Sheriffs of Galloway*, i, 236–37; *Wigtownshire Charters*, nos 37, 134; S.R.O. GD 72/2.
22. *R.M.S.*, ii, nos 11, 31; *Melrose Liber*, nos 525, 526, 534, 545. Fogo also acted as the king's spokesman in parliament in 1433 and attended the council of Basle on his behalf (*Scotichronicon*, viii, 288–291; J. H. Burns, 'Scottish Churchmen and the Council of Basle', in *Innes Review*, 13 (1963), 1–53, 11–12.
23. *R.M.S.*, ii, nos 58, 59, 308; *E.R.*, iv, 419. The grant of Earl Archibald to the king was referred to in a charter of James II in 1450 relinquishing those rights. In 1434 the king had sheep pastured in the Forest (*E.R.*, iv, 576).

24. *H.M.C.*, vii, 728; *Cold. Corr.*, nos cxii, cxiii; *Scotichronicon*, ed. Watt. viii, 290–91; *Melrose Liber*, ii, no. 534. Up to 1426, William Douglas had been styled earl of Angus and lord of Liddesdale. From 1427 he regularly added the title of lord of Jedworth Forest to these, suggesting a change in circumstances (Fraser, *Douglas*, ii, nos 65, 69, 378, 381; *R.M.S.*, ii, no. 111; *H.M.C.*, Milne Hume, nos 5, 583, 631; *Cold. Corr.*, no. cxii).

25. *R.M.S.*, ii, nos 50, 51, 160, 195; *H.M.C.*, vii, 728; Milne Hume, no. 5; Roxburghe, nos 10, 11; Fraser, *Scotts of Buccleuch*, no. 33; *C.S.S.R.*, iv, no. 337. In 1434 Angus, Cavers and Rutherford all witnessed the grant of the earl of Douglas to the Carthusians, the only time any of them appeared in connection with the Black Douglas during the reign (Fraser, *Douglas*, iii, nos 396–97).

26. *E.R.*, iv, 473, 516, 527, 529, 602; S.R.O. GD 119/164, 157/74; *Cal. Docs. Scot.*, iv, 404. Ramsay sought royal confirmations of lands and offices received from the fourth earl of Douglas in 1426 and 1430. His colleague as customer of the west march was Thomas Kirkpatrick, another tenant and councillor of the fifth earl in Annandale. Andrew Kerr, Walter Scott and Thomas Cranstoun all similarly received charters from the earl and attended his council in the late 1420s and early 1430s (Fraser, *Douglas*, iii, nos 390, 391, 392, 393; Fraser, *Scotts of Buccleuch*, no. 25; *H.M.C.*, 11th report, appendix 6, 212, no. 8).

27. *Cal. Docs. Scot.*, iv, nos 949, 1037–38; *Foedera*, x, 487; C. Macrae, 'The English Council and Scotland in 1430', in *E. H. R*, 54 (1939), 415–26, nos iv, v.

28. Fraser, *Douglas*, iii, no. 63; I. O'Brien, 'The Scottish Parliament in the Fifteenth and Sixteenth Centuries' (unpublished Ph.D thesis, University of Glasgow, 1980), appendix I; M. Brown, *James I*, 130–31, 160–61; *P.P.C.*, iv, 296–97; *Scotichronicon*, ed. Watt, viii, 292–93. Angus was first involved in border negotiations in 1430 and in the truce of 1431 was named as one of the march wardens (*Cal. Docs. Scot.*, iv, no. 1032; *Foedera*, x, 487).

29. F. J. Grant, *The Court of the Lord Lyon*, Scottish Records Society (Edinburgh, 1946), 3; *E.R.*, v, 30; Balfour Melville, *James I*, 230; *Scotichronicon*, ed. Watt, viii, 296–99; M. Brown, *James I*, 162–66.

30. Fraser, *Scotts of Buccleuch*, ii, nos 33, 34; *S.P.*, vii, 368; *E.R.*, v, 30. James, Nicholas and George Rutherford all witnessed acts of the fourth and fifth earls of Douglas, George receiving the lands of Chatto and others from the earls (Fraser, *Douglas*, iii, nos 393, 396; *H.M.C.*, Milne Hume, no. 1; *S.P.*, vii, 369). Management of the war over the winter was in the hands of the king's 'lieutenant of the marches', Adam Hepburn of Hailes (*C.S.S.R.*, iv, no. 343).

31. M. Brown, *James I*, 172–201; M. Brown, '"That Old Serpent and Ancient of Evil Days": Walter Earl of Atholl and the Death of James I', in *S.H.R.*, 71 (1992), 23–45; *A.P.S.*, ii, 31; R. A. Hay (ed.), *Genealogie of the Sainteclaires of Roslyn* (Edinburgh, 1835), 90–91; M. Connolly (ed.), 'The Dethe', 64; *Scotichronicon*, ed. Watt, viii, 248–49. The appointment of Douglas as lieutenant took place in 1437 and almost certainly occurred at a meeting of the estates. Rather than the queen's triumphant coronation parliament, which was probably sparsely attended, the meeting of estates at Stirling in May is more likely. I am grateful to Roland Tanner of the University of Glasgow for information about the dating of this council (*E.R.*, v, 12; Fraser, *Melvilles*, iii, no. 31).

32. *A.P.S.*, ii, 31–32, 53–54; Fraser, *Douglas*, iii, nos 404, 406; *R.M.S.*, ii, no. 201. A partial entry in the contemporary Auchinleck Chronicle may well refer to arrangements between Douglas and the queen allowing the latter custody of her son but assigning the government of the realm elsewhere. Douglas certainly took a fee as lieutenant ('Auchinleck Chron.', in C. McGladdery, *James II*, 171; *E.R.*, v, 12, 73, 138).

33. *Scotichronicon*, ed. Watt, viii, 318–19; M. Brown, *James I*, 158–60, 200; Hay, *Sainteclaires*, 90–91; *E.R.*, v, 61; *S.P.*, iii, 172; Robert Lindsay of Pitscottie, *The Historie and Cronicles of Scotland*, Scottish Text Society (Edinburgh, 1899), 24.

34. *E.R.*, v, 84, 86, 87, 88, 89, 166; M. Brown, 'Regional Lordship in North-East Scotland: The Badenoch Stewarts II', 44–46. Alexander of the Isles first appeared as justiciar in

late 1438 and used the title until the end of the following year, suggesting his commission came from Douglas (*Acts of the Lords of the Isles*, ed. J and R. W. Munro, Scottish History Society (Edinburgh, 1986), nos 26–30.

35. Fraser, *Douglas*, iii, nos 75, 301, 302. The new earl of Angus appears to have retained the support of his father's allies in the middle march, Nicholas Rutherford and Douglas of Cavers. James's sheriff of Dumbarton, Colquhoun of Luss, and steward of March, Adam Hepburn of Hailes, seem to have retained their positions (Fraser, *Douglas*, iii, no. 76; *E.R.*, v, 75; *H.M.C.*, Milne Hume, no. 601; 'Auchinleck Chron.', in C. McGladdery, *James II*, 160).

36. Fraser, *Douglas*, iii, nos 301, 401, 403; Hay, *Sainteclaires*, 67–68; S.R.O., AD1/48; *Laing Charters*, no. 117. Egidia's husband had had to deal with local opposition in Nithsdale from men connected to the Black Douglases in 1419. Egidia was the earl's cousin and the mother-in-law of his sister, Elizabeth, and his uncle, Balvenie (N.L.S., Adv. Mss. 22.2.4 fo. 34).

37. M. Brown, *James I*, 156–60, 199–200; C. McGladdery, *James II*, 14–29; 'Auchinleck Chron.', in *ibid*, 160; *A.B. Ill.*, iv, 452; S.R.O., GD 124/1/138; S. Boardman, 'Politics and the Feud in Late Medieval Scotland' (unpublished Ph.D thesis, St. Andrews University, 1989), 159–71.

38. *E.R.*, v, 61–64; Fraser, *Douglas*, i, 419–20; iii, nos 301, 403, 406; 'Auchinleck Chron.', in C. McGladdery, *James II*, 171. The title of Avondale indicates the importance of James's southern lands and, in particular, Strathavon in Lanarkshire (though Abercorn also lay near the Avon in West Lothian). The rewards given to the lord of Gordon by Douglas came via Crichton's influence, while Avondale was one of the lieutenant's advisers during his talks with the lord of the Isles on Bute and may have been concerned to secure control of his own northern estates. Avondale also sat among the auditors of the exchequers (*E.R.*, v, 84, 87; *Copiale*, 152).

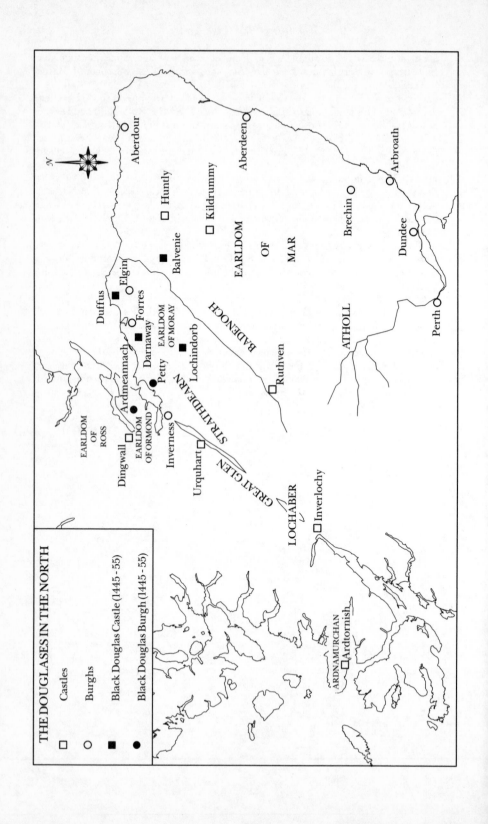

THE DOUGLASES IN THE NORTH

Castles ☐

Burghs ○

Black Douglas Castle (1445 - 55) ■

Black Douglas Burgh (1445 - 55) ●

Aberdour ○

Huntly ☐

Kildrummy ☐

Aberdeen ○

Arbroath ○

Brechin ○

Dundee ○

Balvenie ■

EARLDOM OF MAR

Elgin ○

Duffus ■

Forres ○

Ardmeannach

Darnaway ■

EARLDOM OF MORAY

Petty ■

Lochindorb ■

BADENOCH

Ruthven ☐

ATHOLL

Perth ○

STRATHDEARN

Inverness ○

EARLDOM OF ROSS

Dingwall ☐

EARLDOM OF ORMOND ●

Urquhart ☐

GREAT GLEN

LOCHABER

Inverlochy ☐

ARDNAMURCHAN

Ardtornish ☐

N

CHAPTER TWELVE

A Pride of Douglases

BLACK DINNER

The death of Archibald, fifth earl of Douglas and lieutenant general of Scotland, turned the plague summer of 1439 into a period of intense political competition. For the first time since 1388 this competition would take place without an adult earl of Douglas. The lieutenant's sons, William and David, were in their teens, too young to step into their father's shoes as either lieutenant or active lord of his following in the kingdom. Their only adult kinsman was their great-uncle, James the Gross, earl of Avondale. James desperately needed the power of this Douglas following behind him. Throughout his long career James the Gross had traded on his connections with the Black Douglas affinity to his own advantage, though not always to that of his senior kinsmen. The lieutenancy of Earl Archibald and its alliance of royal and Douglas resources had brought James new status, offices and power, but with the death of Douglas this power seemed precarious. If James the Gross, who was nearing seventy in 1439, was to pass on this power to his pride of sons, he needed to maintain influence with the Douglas earl and his following. This alliance was to be maintained at any cost, even the precipitation of renewed bloodshed over the lands and titles of the family.[1]

James earl of Avondale was not alone in his hopes and fears after the death of the lieutenant. The strength of Earl Archibald's regime had rested on a group of men who combined ties of service and kinship to the earl with successful careers as agents of King James I. Among these were Avondale's brothers-in-law, William Sinclair earl of Orkney and Walter Haliburton of Dirleton, but the other pillar of the lieutenant's council was William Crichton, the newly appointed chancellor. Like Avondale, Crichton's career was built on service to great men. Though he possessed lands and allies, many of these came from his influence as the councillor of, first, King James and, then, Earl Archibald. His direction of royal resources, the lands and revenues accumulated by the old king, flowed from this influence and was the key to his power. This power was, however, dependent on his continued place in the minority regime and, like Avondale, this was made suddenly uncertain by the death of the lieutenant.[2]

TABLE 7. THE DOUGLASES OF DALKEITH

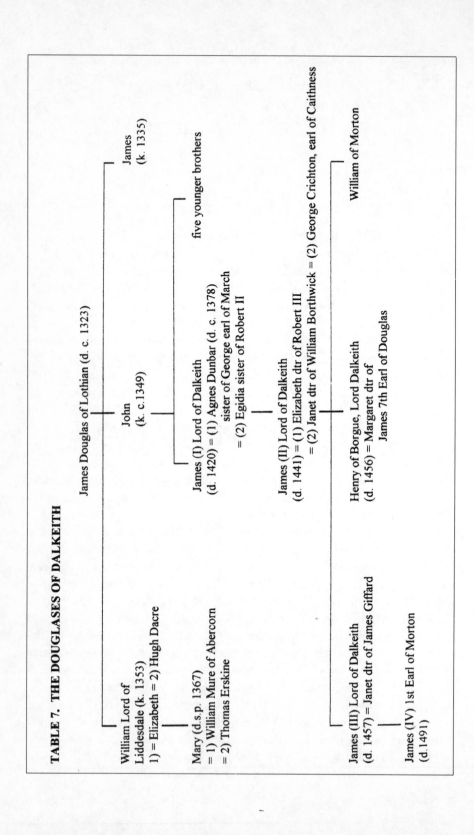

James Douglas of Lothian (d. c. 1323)

James (k. 1335)

John (k. c.1349)

William Lord of Liddesdale (k. 1353)
1) = Elizabeth = 2) Hugh Dacre

Mary (d.s.p. 1367)
= 1) William Mure of Abercorn
= 2) Thomas Erskine

five younger brothers

James (I) Lord of Dalkeith
(d. 1420) = (1) Agnes Dunbar (d. c. 1378)
sister of George earl of March
= (2) Egidia sister of Robert II

James (II) Lord of Dalkeith
(d. 1441) = (1) Elizabeth dtr of Robert III
= (2) Janet dtr of William Borthwick = (2) George Crichton, earl of Caithness

William of Morton

Henry of Borgue, Lord Dalkeith
(d. 1456) = Margaret dtr of
James 7th Earl of Douglas

James (III) Lord of Dalkeith
(d. 1457) = Janet dtr of James Giffard

James (IV) 1st Earl of Morton
(d.1491)

The threat to James the Gross and Crichton came principally from the queen mother, Joan Beaufort. Despite retaining custody of the young James II, Joan had been sidelined by the lieutenant and excluded from her son's council. However, she still held lands granted to her by the king in the mid-1430s, a jointure which included lordships in Perthshire and perhaps also the earldom of Fife. Douglas's death gave the queen the chance to exploit this territorial base to secure control of the minority government. Her first step in this was to extract her son from Edinburgh, and the effective control of Crichton, during July. She took up residence in Stirling, from where she could make contact with potential allies in Perthshire and Fife. The queen's search for support extended to re-marriage. Her new husband was James Stewart of Lorn, a junior member of a baronial family, but despite Stewart's own limited status, the match did have political value. Joan's new brothers-in-law included Robert Stewart, lord of Lorn and of estates in Perthshire, an experienced figure in Scottish politics, and David Murray of Tullibardine, who was already seeking, by force and persuasion, to win the stewardship of Strathearn. Strathearn was the key to lowland Perthshire. The earldom had been left vacant by the execution of Walter earl of Atholl and Strathearn in 1437, and the queen was clearly seeking influence in this power vacuum. Another brother of her husband, Archibald, was made sheriff of Perth, and with her re-marriage backed at the papal *curia* by James Kennedy, bishop of Dunkeld, a protégé of the old king, the queen had drawn together support from across central Scotland in the few short weeks since Douglas's death.[3]

The threat of the queen establishing an alternative council in Stirling would have been unwelcome to Avondale, Crichton and Orkney, who had helped Douglas supplant her in 1437. These men needed to prevent the completion of the queen's plans. On 3 August 1439 Joan and her husband were arrested and imprisoned in Stirling Castle by Sir Alexander Livingston of Callendar. Like the killers of James I, Livingston exploited his custody of the royal residence to strike secretly against the queen and gain control of her son. After years as a figure of local significance, Livingston had seized his chance for greater power, but the arrest of the queen was not a Livingston coup designed to give a minor knight control of central government. Despite control of Stirling, Alexander lacked the means to effect such a revolution. Instead he acted with wider support. The seizure of the queen was carried out, not just by Livingston and his kin, but by Sir William Cranstoun, a royal servant whose father, Thomas, remained a close councillor of the Douglas earls. William's role in the queen's arrest points to the backing of others who shared his links to James I and the Douglases. Within a fortnight of Joan's arrest, these backers were themselves in Stirling. From mid-August, Crichton and Avondale were with Livingston.[4] The links between Sir Alexander and James the Gross which stretched back to 1424 had become an alliance of significance in Scottish politics. While Livingston took the risk of arresting the queen, it was Avondale and Crichton, most under threat from Joan's ambitions, who forced her to accept defeat. Settlement was reached in a general council in Stirling in early September. Though the queen's allies, Stewart of Lorn, Murray and Hugh Kennedy, the uncle of the bishop of Dunkeld, were present, she was kept in custody until she agreed to put her seal to an 'appoyntement' with her captors. To gain her liberty the queen was forced to acknowledge the 'good zeale and motife' of Livingston and Cranstoun, to promise that neither she nor her retinue would seek revenge for her treatment, and to relinquish her son and her pension to Alexander Livingston. Though the bishops who witnessed the 'appoyntement' may have sympathised with the queen, its lay sponsors were not her friends. Crichton, his ally Seton of Gordon, and Avondale's brother-in-law, Walter Haliburton, were the lords who sealed the document. With Crichton's deputy in Edinburgh, William Cranstoun, a kinsman and namesake of the queen's captor, and James Parkle, George Crichton's deputy as sheriff of Linlithgow, representing the burghs, the witnesses to the 'appoyntement' were largely enemies of the queen. The group was completed by William, the new earl of Douglas, who put his signet to the document. William's presence, his first public act as earl, was linked to the ambitions of his great-uncle, James the Gross, earl of Avondale. In the political crisis, Avondale deployed the name and power of the Douglas earl to guarantee his own place in the minority regime.[5]

The support of the young Douglas earl was a valuable weapon, but Avondale's dependence on that support made for insecurity as Earl William approached majority, an insecurity increased by the limited victory won by Avondale and his allies over the queen in 1439. During the autumn the king was returned to Edinburgh, where his council was dominated by Crichton and associates like Orkney, Haliburton and Adam Hepburn, along with Alexander Livingston, in place, though hardly dominant, on the council. The chancellor's influence was reflected by a meeting of the council in his own castle at Crichton in January 1440, but as the new year progressed, his control beyond the Forth seemed in doubt.[6] The council was not the only source claiming to exercise the rights of royal lordship. Despite the 'appoyntement', the queen had not abandoned her ambitions nor had she been deprived of her lands and influence. In late March she was in Falkland with her husband, his brothers and David Murray, and when Bishop Wardlaw of St. Andrews died the following month, her proximity had an impact on the chapter's choice of a successor. The election of James Kennedy of Dunkeld to the see was, despite his absence at the *curia*, a blow to the men round the young king and a mark of the queen's strength. It was not the only sign of difficulty for Crichton and Avondale. The re-emergence of Bishop Cameron as a member of the council cannot have been comfortable for his successor as chancellor and may indicate the end of Crichton's dominance. This became clear when the estates met in early August. There was obvious dissent about the handling of the increasingly violent dispute over the earldom of Mar in the north. The king's council, swollen to include many beyond the chancellor's allies, withdrew support from Crichton's son-in-law, the lord of Gordon, and sought agreement with his rival, Robert Erskine, claimant to Mar. Crichton refused to participate in this agreement, and in late August Avondale and George Crichton met Gordon at Elgin probably to confirm their support, but the events of the council revealed the limits to their control of royal power.[7]

In these circumstances, Crichton and Avondale needed to be sure of their influence with the young earl of Douglas. Avondale's rare appearances at court suggest that he was actively managing the earl's lands and council. Certainly, when Douglas came to Edinburgh in February 1440, he was attended by both core supporters of his family, like Andrew Kerr, the Murrays of Cranstoun and James Rutherford, and men closely associated with William's father only during his lieutenancy. The presence, in particular, of Thomas Cranstoun, the father of the queen's captor, and James Auchinleck, a close adherent of Avondale, points to links between the earl and his great-uncle's allies. However, at Edinburgh Douglas may also have come into contact with a less welcome servant of his father, John Cameron, who was present on the royal council. Any link with Cameron would have been an unwelcome sign of William's growing independence, and during 1440 the earl turned for advice to another man who was no friend of his great-uncle. Malcolm Fleming of Biggar was the son of the man killed by James the Gross at Long Hermiston and a possible rival in Lanarkshire. His association with Earl William would have caused alarm for those dependent on the latter's quiescence. Instead, Douglas seems to have been beginning to exert his own authority. In August 1440 Herbert Maxwell of Caerlaverock appeared at court seeking confirmation of his hereditary rights as the earl's steward of Annandale. Maxwell's action suggests anxiety about the intentions of his Douglas lord and his search for the chancellor's support for his local position.[8]

Any doubts about the ambitions of the young earl of Douglas represented a crisis for Avondale and Crichton. Accounts from the following century speak of Earl William's youthful pride, swelled by the flatterers in his household, and though this smacks of moralising, it would not be surprising if Douglas, now seventeen or eighteen, sought his own council, lands and influence in the kingdom. Nor would it be surprising if Avondale's enemies sought to exploit this to their advantage. In early 1441 Bishop Cameron was accused 'with traitourous magnates' and 'several others of the king's council' of plotting 'to the death' in 'a most treasonable conspiracy against his majesty'. Cameron's crime was probably to move against the chancellor and his allies. The 'conspiracy' may well have been built around Earl William. His assertion of his rights to both the private power and lieutenancy held by his father would, at best, reduce Crichton and Avondale to their place as councillors. The earl's links with Cameron and Fleming coincided with the betrothal of James earl of Angus to Joanna, the king's sister, which suggested that the queen was also seeking allies. Such actions threatened his father's advisers with worse than merely loss of influence. With their control of the royal council slipping, Avondale and Crichton contemplated a grim future unless they took drastic action.[9]

For men whose experience included the fall of the Albany Stewarts, the death of James I, the queen's bloody revenge and Joan's own arrest, drastic action could include the violent removal of the young earl as the best means of securing their hold on both royal and Douglas resources. There had been no open breach between Earl William and his father's chief councillors, and many friends of Crichton and Avondale must have remained in Douglas's entourage. A meeting between the young earl and Crichton in November 1440 was probably not hard to arrange and, as in February, Douglas came to Edinburgh. The earl's great-uncle, Avondale, his uncle, Orkney, Sir William Cranstoun and Sir Alexander Livingston had been with Crichton during early November, while the earl was accompanied by his brother, David, and Malcolm Fleming. It was a gathering of men with strong Douglas connections. It was also a gathering whose purpose was to kill the two Douglas heirs. The contemporary Auchinleck chronicle stated baldly that 'Erll Willam of Douglas, Archebaldis son ... and his brother David Douglas was put to deid at Edinburgh and Malcolm Fleming was put to deid in that samyn place within thre days'. A popular rhyme went further:

> Edinburgh castle, toun and tour
> God grant you sink for sin
> And that even for the Black Dinner
> Earl Douglas got therein.

In the histories from the next century onwards the Black Dinner was embellished with further details. The earl and his brother, dining with the young king, were served with the head of a black bull, a sign that they were condemned to death. Seized and bound, they were dragged from the hall to the castle hill, and without trial or appeal, their heads were hacked off.[10]

Earl William's death was the work of men who had learned their methods from James I, their old master. Douglas's sudden fall, though only the latest in a series of violent coups, was, like the king's murder, capable of inspiring horrified attention in those, in Scotland and beyond, who knew the Douglas name. The Burgundian chronicler, Jehan de Waurin, was one of these contemporaries. He recorded the deaths of the earl, his brother and Fleming as the work of the great men of the king's council who claimed Douglas was conspiring to deprive the king of his rights. The charge was the same as that levelled against Cameron, suggesting it was the justification for the deed circulated by its perpetrators. However, though fear of the earl's intentions may have motivated the king's councillors, the Black Dinner was carried out to secure and extend the power of Avondale and Crichton through the death of Earl William. The executions of both William and David and the failure of the council to forfeit the brothers, as they forfeited Fleming, were designed to smooth Avondale's path to the earldom of Douglas. The legal basis of his inheritance was the family entail of 1342 which stipulated that the principal Douglas lordships should pass to the nearest male heir. In 1440 this was James the Gross of Avondale, great-uncle of the dead earl.[11]

James the Gross followed in his father's footsteps. Just as Archibald the Grim had used the entail to justify the forceful takeover of the earldom from the sister of the second earl in 1388, so James secured the entailed lands over Margaret, the younger sister of Earl William in 1440. As seventh earl of Douglas, James must have hoped that, like his father, the means by which he acquired his new rank would not prevent the acceptance of a new line of earls. However, the events of 1440 were very different from those of 1388. Unlike his father, James the Gross was involved in precipitating the succession crisis. He was at least guilty by association in the violent deaths of his great nephews, whose interests he was bound to protect as close kin. Even within a family not known for its freedom from internal conflict, James's sudden shattering of these bonds of kinship must have left its mark on the loyalties which held the Douglas affinity together.[12] The Black Dinner itself showed that these loyalties were under strain. While Archibald the Grim had rested his claim to the earldom of Douglas and its border lordships on his ability to lead and protect the men of those lordships, Earl James's success was decided by the needs and practice of politics at the king's court. The Black Dinner was the consequence of the efforts of Douglas adherents to control the king and his council and the readiness of these men to kill the Douglas earl to maintain this predomin-ance. For James the Gross, for Crichton, Orkney and lesser men like the Cranstouns, the consequence of James I's reign and Earl Archibald's lieutenancy was to forge an inextricable link between royal and Douglas politics. The careers of these men had been built in service to both kings and earls, and the possibility of conflict between the two was a potential disaster which justified extreme measures. The Black Dinner perpetuated this connection, and for the next fifteen years the fortunes of the crown and the Douglas earldom would remain bound together. But the deaths of Earl William and his brother formed a precedent for both the young king and his ambitious councillors for intervention in the Douglas succession which held clear dangers for James the Gross and his sons.

EARL JAMES AND THE CHANCELLOR

Blows from a headsman's axe had won the earldom of Douglas for James the Gross at a price. To fulfill his immediate ambitions he had to accept only a portion of his great-nephew's estates. Claims based on the 98-year-old family entail entitled him to Douglasdale, the Forest, Lauderdale, and Eskdale. The royal councillors who had connived in the Black Dinner made no objection to James's acquisition of these lands and allowed his claim to the Murray lands of his mother, many of which had already been granted to him. However, James did not receive the whole Douglas inheritance. Galloway and Wigtown were still held as a single lordship by Duchess Margaret Stewart, and with the death of her grandsons, the province was to pass to her granddaughter, Margaret, known to later historians as 'the maid of Galloway'. Similarly Annandale, granted to the fourth earl and his male offspring, now lapsed to the crown. The Douglas lands were threatened with further partition by the existence of two more dowagers after Duchess Margaret. The widow of the fifth earl, Euphemia Graham, had been given a terce which included Bothwell, while, after the sixth earl's execution without forfeiture, Janet Lindsay, his young widow, claimed a share of the remaining lands, presumably backed by her father, the earl of Crawford. This erosion of the power of the Douglas earl must have been anticipated by those who planned the Black Dinner and may have been the price of their support. For his part, Earl James could console himself with the thought that, once again, his father's succession in 1388 provided a precedent for the reconstruction of Douglas power after a divided inheritance.[13]

However, in 1388 the loss of Liddesdale and Jedworth Forest was bal-anced by the union of the Forest and Eskdale with Archibald the Grim's lordship of Galloway and leadership in the west march. Though its centre shifted to the west, the power of the Douglases in the marches increased. By contrast in 1440 James brought no border lordships to compensate for the loss of Galloway and Annandale, and with them the apparent end of Douglas domination in the west march. Instead the new earl's career was built on lands and connections further north and James continued to draw on established allies from Ayrshire and Lanarkshire where he had been sheriff since the early 1430s. While some of these, like the Symingtons, Somervilles and Maxwells of Calderwood, were long-standing Douglas tenants, their prominence in the earl's service and the presence of men like James Dunbar of Cumnock, James Lindsay of Covington, John Wallace of Craigie and James Auchinleck reflected established friendships with James the Gross. His rise had made his power-base in Clydesdale and West Lothian central to the interests of the new line of Douglas earls.[14]

The new earl of Douglas could not accept a major reduction in the power held by his father and brother in the marches and in the kingdom. Earl James spent the rest of his life seeking to recover lands denied to him after the Black Dinner and in establishing his own dynasty, his brood of teenage sons. His first targets were the dowagers and heiresses to whom a significant proportion of the Black Douglas estates had been assigned. Euphemia Graham lady of Bothwell exchanged her rights to a full terce in exchange for the baronies of Carmunnock and Drumsegard in Clydesdale. Earl James's loss of these lands was compensated for by the marriage of Euphemia to the earl's neighbour, James Hamilton of Cadzow, who proved to be the principal ally of the Douglas earls in the coming years. By contrast, Earl James may have denied Janet Lindsay any equivalent share of his lands, risking her father's hostility. The principal aim of the new earl was, however, the rights of the young heiress, Margaret of Galloway. Though outside the earls' direct lordship since 1424, there was now the danger of Galloway being fully separated from the Douglas inheritance. Margaret's marriage beyond the Black Douglas line would deal a fatal blow to the family's power in the south-west, and James the Gross almost certainly exerted pressure during 1441 and 1442 to secure Margaret for his eldest son, William. The marriage, which only occurred after James's death, secured his son's right to Galloway, Archibald the Grim's first marcher lordship and the base of his power.[15]

This recovery of lands and prospects was a source of anxiety for the men associated with James in the Black Dinner, who expected to share in its benefits. For James's brother-in-law, William earl of Orkney, the weakening of Black Douglas power in the south-west would mean the end of the family's interference in Nithsdale, held by Orkney's mother, and plans to renew the lordship of the Douglas earls in Galloway threatened Nithsdale

once again. The recovery of Galloway for Earl James's son was also
unwelcome to William Crichton. Though not from a great family like James
the Gross, Crichton's ambitions stretched far beyond the royal council.
Instead, like his Lothian neighbour, Alexander Seton, who laid the foundations of the great magnate house of Gordon, Crichton sought to build
a position of regional power. As with James Douglas, the Black Dinner
was a stepping-stone to this end which, in the south, focused on the dales
of the Annan and Nith. The interests of Crichton's family in the region
were as long-standing tenants of the lords of Annandale, and the principal
Crichton cadet line were the lords of Sanquhar in upper Nithsdale. Until
the rise of William Crichton, such contacts with the south-west held little
significance beyond the Crichton family, but, with the death of the fifth
earl in 1439, the chancellor used his position in royal and Douglas councils
to build his own influence in Dumfriesshire. In February 1440 William
received lands in the barony of Kirkmichael from the earl of Crawford
and, at the end of April, had an arrangement confirmed whereby he became
heir to his cousin, Robert Crichton of Sanquhar, should the latter's line
fail, in return making Robert his own heir, an act suggesting an attempt
to form a surname-based kindred with significance in the west march. By
the time Herbert Maxwell met with the chancellor in August 1440, Crichton
may already have been considering the potential for influence in Annandale
should Black Douglas lordship be removed, and Maxwell again came to
Edinburgh to meet both William and George Crichton in the aftermath
of the Black Dinner.[16]

The attempt by the Crichtons to widen their influence in the south-west
centred on George Crichton and would lead to major friction with the
Black Douglases. At about the time of the Black Dinner, James, the elderly
lord of Dalkeith, died. The great-nephew of the knight of Liddesdale and
son of the Douglas of Dalkeith who had been Archibald the Grim's close
ally, James had continued his father's friendship with the Black Douglases.
His choice of Janet Borthwick, daughter of the fourth earl's close councillor,
as his second wife tightened these connections, but after 1424 these in-laws,
like many of Dalkeith's Lothian neighbours, turned from the Douglas to
royal service. By 1440 the links between the Douglases of Dalkeith and
their cousins had also loosened. Instead the old king's chief agents in
Lothian, the Crichtons, held sway over the headless house of Dalkeith.
William Crichton received lands resigned by Douglas of Dalkeith in
Dumfriesshire on 26 February 1440, and two days later the elderly lord
gave his wife joint rights in the important lordship of Morton in Nithsdale.
Subsequent events suggest the chancellor's involvement in this grant. By
spring 1441 the old lord was dead and his widow, Janet Borthwick, had

married George Crichton. The new lord of Dalkeith, another James, was declared unfit to run his inheritance and Dalkeith itself was put in the custody of James Gifford, like the Crichtons a servant of the old king.[17] For the chancellor and his kin it was not Dalkeith but the south-western lands of the family which interested them. The scattered estates of the lords of Dalkeith in Galloway, Nithsdale and Annandale represented a potential base for local influence when allied to their influence on the council. For the new earl of Douglas, however, this control over the house of Dalkieth was a clear threat. The lords of Dalkeith were not just another baronial family. They had been key allies in the rise of the Black Douglases whose support had been rewarded with lands which made them the earls' chief vassals in Galloway and Eskdale and neighbours in Annandale and Niths-dale. The danger of these lands falling to the Crichtons was considerable. Allied to the Crichton family's continued links with the Kerrs, Scotts and Cranstouns, their accumulation of a share of the Dalkeith estates suggests an attempt to rival James the Gross in the marches. This attempt was, in turn, threatened by Earl James's plans to recover Galloway.[18]

The result was tension between the architects of the Black Dinner which broke into open conflict during the summer of 1442. The Dalkeith inherit-ance was central to this conflict. James the Gross had found an ally in the younger brother of the mad lord of Dalkeith, Henry Douglas, lord of Borgue in Galloway. Henry had his own ambitions for his brother's lands, ambitions which were fuelled by his marriage to Margaret, Earl James's daughter. With this support, Henry led an attack on his brother's keepers in which Dalkeith suffered 'spuilzie' and 'reif' and which caused 'greit and perowlus strywys' in Lothian. The clash violently split the faction which had engin-eered the coups of 1439 and 1440 and, as we shall see, ended the alliance of Douglas earl and chancellor in the north. Rivalry for influence in Lothian, the marches and Moray was, once more, a struggle for influence on the king's council. In September 1442 it was this council which sought to take direct control of Dalkeith from its feuding claimants. The act aimed, not at pacification, but at partisan intervention by a body now dominated by men opposed to Earl James of Douglas.[19] The chancellor and his close allies were joined by Bishop Kennedy, an ally of the queen who, from 1441, was at odds with Douglas on the issue of the dispute for leadership of Christendom between Pope Eugenius IV and the council of the church assembled at Basle. While Kennedy was active on behalf of the pope, Douglas sponsored Scottish support for the council. His motives were probably opportunistic. Earl James may have been reacting to papal reluct-ance to give dispensation for the marriage of his son to Margaret of Galloway, but he also recalled the enormous influence gained by his brother from his role in the Great Schism. His principal ecclesiastical agent was William Croyser, who had served the fourth earl in these events. Croyser, an avowed enemy of the old king, would also have fuelled Earl James's hostility to Bishops Kennedy and Cameron, James I's clerical adherents. In May 1441 the earl's influence at Basle secured his son, James's, provision to the bishopric of Aberdeen, but in the face of growing opposition to Basle from church and laiety, Douglas turned to violence. In the summer of 1442 he broke up the provincial council of the Scottish church, at which Kennedy and his allies were excommunicating and depriving their enemies. The earl followed this by sending Croyser to Basle to seek his son's translation to Kennedy's own see of St. Andrews.[20]

Douglas's acts were bravado. Unlike his brother he backed the wrong horse in ecclesiastical politics and damaged his family's links with a number of rising churchmen. He also drove Kennedy and Crichton into an alliance with James earl of Angus, a natural rival of the Black Douglases. By the summer of 1442 these men were already co-operating in the east march. The bailiary of Coldingham, lost by the Black Douglases with the deaths of the fourth earl and his local deputy, Alexander Hume, at Verneuil, was a matter of dispute between Alexander's son and namesake and David Hume of Wedderburn, the elder Alexander's brother. Despite their links with the Humes, the Black Douglases were excluded from this rivalry. Instead the steward of March, Adam Hepburn, turned to more recent allies, William Crichton and James earl of Angus, to settle the feud. The support of these men for Alexander Hume secured him the bailiary and Angus obtained the promise that the new bailie would support his claim to renew his family's place as protector of Coldingham. It was only in late 1442 that David Hume, the defeated candidate, turned to the Black Douglas earl. Earl James the Gross travelled to Coldingham as justiciar and, not surprisingly, judged David Hume to be bailie. He condemned the 'partial consale' of the king which had confirmed Alexander's rights, but those present had included Kennedy, Crichton and the earls of Angus, Crawford and Mar, a demonstration of Douglas's exclusion. Soon after Douglas's visit to the priory Alexander Hume gained possession, turning Coldingham into a fortress, and when, in early 1444, he came to terms with his uncle, it was Angus who negotiated the settlement.[21]

By then James the Gross was dead. His end came, fittingly, at Abercorn, his first stronghold, in March 1443. Though he died in the midst of his West Lothian power-base, James's last year had seen a decline in his influence. The general council which met the previous month was dominated by his enemies. However, James had secured enough of the old Douglas inheritance to claim the status of his father, brother and nephew, and the marriage of his daughter, Janet, to Robert, son of the executed Malcolm Fleming, was designed to end the repercussions of the Black Dinner.[22] The power of the family in the marches had proved sufficient for James's predecessors to sit out the hostility of royal governments before 1424. But in the 1440s the new earl was neither willing nor able to accept the domination of the council by Crichton and his allies. William, the eldest son of James the Gross, a youth of eighteen, could not sit back while the chancellor used royal resources to enlarge and entrench his influence. Instead, over the next three years the new earl of Douglas launched a military and political campaign against his enemies which showed both the continued power of the family and a new desire to gain control of the king's council as the means of establishing his own predominance in the south and supporting the ambitions of his brothers in the north.

'IN MYDDIS OF MURRAYE': THE DOUGLASES IN THE NORTH

Earl James the Gross saw Scotland north of the Mounth as a land for younger sons. He himself had been given his family's estates in the region and invited to seek his fortune there by his elder brother. Though he had made only fleeting visits to the north, he saw these lands as a means to provide for his own pride of sons. This attitude was not very different from that of his ancestors who followed their Murray cousins to the north in the 1200s. It reflected the secondary value which the Douglas family placed on their interests in the region compared to those in the marches. Even when William, first earl of Douglas, inherited the earldom of Mar in right of his wife and took care to secure his claims, his priorities still focused on the south. On the death of his son, the second earl of Douglas and Mar, at Otterburn, the two earldoms were separated. The dead earl's sister, Isabella, retained Mar, and her husbands, Malcolm Drummond and Alexander Stewart, exercised control of the earldom.[23]

While Archibald the Grim was denied possession of Mar, he had his own claims beyond the Mounth from his marriage to Joanna Murray. As heiress to both her husband, Maurice Murray, and her cousin and first husband, Thomas Murray, Joanna held the estates won by her ancestors in the twelfth and thirteenth centuries from which they took their name. These included the baronies of Duffus and Petty in Moray, lands on the Black Isle and at Aberdour in Buchan, estates which had provided Andrew Murray with a base from which to act as guardian of the kingdom in the 1330s, but which from the 1360s to 1440s had little effect on the ambition and activities of the Black Douglas lords. This relative neglect was not simply a product of the effort expended by Archibald the Grim and his family in southern war and politics. In the later fourteenth century Moray was another war zone. Effective lordship was not exercised by absentees like the Douglases but, as in the marches, by lords with the influence to extract support from local kindreds. By the early fifteenth century the greatest of these northern warlords were the Clan Donald lords of the Isles, and Alexander Stewart, earl of Mar, the husband of Isabella Douglas, who dominated the region in violent competition. In these circumstances, the northern apanage granted to James the Gross in 1408 offered him limited prospects and wealth.[24] The new lord of Balvenie chose, instead, to base his career in Lothian, but he was not entirely without northern contacts. Chief among these was Alexander Seton, like James a younger son from a Lothian family. Seton inherited the lordship of Gordon and the northern lordship of Strathbogie by marriage and, unlike Douglas, built a career beyond the Mounth as an ally of Mar and, by the 1430s, as the chief agent of the king, a position secured by the marriage of his son, Alexander, to the daughter of William Crichton.[25]

The ambition to build a northern power-base for his sons can only have occurred to James the Gross with the death of Alexander earl of Mar in 1435. The removal of the most powerful lord between the Great Glen and the Mounth had an impact across the north. It sparked local conflict in Aberdeenshire, in which the lord of Gordon opposed the claim of Robert Erskine to the earldom of Mar. It allowed Alexander lord of the Isles to extend his influence into Lochaber, Ross and Inverness, punishing those who had deserted him a decade before. It also aroused the ambitions of James Douglas, lord of scattered lands across Moray. Although, in the lord of the Isles, the earl of Crawford, and the lord of Gordon, there were rivals for Mar's lost influence, from 1437 James the Gross showed a growing interest in the north. Initially this interest was exercised indirectly. As a councillor of the lieutenant, and from 1439 the king, James moulded the council's policy in the north. In 1438, with the northern lord, Walter Ogilvy, he negotiated with Alexander lord of the Isles and earl of Ross on the lieutenant's behalf, securing Ross's support for the council's main northern agent, the lord of Gordon. Gordon's links with Crichton secured him financial and political backing from the council in the struggle over Mar, and James the Gross went north to back Gordon's son and successor directly when the latter's position came under threat in 1440. In building up these contacts beyond the Mounth, both James, now earl of Douglas, and Chancellor Crichton were seeking to establish their families in the region.[26]

These ambitions centred on the earldom of Moray. Since the extinction of the Randolphs the earldom had had an unhappy history, deprived of its highland territories and in the hands of southerners, a junior branch of the Dunbar family. Though pressurised by more powerful rivals, the earls of Moray were still leading lords in the coastal lowlands, and it was this influence which attracted Crichton and Douglas. Since 1430 and the death of James Dunbar leaving two young daughters, there had been no earl of Moray. The dowager countess was the sister of the new lord of Gordon, and the latter may have used his influence with her to the advantage of his allies. It was probably in early 1442 that the marriages of the two Moray heiresses took place. The elder, Janet, married James Crichton, the chancellor's son, while Elizabeth, the younger, became the wife of Archibald, second son of James the Gross. The Aberdeenshire estates of these heiresses were also divided, with James Crichton receiving the barony of Frendraught and Archibald Douglas obtaining the lands of Kintore. However, the fate of the title and lands of the earldom itself was less clear, and may have been one of the issues between the chancellor and the Douglas earl which broke out in open conflict during the latter part of the year.[27] Earl James's northern ambitions were not limited to the earldom of Moray. As well as securing a title for Archibald, he hoped to provide fortunes for his other younger sons in the region. The provision of his son, Archibald's twin brother, James, to the see of Aberdeen and the precentorship of Moray by the council of Basle in May 1441 created the prospect of a Douglas prelate at the head of the northern church. Earl James also laid plans for the division of his own lands in the north between two more sons, Hugh and John. Hugh was to receive his father's lands on the Black Isle, Avoch and Eddirdovar, Duffus in the sheriffdom of Elgin and Aberdour and Crimond in Buchan. John received Balvenie and Boharm in Banffshire, Brachlie and Petty north of Inverness, and Strathdearn, the valley of the Findhorn. Though initially these designs had limited impact in regions dominated by the earl of Ross, they were alarming to Douglas's former allies, Crichton and Gordon, while the attempt to secure the see of Aberdeen alienated David earl of Crawford whose kinsman was James Douglas's rival. In early 1443 Crawford was linked to Crichton and Kennedy in opposition to the Black Douglases.[28]

The revival of the family's northern ambitions after Earl James's death was bound up with the successes won by his eldest son in the struggle to dominate the king's council. Earl William's victories against the queen, Kennedy, Angus and the Crichtons over the next two years allowed the renewal of his father's plans. At the parliament of July 1445 which confirmed the success of Earl William and his allies, his younger brothers appeared for the first time as earls. Archibald was named as earl of Moray, ignoring the rights of James Crichton following the eclipse of his family. For Hugh Douglas a new earldom was created. Hugh took his title of Ormond from the old council hill of Ardmeanach on the Black Isle. The creation of the earldom was a direct affront to Ross, whose chief castle of Dingwall lay close by, and Hugh's promotion suggested that the Douglases were presenting themselves as defenders of Anglicised communities in the north against Gaelic magnates. The title of Ormond was also borne by an Anglo-Irish magnate dynasty who played a similar role in the lordship of Ireland. The adoption of a similar image was directed against Ross, a magnate whose dominance in the north most obviously blocked the establishment of Douglas lords in Moray and the Black Isle. By the summer of 1445 William earl of Douglas and his brothers had apparently succeeded in winning allies who would be valuable in any move against Ross. The family's difficulties with both David earl of Crawford and Alexander lord of Gordon had been resolved, and both men were present at the parliament of July, Gordon having been rewarded by elevation to the earldom of Huntly.[29]

However, if the Douglases hoped to secure Crawford and Huntly as allies, the plan was dashed by the start of open conflict between the two magnates by early 1446. At issue was the dominance of north-east Scotland. This conflict forced the Douglases to seek other ways of securing their interests in the north. In 1446 and 1447, with the backing of the royal council, they sought a negotiated settlement in the region which allowed Earls Archibald and Hugh and John of Balvenie profitable custody of their lands. Part of this settlement was an agreement with Robert Erskine, claimant to Mar, in which he surrendered Kildrummy castle to royal custodians, and in July 1448 the young king himself visited the north-east and stayed in the castle. His presence probably added authority to discussions over lands to the west of Mar. Central to these discussions was an agreement, perhaps even the infamous formal band, between the earl of Ross, the earl of Crawford and the Douglases which was not the aggressive pact later claimed, but a means of ending tensions in the north. Ross, whose experience of the 1420s and 1430s had taught him caution, was keen to secure the acceptance by the king and council of his position in the north. In return for the marriage of his son to the daughter of James Livingston, the king's keeper, Ross was prepared to give ground. By 1448 he had relaxed his hold on Inverness to royal officials, and the earl of Crawford had recovered his lordship of Strathnairn, which had probably been outside his family's control for a decade. The Douglases also gained in the lands round the Moray Firth. In June 1447 Archibald earl of Moray was installed in Darnaway castle, his chief residence, and granted a remission to a local lord for attacks on his brother's barony of Petty, suggesting that the family's rights in both lordships were recognised locally. In practice such recognition could only come from Ross and may have extended to the earldom of Ormond.[30]

Moray was to be the centre of Douglas power in the north. Although Moray had lacked an earl since 1430 and had fallen under the influence of neighbouring magnates, Archibald Douglas gained more than an empty title. He clearly established effective lordship in the heartlands of the earldom in the coastlands around Forres and Elgin and in the lower valley of the Findhorn, districts both rich and relatively easy to govern. He was not just an outsider, but the husband of the Dunbar heiress to Moray, and as earl sought good relations with his wife's kin. Her mother, Huntly's sister, was married to John Ogilvy of Lintrethin, a leading member of an extensive northern family. By 1450 John and several of his kinsmen had been drawn into closer ties with the Black Douglases, a connection built on the dowager's possession of an extensive terce in Moray. Earl Archibald formed similar ties with Alexander Dunbar of Westfield, his wife's bastard half-brother, was emerging as a figure of local significance and probably owed his position as 'our sheriff of Forres' to his brother-in-law.[31] This was not an isolated connection. The principal tenants of the earldom, the Frasers of Lovat, the Grants, the Comyns and Chisholms were, during the early 1440s, brought, willingly or unwillingly, under the influence of the earl of Ross. By 1450 these families had ties of lordship with Archibald earl of Moray, and though some retained connections with Ross, their attitude to Douglas reflected a desire for a lowland lord by men who had supported Alexander earl of Mar's earlier establishment of authority in the region.[32]

Archibald Douglas was not simply an interloper in the north. Though he arranged a marriage and granted lands in his earldom to Walter Haliburton of Gask, a junior kinsman of his family's southern ally, he enjoyed support from his new earldom and drew his servants from the north. The best known of these was Richard Holland, an Orkney man who held a Caithness benefice, and who, by 1450, was secretary of Moray, under whose influence he became precentor of Elgin cathedral. His poem, *The Buke of the Howlat*, was written for his mistress, the 'dow of Dunbar', and celebrated, not just her husband's name and arms, but its setting 'in mirthfull moneth of May, in myddis of Murraye' and specifically in the 'parfyte' forest of Darnaway. Darnaway castle was the centre of Archibald's earldom, and he began a programme of reconstruction of a residence which had probably been neglected since the previous century. His building works signified an intention to put down lasting roots in the north. The formal decision in 1447 that Archibald was younger than his twin brother, James, who was thus next heir to the family's southern lands, confirmed that the earl of Moray's future lay in the north.[33] However, despite such commitments, the roots of Douglas power in Moray were shallow. If Earl Archibald wished for greater success than his Dunbar predecessors, then, like the

Setons of Gordon, he had to remain in the north. Instead, over the few years in which he held Moray, Archibald, like his brother, Ormond, was drawn south to support the head of his family, leaving his northern lands open to rivals like the earls of Huntly and Ross, who had no such distractions and whose attitude to the house of Douglas remained ambivalent. In its end as in its creation, Douglas lordship in the north stood or fell by events in the south and, in particular, at the royal council table.

PARLIAMENT OF FOWLS: THE DOUGLASES AND THE COMMUNITY

When William, eldest son of James the Gross, became eighth earl of Douglas in 1443, this council was dominated by the chancellor, Crichton, and his allies. The control of royal resources which this gave Crichton was used to further the ambitions of his family and friends and to obstruct the efforts of the Douglas earls to secure and extend their own lordship. The young earl of Douglas determined to end this control by force. In the two years from August 1443 William waged a campaign against the Crichtons which showed his mastery of the skills on which Douglas power had been built. The combination of local war, hard bargaining and propaganda recalled the methods of his forebears but aimed, not at regional dominance, but control of royal government.

The offensive which William launched in late August 1443 reflected this goal. He began by securing as allies two men with Douglas connections who had been frequent royal councillors since 1439. On 20 August the earl was at Newark in the Forest confirming the lands of William Cranstoun, while Cranstoun's partner in the arrest of the queen four years before, Alexander Livingston, was also forging new links with Douglas. Four days earlier Livingston had cleared himself on oath of guilt in the death of Malcolm Fleming at the Black Dinner. The oath not only removed grounds for friction between Livingston and Douglas's brother-in-law, Robert Fleming, it also implied that William Crichton was the prime mover in the deaths of the Douglas earl and Fleming in 1440. Despite his father's involvement, William earl of Douglas may now have justified his attack on the chancellor as revenge for the Black Dinner. An alliance with Livingston allowed the earl to justify violence in wider terms. While Livingston sought to eclipse Crichton on the council, Douglas exploited Livingston's custody of the young king in an attack which was launched to the west of Edinburgh, where the lands of Crichtons and Douglases overlapped. On 22 August Earl William with 'ane gret ost' from the marches laid siege to George Crichton's fortified house at Barnton. The attack was presented as more than a private feud. With Douglas were the king and his household, and the garrison was handed royal letters demanding its surrender, and after a short siege, during which the earl flew the royal banner, the defenders capitulated. Royal authority was exploited further. William Crichton was summoned to the king's presence in Livingston's castle of Stirling. When he failed to appear he was condemned by the assembled estates. Though William Crichton retained Edinburgh and waged local war on Douglas and his allies in Lothian, he was stripped of the chancellorship and excluded from the council.[34]

Earl William's success rested on his own armed following allied to claims to act in royal service. Further success was achieved by stepping back from open support of the council of Basle and seeking alliances with Bishops Kennedy, Cameron and James Bruce of Dunkeld. These links and the earl's alliance with Livingston ensured Douglas influence on the royal council, which was maintained by the earl's attendance on the king during early 1444. This influence reversed the situation faced by his father in 1442. Douglas now held the advantages of royal support. In the summer of 1444 the earl at last received a papal dispensation for his marriage to his cousin, a product of influence at both royal and papal courts. This influence was also used to enlarge the orbit of the Douglas family. In September Earl William was with the royal council at Dalkeith, intervening in the feud over the lordship. Crichton's ally, Gifford of Sheriffhall, was removed as keeper of Dalkeith castle, and Earl William's brother-in-law, Henry Douglas, received formal rights in the family lands in place of his feeble elder brother. Earl William also intruded his influence into the south-east. Though Douglas did not overturn the *status quo* achieved against his father's wishes, he was in a position to forge new links with the rival branches of the Hume family and the Hepburns, and thus renew his family's influence in a region from which it had been excluded since 1424.[35]

Though Earl William's success owed much to his own ability and energy, it was also a sign of the increasing importance of royal lordship in Scotland. Douglas and the Livingstons maximised this importance, and in a general council in November 1444 the fourteen-year-old king was declared of age to rule. Control of the nominally adult James II gave greater power to his keepers, enabling them greater control of royal resources and greater powers to condemn opponents as rebels. The action was a preliminary to further attacks on these opponents, a group which included the king's own mother, his cousin, Angus, and the former chancellor, Crichton. It also included Bishop Kennedy, who had written with the queen against 'tha persownis that nw has the kyng in governance'. The bishop and the queen were first targets of these 'persownis'. In late November 1444 Methven castle, near Perth, which was probably held for the queen, was surrendered to a force which, like that at Barnton, included the king and his council. Two months later 'ane richt gret herschipe', a massive raid, was launched on Fife where both the queen and Kennedy held lands. The raiders included the Livingstons, Douglas's ally, James Hamilton, and the earl of Crawford, soon to be a supporter of the Douglases in the north, and the attack aimed at neutralising the council's opponents north of Forth. By the summer the rest of these opponents came under attack. For nine weeks in June and July, William Crichton was besieged in Edinburgh castle, and during the blockade a raid was led on Angus's lands in East Lothian, and finally, in August, Dunbar castle was blockaded and surrendered following the death of Queen Joan within its walls.[36] Though political violence was nothing new, the ten months from November 1444, and especially June, July and August 1445, represented a civil war of a new type. Unlike the conflicts of 1388–89 and 1397, unlike even the struggles for control of royal government in 1401 and 1437, the war of 1445 was a sustained clash between coalitions of magnates who recognised control of James II's council as the means to regional dominance.

The fruits of victory in this war were distributed even as the Douglases and their allies besieged Edinburgh castle. They sat as a parliament whose purpose was to confirm the gains of the previous months and obtain new rewards. The parliament marked the first appearance of the Douglas brothers as earls of Moray and Ormond and of their brother-in-law, Henry Douglas of Borgue, as lord of Dalkeith, supplanting the senior line of his family. Earl William's ally, James Hamilton, had his lands united into a single lordship, while his councillor, James Auchinleck, received new lands from the king. Control of parliament was also used by the king's keepers to exert further pressure on their opponents. James earl of Angus, who had failed to answer a summons to attend, was threatened with forfeiture unless he submitted to the king in parliament. Though Earl William must have been tempted to seek the downfall of his Red Douglas rivals, the aim of this threat was limited. Crichton submitted at the end of July and after the fall of Dunbar Angus must have also sought terms. Though neither was forfeited, Angus had to accept the end of his betrothal to the king's sister and Crichton lost Edinburgh. Douglas and the Livingstons were satisfied with a settlement which confirmed their gains without forcing a fresh conflict.[37]

The civil war of 1445, though it ended in compromise, revealed the extent to which Douglas power still rested on the coercive force of the earl. Military action secured the Livingstons as custodians of the king, his lands, castles and revenues. It also allowed Earl William and his brothers to use the crown as a source of moral and physical support to their own ends. Even in the marches, where Douglas lordship was strongest, links to the royal council were a major element in the expansion of the family's influence after 1445. In the south-east Earl William's success contrasted with his father's lack of influence. Earl James's local efforts had been blocked by his enemies on the council. Earl William dealt with those enemies, Crichton, Angus and Bishop Kennedy, and in the process showed his ability to dominate the region in the civil war, which ended in the capture of Dunbar from Adam Hepburn. In early 1446, Adam's son Patrick sought to reverse this defeat by occupying Dunbar 'as a violent intruder', using it as a base for attacks on his local enemies. Chief among these enemies were the Humes, both Alexander and his uncle, David, who had built links with Earl William during his rise to power in 1443 and 1444. The Humes responded to Hepburn's attacks by waiting on Douglas at Stirling, and at the end of the year the earl exploited the capture of Hepburn's castle of Hailes by Archibald Dunbar, son of the last earl of March. Dunbar's aims and backing are unclear but he 'cowardlie' surrendered his conquest to James master of Douglas, the earl's brother. Though local violence rumbled on, the Douglases, with allies like the Humes and John Haliburton, had established themselves as major lords in the east march. This power, although built on local support, also rested on influence derived from the earl's place on the king's council, and allowed William to sideline the new earl of Angus, George, who had succeeded his brother in 1446 and was forced to abandon ambitions and allies to his Black Douglas kinsmen.[38]

In the south-west, too, influence at court was vital to Earl William's interests. Marriage to his cousin, Margaret, was secured through this dominance, but although William began to call himself lord of Galloway in 1444, it was after his victory the next year that he secured effective control of the province. In a deal only formally recognised in 1450, but probably dating from 1447, the earl extracted the resignation of Galloway from his wife's aged grandmother, Margaret duchess of Touraine. By August 1447 William was in Threave, her chief residence, granting lands in Galloway without reference to either his wife or her grandmother. He was accompanied by Alexander Mure, the latter's long-standing servant who, the next year, was the earl's justiciar in the province. For Mure, the opportunities of service to a powerful lord clearly outweighed any doubts about the fate of his old mistress, now confined within Lincluden priory.[39]

For Earl William, control of Galloway was only part of wider ambitions in the south-west. As lord of Stewarton and Dunlop, Douglas inherited his father's interests and allies in central Ayrshire as well as older connections between Galloway and the neighbouring earldom of Carrick. The desire to secure Galloway led the earl to bring his influence to bear in Carrick.

At the same time as William gained control of the province, the castle of Loch Doon, just across the border in Carrick, was surrendered to a force which included Alexander Livingston. As at Barnton, royal councillors were present to justify an attack against Douglas's enemies. Loch Doon was held by men 'of the name MacLellan', suggesting that men from Galloway opposed William's growing regional power. From 1446 this power may have extended into Carrick, where politics was dominated by the feuds within the Kennedy family and the attempt of Gilbert Kennedy of Dunure to bring his kinsmen into his lordship. Despite being Bishop Kennedy's brother, Gilbert had received lands from Douglas, and his recovery of the office of bailie of Carrick in 1447 suggests his good relations with Earl William and the royal council.[40] The takeover of Galloway also allowed William to renew Douglas leadership of the west march. Since the Black Dinner, Annandale and Nithsdale had seen a number of disputes connected with the efforts of the Maxwells, now royal stewards of Annandale, to extend their lands. With the success of Douglas and his allies in 1445 there were even signs that the earl intended to renew his family's interest in Annandale. Even without Annandale, Douglas, as lord of Galloway, had the means to exercise lordship in the region, based both on ties of land-holding with local men which had been fostered by the Black Douglases since the days of Archibald the Grim, and on Archibald's wardenship of the west march. In the late 1440s Earl William made it clear to rivals like the Maxwells and Crichtons that the Douglases still claimed to be principal defenders of the western borders.[41]

With the outbreak of open war in the marches in late 1448, the family were given an opportunity to put these claims into practice. The war owed something to the efforts of Earl William to renew his family's European status. Douglas's influence on the council was accompanied by increased Scottish diplomatic activity which culminated in treaties with France, Brittany and Burgundy in 1448 and 1449. William took the opportunity to seek the restoration of at least part of the duchy of Touraine from Charles VII by cynically championing the rights of Duchess Margaret, recently deprived of her Scottish lands. The Douglases were at the forefront of Scottish negotiations with Burgundy. When, as part of these, a group of Burgundian knights sought Scots opponents in a joust, it was to James, master of Douglas, that they directed their challenge. The combat at Stirling before King James gave the Black Douglases the chance to pose as the armed defenders of Scotland's honour.[42] A more serious display of this claim came later in the year when Scottish raids in both east and west marches provoked a counter-attack. In October an English force led by Thomas Percy, the younger son of the earl of Northumberland and grandson of Hotspur, crossed into Annandale. It was met and defeated by Hugh Douglas, earl of Ormond, at Lochmabenstone near the mouth of the river Sark. The next year, in retaliation for English attacks on Dunbar and Dumfries, Earl William himself led raids which burned Percy's towns of Alnwick and Warkworth.[43] Douglas attached symbolic as well as military significance to the war. The return of war and Douglas leadership was emphasised in December 1449 when Earl William assembled a council of 'lordis, frehaldaris ... and bordouraris' at Lincluden to codify statutes of the marches ordained 'in Blak Archibald of Douglas dais'. The earl's appeal to the authority of his grandfather emphasised that for nearly a century the Black Douglases had laid down the rules of war and defence in the march and that their new rivals and even the crown owed lands and survival to this leadership. The laws repeated the statutes already included by James I in his 'item on the marches' in 1430. By reclaiming them as Douglas ordinances, William was reclaiming the role of his dynasty as leaders of a special community within the kingdom.[44]

It was in this atmosphere that Richard Holland produced *The Buke of the Howlat* for his Douglas patrons at the end of the 1440s. Holland followed earlier tales which represented human hierarchy as species of birds gathered in council. His poem described the appeal of the howlat, the owl, who complained of his ugliness and begged help from his fellow birds. He was given a feather by each but, in this borrowed plumage, the howlat claimed to be greater than princes, rebuking those who refused to bow before him. The birds responded by stripping the howlat, leaving him once again 'hidowis of hair and hyde'. The moral was a typical medieval attack on the unnatural pride of those raised 'fro poverte' with riches which were 'othir mennis and nocht thi awne'. Set against this was Holland's description of the fame and achievements of the Douglases who 'blythit the Bruse', defended 'Scottis blud' and whose lands were not borrowed but won in war. Earl William and his brothers were shown as the heirs to the honour and rights of their ancestors. As in the 1350s and 1390s the family's propaganda smoothed over the change of Douglas lord by drawing on traditional duties and images of Douglas power, chief amongst them the bloody heart. Holland went further than this. The unique status and role of the Douglas earl in the kingdom was clearly expressed and contrasted with the borrowed power of other families.[45]

In the *Howlat* and elsewhere there were clear signs of Earl William and his brothers laying claim to special powers born of the achievements of the Good Sir James, Archibald the Grim and the Douglas dynasty. Such claims were made in the face of changes in Scottish political society in the previous quarter-century, and the *Howlat* and the Lincluden statutes suggest that the

Douglas magnates were consciously defending what they regarded as their rights against the encroachment of the crown and new aristocratic rivals. These rivals, quite probably the target of the *Howlat*, were baronial families who, by the 1440s, formed the top rank of the nobility in a kingdom stripped of its provincial earls and regional lords by royal hostility and extinction. Though almost entirely from established Scottish lines, these rising men sought formal status to match their new influence. James II's council recognised this by giving out the title of lord of parliament. The recipients of this, men of growing significance on the council and in the kingdom like William Crichton, Herbert Maxwell, Alexander Forbes, John Stewart of Lorn and Alexander Montgomery, reflected the dispersal of aristocratic power in Scotland among a wider group of men. In both the theoretical obligations of their new rank and in practice, these new lords were bound to the crown which, as the holder of a growing number of major estates, was the potential arbiter of their local power. Though 'ever servabile' to the crown, the Douglases claimed, by contrast, to have earned the blessing of the royal house and to belong, in this and other ways, to a different rank from lesser nobles. At the parliament of July 1445 three of the five earls present were Black Douglases, while lordships of parliament were conferred on four of their allies, Laurence Abernethy, William Somerville, Henry Douglas of Dalkeith and James Hamilton, seeming to show the distinction between the family and its peers and its powers to dispense rewards amongst its satellites.[46]

The display of this special place in the aristocratic elite of the kingdom also obscured shifts in the power of the Douglas family in the decade from 1439. Appeals to older traditions of Douglas lordship were made for the benefit of the brothers and the men they looked to for support, but they could not alter the reality of changes in that lordship or in the family's place in the kingdom. Earl William and his brothers were sons of a man whose rank and resources had been on a par with those made lords of parliament in the 1440s. While he had been raised to an earldom in 1437, James the Gross was not a great magnate and his family's interests in Clydesdale and Ayrshire and in the north reflected different aims from those of previous earls. Despite the propaganda, the Black Dinner had broken the continuity of Douglas power. Though it allowed Earl James to claim the bulk of the Douglas estates, the recovery of the influence held by previous earls and the quest for power in the north were part of a process which saw a series of struggles to succeed to the regional interests of other extinct or dispossessed magnate lines, of Albany, Atholl, Mar and March. The success of James and his sons in acquiring Galloway and Moray and in extending their influence in the kingdom concealed flaws in the fabric of this lordship. The first of these flaws was dynastic. In August 1447, at Edinburgh, a formal indenture was concluded in the presence of Earl William and his mother, Countess Beatrice, which named James as older than his twin, Archibald earl of Moray, and thus next in line to inherit the Douglas lands. By 1449 James was consistently referred to as master of Douglas. This designation, when his brother could be expected to produce heirs, suggests an anxiety that Earl William's marriage was likely to remain childless. Beyond the normal problems this created, the lack of offspring left future Douglas possession of Galloway in doubt and perpetuated the insecurity which had affected the province since 1440.[47]

There was uncertainty, too, over the Douglas earl's hold on the loyalty of his adherents, especially in the marches, the heartlands of Douglas power before 1440. Though Earl William's councillors included men from border families, the extent to which the earl could draw on the middle and west march as a source of support was uncertain. At Lochmabenstone in 1448, Ormond met the invading English with a force which, though it included the Johnstones and Stewarts of Dalwswinton from the west march, was dominated by 'gentillis of the westland' like Wallace of Craigie and John, son of Lord Somerville. It was men from the 'westland', Clydesdale and Ayrshire, who also provided the core of Earl William's following, and the assembly at Lincluden may have been the attempt to extract support from a community which had shown only limited enthusiasm for the Douglases in recent war. The appeal to traditions of regional lordship which William made challenged the ambitions of baronial families and was itself threatened by the new title of lord of parliament and its implications of direct loyalty to the crown. To many men in the south, from royal councillors to free-holders, the efforts of Earl William to rebuild the dominance of his predecessors was unwelcome and not unresisted. In April 1449, with the earl at the height of his power, James Auchinleck, one of his closest adherents, was slain by Richard Colville, a member of the family which held the neighbouring barony of Ochiltree. Douglas responded by besieging Colville and, on his surrender, beheading him with three of his accomplices. Despite his success the incident contained warnings for Earl William. The killing of Auchinleck was not simply a local feud. It was also the murder of a man at the heart of the new Black Douglas affinity by one of a kindred who had been followers of the earls from the 1380s to the 1420s, valued for their influence in Teviotdale. William had again demonstrated that his principal adherents were from the 'westlands', not the marches, where his dominance was no longer automatic. If this marked a departure from the past, in his readiness to use force against his Scottish enemies William was drawing on the methods of his ancestors and confirmed the means by which he had built his power since 1443. Such displays of coercive lordship were not just limited to secular rivals and tenants but extended to the church. Earl James's break-up of the provincial council was not repeated, but his son sought to maintain his family's grip on the offices and institutions of the church in their lands by physical force and political influence as before 1424. Yet this coercive approach to lordship, like the claims to regional power, fitted less easily into the Scottish realm of the late 1440s, a realm which contained many alternatives to the house of Douglas as sources of leadership and protection. The greatest of these, and the one to which Earl William would now have to answer, was James II, the young King, nineteen years old in October 1449 and ready for power.[48]

NOTES

1. The new earl, William, was said in the contemporary *Auchinleck Chronicle* to have been eighteen years old in November 1440 but the dispensation for his parents' marriage was only granted in 1423, suggesting he may have been slightly younger than this ('Auchinleck Chron.', in C. McGladdery, *James II*, 171; *C.S.S.R.*, ii, 94).
2. *R.M.S.*, ii, nos 201, 202; Fraser, *Douglas*, iii, nos 403, 404, 406.
3. *H.M.C.*, vi, 691–92, nos 17, 20, 22; *E.R.*, v, 483; *C.S.S.R.*, iv, no. 513; S.R.O., GD 20/301; *R.M.S.*, ii, nos 202, 203; 'Auchinleck Chron.', in C. McGladdery, *James II*, 160; *Calendar of Papal Letters*, viii, 255–56; *S.P.*, v, 2–3; *E.R.*, v, 110; S. Boardman, 'Politics and the Feud in late Medieval Scotland' (unpublished Ph.D thesis, University of St. Andrews, 1989), 159–62.
4. 'Auchinleck Chron.', in C. McGladdery, *James II*, 160; *A.P.S.*, ii, 54–55; Fraser, *Douglas*, iii, nos 392, 398; *H.M.C.*, xi, appendix 6, 212, no. 8; *R.M.S.*, ii, no. 203.
5. *R.M.S.*, ii, no. 204; *A.P.S.*, ii, 54–55. Hugh Kennedy, provost of St. Andrews, had resumed an ecclesiastical career after two decades as a mercenary in France, during which time he had served at Baugé and with Jeanne d'Arc. His presence may suggest a military force had been assembled by the queen's allies. Parkle was with Avondale and George Crichton at Lanark in December 1439 (*R.M.S.*, ii, no. 246).
6. *R.M.S.*, ii, nos 206, 208, 209, 210, 211, 212.
7. S.R.O., GD 20/300, 301; *Calendar of Papal Letters*, ix, 129; A. I. Dunlop, *The Life and Times of James Kennedy Bishop of St. Andrews*, 39; *R.M.S.*, ii, nos 212, 223, 224, 370; *Illustrations of the Topography and Antiquities of the Shires of Aberdeen and Banff*, Spalding Club, 4 vols (Aberdeen, 1847–69), iv, 192.
8. Fraser, *Douglas*, iii, no. 303; *R.M.S.*, ii, nos 226, 242; 'Auchinleck Chron.', in C. McGladdery, *James II*, 171. David earl of Crawford, Earl William's father-in-law, was also present at Edinburgh in February 1440 (*R.M.S.*, ii, no. 213).
9. Pitscottie, *Cronicles*, 24–25, 28–29, 40; *C.S.S.R.*, iv, no. 748; Fraser, *Douglas*, ii, 42.
10. Dunlop, *Bishop Kennedy*, 397; *R.M.S.*, ii, nos 247–51; C. McGladdery, *James II*, 171; Pitscottie, *Cronicles*, 41–46. Though Avondale was in Edinburgh in the weeks before his great-nephew's death, he was probably absent from the Black Dinner itself in a deliberate effort to distance himself from the event.
11. Waurin, *Chroniques*, v, 213–14.
12. The confirmation by the royal council of a number of Douglas grants in the aftermath of the Black Dinner suggests a sense of insecurity amongst certain of the family's tenants (*R.M.S.*, ii, nos 252, 254, 255, 256).
13. *R.R.S.*, vi, no. 51; Hume of Godscroft, *House of Douglas*, i, 156–57; M. G. Kelley, 'The

Douglas Earls of Angus: A Study in the Social and Political Bases of Power of a Scottish Family from 1389 until 1557' (unpublished Ph.D thesis, Edinburgh University, 1973), 28–32; *R.M.S.*, ii, no. 19; *H.M.C.*, Hamilton, no. 131; S.R.O., RH2/321.

14. *H.M.C.*, Hamilton, no. 131; *E.R.*, iv, 670. The Somervilles, Auchinlecks and Dunbar of Cumnock were associated with James before he became earl of Douglas (*R.M.S.*, ii, nos 246, 370).

15. *H.M.C.*, Hamilton, nos 16, 131; Fraser, *Douglas*, i, 421; *C.S.S.R.*, iv, nos 745, 1045. The marriage of William and Margaret was said to have been to preserve concord amongst their kin and friends.

16. S.R.O., GD78/1; Fraser, *Carlaverock*, ii, no. 24; *R.R.S.*, vi, nos 81, 282; *R.M.S.*, ii, nos 226, 234, 234, 242, 258.

17. *R.M.S.*, ii, nos 224, 226; S.R.O., GD150/109; *Morton Reg.*, ii, no. 219; *E.R.*, v, 147; *C.S.S.R.*, i, 236; Dunlop, *Bishop Kennedy*, 58–59.

18. Fraser, *Scotts of Buccleuch*, ii, nos 21, 36, 41; *H.M.C.*, Roxburgh, no. 10.

19. S.R.O., RH2/307; Fraser, *Douglas*, i, 445.

20. Burns, 'Scottish Churchmen at the Council of Basle. Part Two', in *Innes Review*, 13 (1963), 157–83; Dunlop, *Bishop Kennedy*, 37–48; *Copiale*, 313–21, 322–24. Douglas's councillor Adam Auchinleck was also to be provided with the archdeaconry of Aberdeen. The news of Earl James's action against the council was relayed to Basle by Croyser.

21. *Cold. Corr.*, nos cxxxvi-cxl, cxlii, cli, clii, cliii, cliv, clvi, clviii, clx, clxi; J. Raine, *North Durham* (London, 1852), appendix 99, no. dlxvii; *H.M.C.*, xii, appendix 8, no. 296.

22. 'Auchinleck Chron.', in C. McGladdery, *James II*, 161; *Wigtown Charters*, nos 23, 24, 26; Fraser, *Douglas*, i, 446.

23. Fraser, *Douglas*, iii, nos 27, 293; Boardman, *Early Stewart Kings*, 81–82, 260–66.

24. *R.M.S.*, ii, nos 43, 49; Fraser, *Frasers of Philorth*, ii, nos 16–19. For northern politics in the early fifteenth century, see Brown, 'Alexander Stewart Earl of Mar'.

25. S.R.O. GD 44/4/3. James Douglas had confirmed a grant to Seton's brother-in-law, John Gordon, in 1423, a grant witnessed by the lord of Gordon, and had also used Seton's seal in 1430 (*A.B. Ill.*, ii, 378; Fraser, *Frasers of Philorth*, ii, no. 21).

26. Brown, 'Alexander Stewart Earl of Mar', 46–47; J. and R. W. Munro, *The Acts of the Lords of the Isles*, nos 23–31; *E.R.*, v, 84, 87, 191, 265; *Highland Papers*, ed. J. R. N. MacPhail, vol. 1, Scottish History Society (Edinburgh, 1914), 46; *R.M.S.*, ii, no. 370.

27. Fraser, *Douglas*, iii, no. 410; S.R.O. GD 52/404. The plan to marry the Moray heiresses to sons of Crichton and Douglas may have been formulated in 1440 when James the Gross met Gordon and a number of local men at Elgin, within the earldom (*R.M.S.*, ii, no. 370). James Crichton first appeared as lord of Frendraught in about 1445 but quite probably received the title in 1442 (Fraser, *Scotts of Buccleuch*, ii, no. 21). In 1436, the lord of Gordon had custody of Kintore by reason of James Dunbar's death (*E.R.*, v, 8).

28. Burns, 'Scottish Churchmen', 168–69; S.R.O. GD 1/177/1; *A.B. Ill.*, iii, 118; *E.R.*, vi, 212, 265, 648. Much of the evidence for the division of Douglas lands in the north comes from the records of the royal administration after 1455.

29. *A.P.S.*, ii, 59–60; Fraser, *Douglas*, iii, no. 413; *H.M.C.*, Hamilton, no. 12; *E.R.*, vi, 648.

30. *A.B. Ill.*, iv, 196–201, 393; *E.R.*, v, 306; Munro, *Lords of the Isles*, no. 45; A. Grant, 'The Revolt of the Lord of the Isles and the Death of the Earl of Douglas', in *S.H.R.*, 60 (1981), 169–74; S.R.O., GD 1/177/1; GD 52/1044. A royal keeper held Inverness and Urquhart from 1448 and John had to retake both strongholds by force in 1451 (*E.R.*, v, 380; 'Auchinleck Chron.', in C. McGladdery, *James II*, 169).

31. S.R.O. GD 16/1/3; GD 40/4/61; *Rot. Scot.*, ii, 346; *E.R.*, vi, 219, 221, 270; Fraser, *Douglas*, iii, nos 79, 80.

32. Fraser, *Douglas*, iii, no. 80; W. Fraser (ed.), *The Chiefs of Grant* (Edinburgh, 1883), no. 29; *E.R.*, vi, 217.

33. M. Stewart, 'Holland of the Howlat', in *Innes Review*, 23 (1972), 3–15; *Buke of the Howlat*, 84, lines 989–1000; *E.R.*, vi, 220, 453; *R.M.S.*, ii, no. 301; Fraser, *Douglas*, iii, no. 306.

34. *Wigtown Charter Chest*, no. 29; *Laing Charters*, no. 122; 'Auchinleck Chron.', in C. McGladdery, *James II*, 161. Fleming had earlier protested to James the Gross as justiciar about his father's death and received the latter's support for his recovery of his lands, an event associated with his marriage to James's daughter. Crichton as chancellor had, however, refused to give formal sasine of them in early 1442 (*Wigtown Charter Chest*, nos 23–28).

35. *C.S.S.R.*, iv, no. 1045; *R.M.S.*, ii, nos 272, 273, 515; *E.R.*, v, 146, 150, 182; Fraser, *Douglas*, iii, nos 81, 412.

36. *H.M.C.*, volume 12, appendix 8, no. 85; *Extracts from the Council Register of the Burgh of Aberdeen*, Spalding Club (Aberdeen, 1844), i, 399; *R.M.S.*, ii, no. 283; *E.R.*, v, 186, 187, 230; 'Auchinleck Chron.', in C. McGladdery, *James II*, 162; S.R.O. GD 1/479/1; GD 52/1042; Fraser, *Douglas*, iii, no. 414.

37. *A.P.S.*, ii, 59; *H.M.C.*, Hamilton, no. 12; Fraser, *Douglas*, iii, no. 413; *E.R.*, v, 259, 305. Angus's betrothed, Princess Joanna, was sent to France with her sister, Eleanor, in search of a continental husband (*E.R.*, v, 225; Dunlop, *Bishop Kennedy*, 62).

38. *Cold. Corr.*, nos 168, 169, 170; Raine, *North Durham*, appendix 22, no. 96; *Laing Charters*, no. 122; Fraser, *Douglas*, iii, nos 81, 412, 415; 'Auchinleck Chron.', in C. McGladdery, *James II*, 162. The king's council had been at Dunbar in late August 1445 (S.R.O., GD 98/xiv/1). The feud between the Humes and Hepburns was settled in 1449 (Fraser, *Scotts of Buccleuch*, ii, no. 44).

39. S.R.O., GD25/33; Fraser, *Douglas*, iii, no. 81; *Laing Charters*, no. 122; *R.M.S.*, ii, nos 309, 383; Fraser, *Carlaverock*, ii, no. 38.

40. Fraser, *Douglas*, iii, nos 81, 412; *E.R.*, v, 261, 266, 328; S.R.O., GD 25/33, 34, 35, 37, 55; H. L. MacQueen, 'The Kin of Kennedy, 'Kenkynnol', and the Common Law', in Grant and Stringer (eds), *Medieval Scotland*, 274–96. Loch Doon was placed in the custody of Edward Mure, probably a kinsman of Douglas's justiciar of Galloway (*E.R.*, v, 261).

41. *H.M.C.*, Drumlanrig, i, nos 84, 129; N. R. A., no. 776, Earl of Mansfield, no. 19; *E.R.*, v, 357. In October 1445 Douglas promised David earl of Crawford that he would help Crawford's daughter, the widow of the sixth Douglas earl, to recover her terce in Annandale (S.R.O. RH6/321). The Maxwells sought the verdict of the earl in a dispute over lands in Galloway with Sweetheart abbey in 1448 (Fraser, *Carlaverock*, ii, no. 38).

42. R. Griffiths, *The Reign of King Henry VI*, 408–409; Beaucourt, *Charles VII*, iii, 320; iv, 180–81, 365–70; Bibliothèque Nationale, Ms Latin, no. 10187; 'Auchinleck Chron.', in C. McGladdery, *James II*, 164; *Chronique de Mathieu d'Escouchy*, ed. G. du Fresne de Beaucourt, 2 vols (Paris, 1863–64), ii, 148–53.

43. C. McGladdery, *James II*, 164, 173; R. Griffiths, *The Reign of King Henry VI*, 402–11; R. Storey, *The Fall of the House of Lancaster* (London, 1966); J. Stevenson, *Letters and Papers of the English in France*, i, 491.

44. *A.P.S.*, ii, 715.

45. *The Buke of the Howlat*, 46–84; M. P. McDiarmid, 'Richard Holland's *Buke of the Howlat*: an interpretation', in *Medium Aevum*, 38 (1969), 277–90; M. Stewart, 'Holland's *Howlat* and the Fall of the Livingstones', in *Innes Review*, 23 (1975), 67–79.

46. *Maitland Misc.*, i, 383, 393; *A.P.S.*, ii, 59. For the introduction of parliamentary peerages see A. Grant, 'The Development of the Scottish Peerage', *S.H.R.*, 57 (1978), 1–27.

47. Fraser, *Douglas*, iii, no. 306; *R.M.S.*, ii, no. 301.

48. 'Auchinleck Chron.', in C. McGladdery, *James II*, 164, 171.

CHAPTER THIRTEEN

The Fall

'EVER SERVABILE'

At the end of the 1440s the Douglases pronounced themselves 'ever serv-abile', ever ready to serve their lord, Scotland's king. Such service was at the core of the Douglases' self-image, but as the *Howlat* made clear, it was delivered by a family whose place was securely 'next the soverane', whose status was enhanced, not restricted, by their links to the crown. The honour, fame, lands and leadership held by the Douglas earl were the fruits of royal patronage, but they were also rewards which his ancestors had justly won by their own efforts in defence of Scottish king and realm. If the family's ideal remained the loyalty of James Douglas to Bruce, which the *Howlat* described at length, the heirs of Bruce should honour the debt owed to Douglas's descendants for his service, acknowledging their place 'next' to the king and their regional power. As well as Holland's verses, the joust at Stirling, the victory on the Sark and the statutes of Lincluden all recalled earlier achievements of the dynasty and the power of the family which had reached its height in the days of Earl William's uncle, the fourth earl. By adopting his uncle's title of 'great guardian' whilst on the continent in 1450, William was linking himself with the fourth earl's claim to exceptional status and powers in the management of war and peace on the marches.[1]

These claims made little allowance for royal lordship in the marches and implied limits to royal authority over the house of Douglas. Yet it had been only a dozen years since King James I had sought both, and in 1449 his son and successor was nineteen and showing a readiness to grasp the reins of power. The burst of Douglas propaganda, of war leadership and political muscle-flexing, may have been directed by Earl William and his family at their young king, protesting their loyalty, showing their indispensability and demonstrating their rights and rank. However, in this aim too, the Douglas brothers consciously concealed the means by which they had built their power since 1440, stressing instead the continuity with predecessors whose careers indicated restrictions on the crown's real authority in the south. By contrast, the successful usurpation of Douglas lands and lordship and the subsequent extension of their influence by James the Gross and his sons

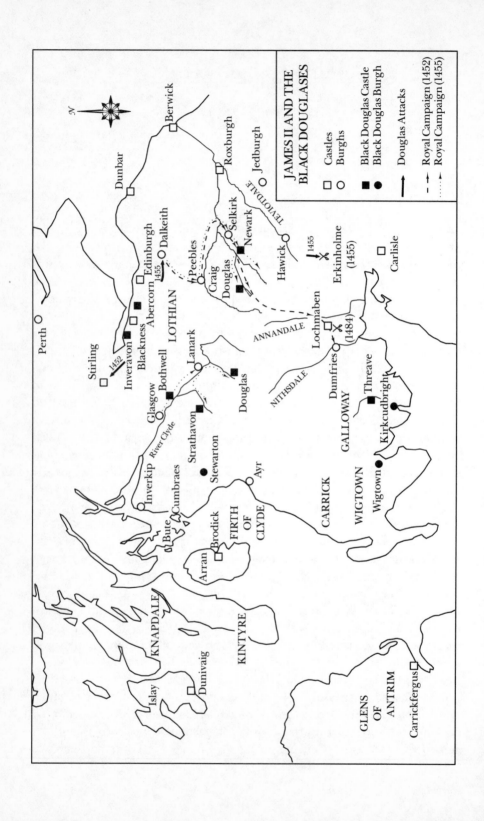

JAMES II AND THE
BLACK DOUGLASES

- □ Castles
- ○ Burghs
- ■ Black Douglas Castle
- ● Black Douglas Burgh
- → Douglas Attacks
- ⤏ Royal Campaign (1452)
- ⤏ Royal Campaign (1455)

N

Perth ○

Stirling □

Inveravon ■   1452

Blackness ■

Abercorn ■

Dunbar □

Berwick □

Edinburgh □

Dalkeith ○   1455

LOTHIAN

Peebles ○

Roxburgh ■

Jedburgh ○

TEVIOTDALE

Selkirk ○

Newark ■

Craig
Douglas ■

Hawick ○

1455   Erkinholme (1455) ✗

Carlisle □

Glasgow □

Bothwell ■

Lanark □

Douglas ■

Annandale

Lochmaben □   (1484) ✗

River Clyde

Strathavon ■

Stewarton ●

NITHSDALE

Dumfries ○

GALLOWAY

Threave ■

Kirkcudbright ●

Inverkip ○

Cumbraes

Bute

Brodick □

Arran

FIRTH
OF
CLYDE

Ayr ○

CARRICK

WIGTOWN

Wigtown ●

KNAPDALE

KINTYRE

Islay   Dunivaig □

GLENS
OF
ANTRIM

Carrickfergus □

was a product, not of local dominance, but of their ability to use the authority of the crown to their advantage. Earl James had made his political breakthrough as James I's councillor and agent, and despite the long minority he and his sons continued to seek power by this route. In the 1440s the prestige and resources of the crown, which had seemed so threatening when wielded by James I, formed a vital source of support. Earl William used royal imagery and prerogatives to his and his allies' advantage. He had the young James II declared of age, bore the king's banner in sieges and claimed to be the agent of royal authority, all to give weight to attacks against enemies on and around the royal council. Unlike his predecessors before 1440, he saw the royal council as the principal forum of Scottish politics, recognising, and seeking to exploit, the achievement of James I as king. Earl William's appearances as a councillor of the king from 1444 to 1449 were designed to secure and safeguard his personal ascendancy in the minority regime. Though neither lieutenant nor keeper of the king's person, Douglas was able to direct royal resources and authority to support his family's interests in the north, Galloway and the marches. The prospect of James II's assumption of power limited the chances of continued dominance, and it was this that produced a change in the political language, though not the political objectives, of the Black Douglases in the late 1440s. Earl William and his brothers claimed a place at the royal council table as the king's greatest subjects, his 'natural' advisers, the heirs, not of James I's servant, but of Robert I's right hand.[2]

Such claims, though expressed in loyalist language, raised possible tensions with James II. They could be presented as challenging the authority of the young king, and James could not be unaware that the history of his family as Scotland's rulers was dominated by rivalries between the royal line and magnates who had built their regional ambitions on influence at court. Though he lacked the royal blood of Albany and Atholl, Earl William may also have appeared to be a threat to the king. This threat was exacerbated by the personal relationship between the king and the earl, who had known each other from James's baptism in 1430, at which William was knighted as a 'fellow soldier' of his future lord. While the conflict between the two men which began in 1450 grew out of the antagonisms of the minority, it was transformed into a personal rivalry. At issue in what became a struggle for ascendancy between the crown and its greatest subjects were different perceptions of the Scottish realm and the place of king and magnate both within and beyond Scotland. The chief battleground and prize of this struggle was local lordship, the efforts to win support from lesser men in the communities essential to Douglas power.[3]

In late September 1449 James II demonstrated his emergence as a political force by casting down the Livingston family. The leading members of the family were executed, imprisoned or exiled, stripped of their offices and deprived of the royal castles which they had held. It was an act with clear echoes of the methods of the king's father and seemed to provide Earl William with a warning of James II's approach to kingship. The king was motivated by the desire to recover full rights in the crown's lands from the family which controlled them, not just as a direct source of revenues, but to provide his new queen, Marie of Gueldres, with her promised dower lands. Like his father, James II was driven to enlarge the royal domain and was prepared to use force to this end. The fall of the Livingstons was the fall of a family whose rise since 1439 had occurred in alliance with James the Gross and his sons. The king was now in closest contact with men who had reasons to hope for the reduction of Douglas power, among them William Crichton, once again chancellor, and William Turnbull, bishop of Glasgow and a constant presence at James's side in 1449 and 1450. Turnbull's career since 1440 had been dogged by local friction with the earls of Douglas, and his rise, with strenuous papal backing, was a mark of the reduced influence of the earls at the curia, the legacy of ill-judged support for the council of Basle. The family's difficulties with Turnbull may, however, have gone back to the Black Dinner, which severed bonds within the ecclesiastical as well as the temporal affinity of the Douglas earls. By early 1450 Turnbull and two other former Douglas servants, John Railston, bishop of Dunkeld, and Thomas Spens, bishop of Galloway, combined high church office with influence on the king's council. When the king confirmed the liberties of the church in 1450, his act seemed to have special significance for relations between the greatest lay and ecclesiastical magnates south of Forth, the earl of Douglas and the bishop of Glasgow.[4]

Yet any impression of gathering tension leading to the king's first clash with Douglas is misleading. To Earl William and others the attack on the Livingstons was the latest in a series of turns of fortune since 1437, not a major change of the political rules. The king's determination to recover his possessions was natural, and Douglas may cynically have written off his allies as scapegoats for the acts of minority regimes. The earl showed no reluctance to participate in the Livingstons' forfeiture and was present at the siege of Dundas, only two miles from Abercorn, held by Livingston partisans. During the siege William received the lands of its owner, James Dundas, and when the tower fell in April the goods it held were divided between the king, Douglas and William and George Crichton. To these latter three the episode was another stage in the complex relationship between their families in their West Lothian heartlands. Allies before 1442, the Crichtons and Douglases of Abercorn may have moving towards better relations since 1448, and appeared, once more, to recognise the advantages of co-operation on the king's council; and in March 1450 the chancellor also sat as a councillor of Earl William.[5] By supporting the royal attack on the Livingstons, Douglas had secured concessions which appeared to be of greater value than this alliance or additional lands in West Lothian. At the parliament of January and February 1450, Douglas formally resigned the Forest and Galloway beyond the Cree, receiving them back from the king, who gave up the rights established by his father in the Forest and recognised the duchess of Touraine's resignation of Galloway. The king also granted William all rights from his marriage to Margaret of Galloway and confirmed his brother, James, as his heir, acts which seemed to remove the earl's anxieties should the match remain childless. James II's acts seemed to show a king readily giving concessions to the Douglas earl out of a stated regard for William's 'singular favour, zeal and love'. Royal generosity in confirming the gains won by Douglas before 1449 suggested, not the mistrust of the king's father for great lords, but the renewal of the close relations between Black Douglas lords and their sovereigns.[6]

It was in these circumstances and probably with 'the licence and blessing of the king of Scots' that William earl of Douglas departed on pilgrimage to Rome to celebrate the papal jubilee declared by Nicholas V. Douglas himself left by ship in early October 1450, sailing to the Low Countries for an audience with Duke Philip of Burgundy at Lille. He was probably joined there by an escort which had travelled through England and which included many of his close allies, James master of Douglas, Lord Hamilton, Alexander Hume, John Ogilvy of Lintrathen and William Cranstoun, and long-standing adherents of the Douglases like Andrew Kerr from Teviotdale and Charles Murray of Cockpool from Annandale. With this 'great and honourable household' Douglas travelled on to Rome, reaching the holy city in early January 1451, where 'the pontiff commended them above all other pilgrims'. The earl's obvious purpose was to restore credit at the papal curia lost by his father's association with Basle, but his journey had the wider goal of recovering the influence of his family across western Europe. In taking the well-trodden route of previous Douglas earls into international politics, William was again drawing links with these predecessors and seeking a place amongst the princes of the west. In his six months away from Scotland, the earl visited the courts of Philip of Burgundy, Charles VII of France and Henry VI of England, all of whom had reason to recall the power of the earls of Douglas. The timing of these visits was not dictated by Scottish considerations alone. As the war in the marches was brought to a temporary close by a truce which James II ratified in June 1450, the war in France reached its decisive stage. During the summer of 1450 the English position in Normandy collapsed, and the French began the reduction of the Plantagenets' lands in Gascony. Like Philip of Burgundy, William may have seen England's emergency as his opportunity to make the kind of deals which the fourth earl had struck with Charles of France after 1419. In late February 1451 the English king and council were anxiously awaiting Douglas's arrival.[7]

If Earl William entered England with hopes of finding a role in European politics, the news which reached him there was of a challenge to his lordship at home. The challenge came from James II, but, as through the 1440s, it was the men on his council, led by William bishop of Glasgow and the Crichtons, the chancellor and his cousin, George, now admiral of Scotland, who guided the king's hand. The chancellor had not forgotten the importance of controlling the council, and he and his allies exploited Douglas's absence to their advantage. Despite their apparent reconciliation with the earl, the Crichtons saw the chance to recover ground lost to their rival in the north, in the marches and, above all, in the south-west. Twenty years before, James I had encouraged William Crichton to seek allies in the marches, now the Crichtons encouraged James's son to seek to renew royal lordship in the region. With symbolic significance James II travelled to Melrose in December 1450 accompanied by the Crichtons, Bishop Turnbull and an entourage of Lothian knights. The king's presence at the abbey recalled his father's interventions in the middle march at the expense of the earl of Douglas. He was joined there by George earl of Angus, the son of James I's principal lieutenant in the marches who, like his three Red Douglas predecessors, found it hard to defend his local position without allies against his Black Douglas cousins. Despite this, both Angus and the Crichtons retained friends in the middle march, men and families who had sought James I's lordship in the 1420s and 1430s, among them Walter Scott of Buccleuch and Archibald Douglas of Cavers, the sheriff of Roxburgh. The support of these men, crucial in the coming years, was confirmed by the king in his first progress through the marches.[8]

From January 1451 there was a sudden change in the intensity and direction of the king's interference in the Black Douglas lands. This change may have been the consequence of the death of the king's aunt, Margaret duchess of Touraine, dispossessed lady of Galloway. She had received Galloway from her brother, James I, for her lifetime, and despite having recognised her resignation to Earl William only a year earlier, James II was presented with the opportunity to claim part or all of the province, basing his right on the temporary nature of his father's charter. For the king, in search of new royal lands, such an opportunity was tempting.

Royal intervention in Galloway allowed the Crichtons to renew their own ambitions in the dales of the Nith and Annan. The chancellor was already forging new links with men from Annandale and was with James II when he visited Lochmaben in January 1451. The strategy of king and council was to win over the leading men of the communities of the south-west. From Annandale the king went to Ayr in February to meet Gilbert Kennedy and confirm his dominance in Carrick. From there the royal household went to Lanark, a centre of Earl William's influence. At Lanark James secured the adherence of Robert Colville, whose kinsman had been executed by Douglas the previous year, and who was now married to the daughter of the chancellor's kinsman, Crichton of Sanquhar. The royal offensive consisted of more than the display of the king's person and banner across the south. As through the previous century and a half, the extension of lordship in the region rested on force or the threat of force. James's progress from Lochmaben to Ayr to Lanark was a military as well as a political expedition. The king's aim, like that of Douglas lords before him, was to force 'the free tenants ... into his peace under oath'.[9]

These displays of royal force put pressure on Earl William's allies. Especially in the south-west, they provided an opportunity for those unhappy with Douglas's newly won dominance. Despite his tacit alliance with the earl, Gilbert Kennedy was one of these, and in his following was Andrew Agnew, the protégé of Duchess Margaret, Kennedy's aunt. Agnew may have left Galloway after her fall, and with the MacLellans, who had also opposed Earl William's usurpation of her rights, probably welcomed royal intervention in the province. Elsewhere, too, the fragility of Earl William's connection was revealed in his absence. In Clydesdale, Lord Somerville was drawn into the king's orbit, while in Lothian, the earl's principal ally, John lord Haliburton, forged a link with the earl's enemy, Patrick Hepburn, through the marriage of his heir to Hepburn's daughter. After the king's successes, Haliburton, whose family's connections with the Douglases of Abercorn went back fifty years, was pushed into seeking his own security above such personal ties.[10]

However, James II's success proved equally fragile. Douglas's absence with many of his adherents weakened his party, as Verneuil had done in 1424. Unlike Verneuil, though, Earl William was coming back, bringing with him lords and men untouched by the king's success. By early March, Douglas's servant, William Lauder of Hatton, was back in Scotland as his lord's forerunner and in contact with the men around the king. In early April, after several weeks at the English court, the earl of Douglas and his escort returned to Scotland. They were met with a fresh demonstration of royal hostility. The king sent a force to raid Douglas's lands in the middle march. It penetrated deep into the Forest where the small castle of Craig Douglas was taken and destroyed. The raid was designed to force Earl William to submit rather than initiate open conflict, and by mid-April Douglas was apparently in the king's peace and said to have put his seal to negotiations with England.

However, the earl was planning his own show of strength. The need for a new English truce suggested that the king feared the breakdown of the recent agreement. As Douglas had just come from England and his consent had been specifically sought for talks, the king's fears probably focused on the earl. Like previous Scottish rulers at odds with Douglas magnates, James recognised the dangers of English support for the most powerful magnate in the marches. When the king sent Snowdon Herald to London 'to bynd up the trewis', Douglas secretly dispatched his brother, James, to the English court where 'he was meikle maid of'. The Douglases were ensuring that English backing was still available. This alliance did not damage Earl William's position in Scotland. A partial entry in the contemporary Auchinleck chronicle, apparently indicating that Douglas's supporters 'cryit him luftennent', acclaimed him as lieutenant in challenge to the king who had shown open hostility to the earl, and that 'sone efter thai worthit als strange as ever thai war' (soon after the Douglases became as strong). In late April Douglas was at Jedburgh, a base convenient for English aid and perhaps indicating Angus's expulsion from his nearby lordship. With the earl was an impressive following including his brothers and his leading southern companions from his pilgrimage. The continued adherence of the Humes, Kerrs, Pringles and Murray of Cockpool showed that Douglas had held together much of his connection in the marches through personal leadership and long-established ties. In mid-May Douglas obtained a licence for this retinue to enter England. His plan was not flight but open access to English bases and support. His aim was to increase pressure on the king.[11]

In the face of Douglas's resistance, the king sought a compromise. By early June, the outlines of a settlement had been drawn up. The king commissioned Andrew Agnew as sheriff of Wigtown, indicating his intention to retain the earldom while Douglas was in the king's 'respite', his temporary peace. The public mood by late June was one of reconciliation, and as parliament assembled at Edinburgh, King James granted lands to Douglas's close supporters. The principal business of the parliament was the formal peace between king and earl:

> Erle William of Douglas ... put him body landis and gudis in the kingis grace and the king resavit him till his grace ... and grantit him all his lordschipps agane outtane the erledome of wigtoun ... and stewartoun.

The king also 'gaf him and al his a fre remission of all things'. Douglas
had come through a sustained royal attack with a full pardon and a series
of eighteen charters issued by the king before parliament which confirmed
Douglas's rights to lands and offices, including a hereditary grant of the
wardenships of middle and west marches. In return James II retained
Wigtown and Stewarton in Ayrshire. However, the display of royal authority
was not wholly convincing. Confronted by James's attack, Douglas had
faced down the king, retaining crucial allies, exploiting the English threat
and avoiding any direct clash. He counted successfully on claims to be
acting in defence of his inherited rights against the aggression of king and
councillors, and on the personal bonds and stature he had forged since
1443. Continued resistance fuelled the desire of the community to avoid a
drawn-out conflict and 'gud scottismen war rycht blyth of that accordance'.
Peace, even at the price of territorial losses, restored, perhaps even raised,
Douglas's standing in the kingdom, and it was James, not the earl, who
now seemed vulnerable.[12]

BREAKING THE BONDS

King James's readiness to compromise was not simply a mark of Earl
William's strength, but of wider concerns. In early 1451 the most pressing
of these was, not English intervention, but the growing crisis in the north
of his own kingdom. As in the south, the king's attempt to demonstrate his
authority had provoked a violent reaction. In March, John, the new earl
of Ross and lord of the Isles, launched a direct attack on the crown's lands
in the north. John was incited to make war by his father-in-law, James
Livingston, who had escaped to the north from royal custody. Ross complained that the 'kingis awne person gart him marry [Livingston's] douchter
and hecht [promised] him gud lordschipe the quhilk he had nocht gottin'.
The king's purge of the Livingstons severed the personal alliance which
linked Ross to the royal council and which had secured his father's settlement with Crawford and the Black Douglases. The response of Ross to
the loss of this 'gud lordschipe' was to unleash his adherents in attacks
which captured Inverness and Urquhart and destroyed the castle of Ruthven in Badenoch. Though Ross directed his efforts at royal estates, his acts
contained a clear threat to his other neighbours. At the end of April two
of these, the earls of Huntly and Crawford, were at court seeking the king's
advice. James's response was dictated by the influence of the Crichtons.
Huntly, the chancellor's son-in-law, was granted the lordship of Badenoch
and probably named the king's lieutenant in the north. Crawford, 'a
rigorous man', was Huntly's enemy for influence in the north-east and was

alienated by the king's actions. Fear of widening conflict in the north dictated James II's attitude to Douglas through the second half of 1451.[13] While they remained at odds with the king, Douglas and his brothers, Archibald earl of Moray and Hugh earl of Ormond, had strong reasons to ally with Ross. The king hoped that his reconciliation with Douglas would end this possibility and improve his relations with Crawford, who witnessed Douglas's reinstatement and received a share of the forfeited Livingston lands in July. The king also sought the active support of Douglas, Moray and Ormond for moves against Ross. The Douglases had no love for Ross and much to gain from his exclusion from the Moray coast, but they were slow to back a king who had recently launched an assault on their own lands. To win their help James restored Wigtown and Stewarton, his gains of the previous winter, to Douglas. However, his efforts did not ease the situation in the north. Ross retained Inverness, Crawford remained hostile to the king's lieutenant and Douglas continued to resist James's efforts to secure his support against them.[14]

Instead Earl William treated his restoration as a licence to strengthen his authority in Galloway and the marches. Sixteenth-century chroniclers give lurid accounts of the earl's arbitrary and violent treatment of his tenants in this period. Though designed to paint a picture of aristocratic lawlessness and inaccurate in detail, the stories of Douglas persecuting men from the families of Herries and MacLellan are not implausible in essence. Both families were vassals of the Douglas earls in Wigtown, and neither had resumed their ties with their lords since the Black Dinner. They were likely targets for Earl William's hostility to those who had welcomed the royal takeover of Wigtown and Douglas, who had beheaded Colville two years before, may well have treated such men in similar fashion. The earl certainly withheld lordship from men he distrusted. Douglas refused to give possession of the barony of Hawick to William Douglas of Drumlanrig, the heir to the barony. He cited the fact that he was in the king's respite but, in reality, the earl delayed handing over lands to one of a number of disaffected men in the middle march. Such partisan lordship had always been a staple means of controlling local communities. Now, however, Drumlanrig had allies to support his cause. In June 1451 he had the earl's refusal to give him Hawick recorded before witnesses who included Patrick Hepburn, George Crichton and Robert Crichton of Sanquhar, all enemies of Earl William. During the same period the earl may deliberately have stirred up friction with England. Men who 'belong to rewle and governance of th'Erle Duglas' were reported as having broken the truce by early 1452 in a probable attempt to increase the king's insecurity.[15]

But in acting as a lord in the manner of his forebears, Earl William overplayed his hand. He underestimated the hostility of king and council to him and his own difficulties. Despite the tension between them, the earl attended King James at Stirling in October and November and at Edinburgh in January 1452. Even whilst he was absent, Douglas kept in contact with James. Among the king's councillors were Douglas's own adherents, Simon Glendinning, William Cranstoun, Alexander Hume and William Lauder of Hatton. However, the earl's meetings with the king only added to tensions. In this atmosphere of mistrust, contacts between Douglas's servants and the king's council had ominous parallels with events a dozen years before. Proximity to royal power, though a feature of the earl's career, had proved fatal to another Earl William Douglas at the Black Dinner. By February 1452 Douglas was not immune to such fears. When James wanted the earl to come to Stirling, he was forced to issue a 'speciale assourans' to Douglas which he and his councillors sealed. James employed William Lauder as his messenger, perhaps to prove his good faith, and met the earl on his arrival with a show of friendship. The next day, 22 February, Douglas 'dynit and sowpit' with the king in Stirling castle, but at seven in the evening the atmosphere suddenly soured. James spoke of an agreement, a 'band', between Douglas, Crawford and Ross, and demanded that William break it. The earl replied that 'he mycht nocht nor wald nocht, than the king said, fals traitour sen yow will nocht I sall'. James drew his knife and stabbed his guest in the neck. The other men present gathered round the earl, one with a pole-axe which 'strak out his branes', the rest striking with daggers and leaving William's corpse covered with wounds.[16]

Douglas was killed for refusing to break his bond with the king's northern enemies. The alliance with Ross and Crawford had secured Douglas interests in the north during the late 1440s, but was not the only means to maintain those interests and of limited direct importance to William's own lordship in the south. But Douglas's defiance of the king was not an empty gesture born of over-confidence. He 'mycht nocht nor wald nocht' break the bond because of what it represented, not just in the north but in the whole kingdom. The bond was a private agreement between great men which, like the indenture of 1409 between Albany and the fourth earl of Douglas, defined and regulated the exercise of lordship in part of the kingdom. It was through such 'bands and ligues' that Scottish political society had functioned for most of the period since 1330. In essence, though, they represented a challenge to royal authority of the kind to which James's father had aspired. As part of his campaign against the Albany Stewarts, James I had forbidden such magnate leagues and repeatedly stressed his public authority as king, his right to rule and demand loyal service regardless of the private obligations of his subjects. In 1452 Douglas refused James II

service against rebels because of a private bond. His action was the calcu-
lated defiance of the king in a period of weakness by a magnate who was
aware of the nature and ideology of royal power. Unlike the fifth earl in
the 1420s, Earl William was not prepared to fit into a new royal establishment, and the king was faced by a magnate who was caught up in the
politics of court and council but who showed no respect for their authority.
The frustration of the situation produced the explosion of violence, with
the king plunging his dagger into a protected guest. Though renewed
crown-Douglas conflict was probably inevitable, the earl's murder had no
precedent in the frequent conflicts of the previous reign. It surpassed even
its closest contemporary parallel, the murder of Duke John of Burgundy
by the councillors of the dauphin, the future Charles VII, in 1419. Like this
event, the murder of Douglas reflected the inability of a prince to deal with
a great magnate by any other means, direct attack and bribery having
already been tried. Like Burgundy's murder, too, the killing of Douglas
risked turning a crisis into a catastrophe. James II had shown a disregard
for the law and authority he claimed to represent. In reacting with violence
to his failure to win Douglas's support against his enemies, he had ensured
the violent hostility of the murdered earl's brothers and allies and launched
fresh conflict.[17]

However, James's assault on Douglas also exposed flaws in his victim's
position. As in the case of his cousin at the Black Dinner, Douglas's killers
included his own men. William Cranstoun and Simon Glendinning, who
joined in the savaging of the fallen earl, were regular members of Douglas's
council and were probably present as his advisers. If so, their part in his
death sprang from a desire to avoid the same fate but perhaps also from
contacts with the royal council during the previous six months which fuelled
hopes of the king's gratitude. Though Cranstoun was a veteran of such
coups, Glendinning came from a line of Douglas loyalists. His defection
suggested that even amongst such ancestral border adherents there was a
loosening of these ties of lordship. The killing of Earl William severed
personal friendships, like that between the earl and Alexander Hume, and
gave the king the opportunity to weaken the family further.[18] A week after
the murder James was in the marches. On the last day of February he was
at Jedworth before moving through Eskdale, where his new adherent,
Glendinning, was the leading tenant, to Lochmaben and then Dumfries.
As in the previous year, the king was displaying royal lordship in the absence
of the Douglas earl, and in his journey James was escorted or joined by
significant local men including Glendinning, Lord Maxwell, Scott of Buccleuch and the Humes. Others, the Pringles from the Forest and the
Rutherford from Teviotdale, came in search of reward or security from

the king. Once more, though, it was the Crichtons who exercised the key influence on the king. The family had been conspicuously absent from Douglas's murder, but in the aftermath of the deed, the chancellor exploited his place at the king's side. James's presence in Annandale and Nithsdale was used again to support George Crichton's rights in the Dalkeith estates. On 8 March an assize of local men at Dumfries recognised George's claim, on his wife's behalf, to the lands of Preston and Buittle in Galloway. On the same day James II moved up Nithsdale to George's castle of Morton where he showed his clear intention to annexe Wigtown once more. Earl William's murder had cut the link between Galloway and the Douglas earldom, and despite several royal charters which named William's brothers as heirs to all his lands, the king and the Crichtons were again seeking to dismantle Douglas lordship in the south-west, dividing the spoils between them.[19]

Yet though king and council reacted quickly to take advantage of the gap in Douglas lordship and win over a number of the family's partisans, the King's presence in the south was not a killing blow. He had, moreover, placed himself in danger. His journey northwards from Morton in Nithsdale took him past Douglasdale where James master of Douglas, now earl in succession to his brother, gathered allies. By mid-March Earl James had assembled a force which included his brothers, Ormond and John of Balvenie, and Lord Hamilton and Andrew Kerr of Attonburn. The king had to pass through hostile terrain with an escort which included men whose loyalty to him was new and untried. The royal entourage did not linger in Clydesdale but the new earl of Douglas did not allow the king to withdraw in peace. Perhaps in pursuit of the king, Douglas led a band of 600 men in an immediate counter-blow. On 17 March they burst into the burgh of Stirling, where the royal household had been a few days before, plundering and burning the town. The raid was no empty gesture, but employed the family's traditional methods to spread fear and demonstrate the continued power of the Black Douglases. His ability to raise and lead 600 'gud men' to devastate one of the centres of royal power showed Earl James's defiance of the king and his ability to carry on the rights and role of his dynasty. By contrast the king's absence from Stirling did not reduce the effectiveness of Douglas's gesture but showed royal inability to defend the burgh. Though later stories of planned royal flight were exaggerations, the king had been thrown onto the defensive. The murder of Earl William brought no easy victory, and it would be four months before the king would return to the Douglas heartlands.[20]

King James used this time to strengthen his shaky position in Scotland and his reputation in Europe. On 12 April he wrote to the French king

informing him of Douglas's death and seeking Charles VII's confidence
and support. Within Scotland the king turned his attention to Douglas
lands and influence in West Lothian, probably the base for the attack on
Stirling. In early April the king wheeled out his father's guns to lay siege,
not to the Douglas stronghold of Abercorn, but to an easier target, the
tower house of William Lauder at nearby Hatton. The house was taken
before Douglas, who was only thirty miles away at Bothwell, could intervene. Lauder was killed or executed in a success which signalled recovering
royal fortunes. The king sought to exploit his success to secure new defectors
to his camp through the rapid distribution of the dead man's lands amongst
former Douglas adherents like William Cranstoun, John and Andrew
Rutherford and James, brother of Andrew Kerr. Over the next few months
the king continued to employ his powers of patronage, to extend or confirm
lands and rights, to strengthen his support. Among the beneficiaries were
old Douglas partisans, Murray of Cockpool and Newton of Dalcove, and
key royal supporters, Angus, Bishop Turnbull and Douglas of Cavers. The
process reached its peak at the parliament which met at Edinburgh in June,
where the 'kingis secret counsall' gave 'sundry landis' to 'sundry men', who
included the Humes and James Kerr, and created new lords of parliament,
among them the Douglas brothers' enemy Hepburn as lord Hailes and
their brother-in-law as lord Fleming. The Crichtons received the greatest
rewards. George Crichton was made earl of Caithness, and James, the
chancellor's son, was recognised at last as earl of Moray.[21]

The gift of the two northern earldoms reflected the improved fortunes
of the king and his allies in the region. In March James II had been reduced
to negotiating with the earl of Crawford, now openly in conflict with Huntly,
in an effort to prevent his support for the Douglases. However, in May
Huntly defeated Crawford in battle at Brechin. The fight was hard and
Archibald Douglas, earl of Moray, may have used the opportunity to
plunder Huntly's Aberdeenshire lands, but after Brechin, the king felt able
to forfeit Crawford and grant Moray to James Crichton, Huntly's brother-
in-law. The power to enforce both judgements was not, however, possessed
by the king in June. King James was preparing instead for renewed conflict
with Earl James Douglas. His efforts were not limited to material patronage.
It was equally important for him to recover his full rights and status as
king both of which were still clouded by his murder of a subject under his
protection. To this end parliament formally absolved the king of his guilt.
The estates justified the murder by claiming that Douglas had renounced
royal protection the day before his death, that he had entered into private
bonds, had 'perpetrated oppressions' and had refused to help the king
against rebels. Though the earl's deeds were reported with some accuracy,

the key to the king's absolution was Douglas's unlikely renunciation of royal protection. Truth was less important to an assembly dominated by the king's allies than a formal statement allowing James II to treat the Douglases as rebels and enemies. But, despite this exoneration, the king was not able to proceed at once to the forfeiture of the murdered earl's kinsmen. The king's authority remained weakened by his own murderous actions.[22]

Earl James did his utmost to declare publicly that these actions put the king beyond the pale. During his attack on Stirling, Douglas 'blew out twenty-four hornis attanis apon the king ... for the foule slauchter of his brother'. He had the king's protection dragged through the town behind a horse and spoke 'rycht slanderfully of the king'. In June, a letter bearing the seals of Douglas, Ormond and Hamilton was nailed by night to the door of the house in which parliament was sitting. In it the Douglases 'declynand fra the king sayand that thai held nocht of him nor wald nocht hald with him with mony uther sclanderous wordis calland thaim traitours that war his secret counsall'. Such statements were designed to justify resistance to a king who was identified as unjust and surrounded by traitors. As his father had been killed following just such defiance, James II could hardly regard the charges as having no force. The formal renunciation of bonds of lordship, that Douglas 'held nocht of him', justified his dealing with other enemies of the king and even seeking another lord. Like his brother the previous year Douglas looked to England as a refuge and a source of support. In early June safe conducts were issued allowing access to England for the earl's mother, Countess Beatrice, and her widowed daughter-in-law, the maid of Galloway, whose custody was vital for the Douglases' efforts to retain her lands. At the same time, Douglas servants carried gold and belongings into England by ship. The exodus secured the family's assets in preparation for renewed conflict. The English government, fresh from crushing the rebellion of the duke of York, was led to hope for even closer links with the Douglases. A commission led by the march wardens, Northumberland and Salisbury, was appointed to negotiate with the earl of Douglas for his entry into English allegiance. Mistreated by the Scottish king, the Douglases threatened to withdraw from his service and seek the English king's protection, an act which would turn their defiance into war.[23]

King James was ready to treat the Douglases as enemies of the realm. Despite the lack of any formal proceedings against the family, the parliament of June may have been aware of Earl James's contact with England. The king may have used this to justify the renewed confiscation of Stewarton and Wigtown from the family and to issue a general summons to the royal host, which would assemble on Pentland Muir south of Edinburgh in July.

The king acted as the community's war leader in raising the host for the first time since his father's siege of Roxburgh in 1436. Like his father, he was able to raise an impressive force, numbered by the Auchinleck chronicle at 30,000 men, but precedent also suggested that it was an army whose size and political complexity were not ideal for what remained an internal conflict against fellow Scots. So it proved. The army crossed the Pentlands to Peebles, moved down the Tweed to Selkirk, then through the Forest to Corhead in Moffatdale, and finally down Annandale to Dumfries. The route was chosen to inflict maximum damage on Douglas lands and allies in the marches, but the host proved too crude for what was a political campaign. The king's army 'did na gud bot distroyit the cuntre richt fellonly and heriit mony ... that war with himself'. By employing a large force from beyond the south, the king inflicted indiscriminate damage in the marches and undid much of his careful wooing of local men. Against a strategy reminiscent of that used by English kings since the 1290s, Douglas responded, after his family's custom, by maintaining a small but effective retinue and striking elsewhere. It was probably at this point that the earl plundered his own enemies in Galloway, where, once again, there was hostility to the sons of James the Gross and support for the king, perhaps encouraged by the efforts of Gilbert Kennedy, who was now firmly in the king's camp. However, Douglas remained active, at large and in arms. His network of major castles, from Abercorn on the Forth, through Bothwell and Douglas in Clydesdale to Newark in the Forest and Threave in Galloway, was untaken, his links with England were unbroken. As the royal host dispersed after a fortnight-long campaign, the king and his council were left to ponder their inability to defeat the Douglases or force their submission.[24]

On his return to Edinburgh the king probably already recognised the need to seek another settlement with Douglas. After the violence of the previous months such a reconciliation was more difficult, and it was only on 28 August that it was finally agreed. Earl James and Lord Hamilton put their names to an 'appoyntement' between them and the king at Douglas. In it the earl agreed to relinquish his claims to Stewarton and only to seek to recover Wigtown with the permission of Queen Marie, who had received the earldom from her husband. He also released the king from the damages done to his own lands and promised to repair the damage he had committed in Galloway. He would renounce all bonds 'contrare' the king, keep order in the marches, not seek to dispossess his tenants and, finally, forgive the king for the murder of Earl William. Though Douglas recognised the issues which had led to conflict between his brother and the king and promised not to repeat them, the document did not mark the victory which James had sought in July. After months of military and politcal conflict, peace took the form of Douglas's promise to offer better lordship and to forgive the king for 'arte and parte of the slaughter' of Earl William. Douglas, who had withdrawn his allegiance from the king and spoken strongly and publicly against him, was received back into royal peace without major sanctions. Despite his exoneration in parliament, it was the king who appeared to be freed from the stigma of the year's events. Once again the crown had been defied by the Douglas earl and it would be King James who, again, had to make concessions to maintain the earl's loyalty.[25]

CONSPIRACY, TREASON AND REBELLION

James II had killed Earl William over the place of private leagues and bonds in his kingdom. He was now forced into such bonds himself. The 'appoyntement' of August 1452 was followed, five months later, by a second agreement between the king and the Douglas earl at Lanark. The Lanark bond made even more apparent the king's inability to use his public authority to extract obedience from his greatest southern subject. Instead James II sought to rebuild relations with Douglas in terms of private lordship, with the latter entering into bond of manrent, of personal service, to the king. James II would be forced to pay a high price for even this relationship. In the 'appoyntement', Douglas had agreed to the loss of Stewarton and to recognise the queen's prior claim to the earldom of Wigtown. In the new bond the king promised to restore the earl to both estates and assist Douglas's efforts to obtain papal permission for his marriage to his brother's widow, Margaret of Galloway. The marriage, which took place in March 1453, renewed the link between the sons of James the Gross and Galloway, and must have strengthened the earl's hand.[26]

As in 1451, the second royal attack on the Black Douglases had ended with James II relinquishing his hard-won advantages. The Lanark bond was a capitulation which could only be the product of the king's weakness. He had used the powers of the crown to seek a decisive victory over Douglas but had failed, not just due to the resistance of his enemies, but also because of his inability to maintain support from the rest of the community. During the spring and summer of 1452 the king had gained the backing of men who anticipated a quick royal victory allowing them to enjoy the fruits of his generous patronage. These men had been prepared to exonerate the king for Earl William's murder, although they had not backed any call for his family's forfeiture. The July campaign led to a quick unravelling of the king's position. His use of the host antagonised the borderers without forcing

Douglas to submit. According to the Auchinleck chronicle, with the earl at large, 'men demyt' that the king's grants of land 'wald nocht stand'. In the case of the largest displays of patronage, the grants of Wigtown to the queen, of lands in Stewarton to Gilbert Kennedy, of Moray to James Crichton and of lands in Galloway to George Crichton, such fears proved accurate. All were reversed in the king's capitulation. Like David II in the 1340s and Robert II in the 1370s, James II experienced the difficulties of challenging the Douglases directly. Whether or not he remained determined to destroy Black Douglas power in the long term, James II was forced to avoid immediate conflict and sought to restore ties of lordship with the earl by any means. It was to display renewed trust in Earl James Douglas that the king gave him control of negotiations for a fresh truce with England. Though the king's steward, Robert Liddale, was associated with Douglas, it was the earl, accompanied by his brothers and a large retinue, who led the embassy. His role bound him to the terms of any truce and showed him in King James's service, but given the recent dealings of his family with England and the Douglas dynasty's long-standing influence in relations between the two kingdoms, the earl's prominence emphasised his continued power in the marches.[27]

Yet Douglas power in the south did not emerge unscathed from two years of conflict with the king. Twice King James had progressed across the south seeking and rewarding adherents from the Douglas lordships. Twice he had launched attacks on the earl's tenants and allies which had caused indiscriminate damage. Twice he had abandoned his supporters to the resurgent earl's hostility. The result was an atmosphere of instability even in marcher communities used to war and shifting loyalties. However, this uncertainty was clearest in the further draining away of support from the house of Douglas. Among those who severed links with the Douglas earls were rising barons like lords Somerville, Fleming and Haliburton and Alexander Hume, and ambitious churchmen such as James Lindsay of Covington, provost of Lincluden, men who saw royal service and the fall of the Douglases as in their interests. For example, Lindsay of Covington, the secretary of Earl William Douglas and successively rector of Douglas and provost of Lincluden through his master's patronage, followed the example of John Cameron to become a regular royal councillor from late 1452, and keeper of the privy seal. More worrying, though, was the way the family's border tenants had sought the king's peace. While the king's committed adherents in the marches, George earl of Angus, Scott of Buccleuch and Douglas of Cavers remained fixed in their attitudes, the Pringles, Rutherfords and others had not backed Earl James in 1452 as they had supported his brother the previous year, but had submitted to the king. In Galloway this collapse of lordship may have been striking, with even the Douglases' justiciar, Alexander Mure, attending on James II and the Crichtons in March 1452. With the exception of Andrew Kerr, Earl James did not receive major support from his family's traditional heartlands. Instead, though he retained a retinue of minor men from across the south, Douglas relied on his father's connection from Clydesdale and Ayrshire, James Hamilton, John Auchinleck and John Dunbar of Cumnock, and used Clydesdale as his principal base for raids on Stirling, Galloway and the marches. The differing responses of men in Galloway and the borders and those in Clydesdale to their Douglas lords during 1452 suggests lingering antipathy to the usurping line from old adherents of the dynasty.[28]

These attitudes also suggest that the king's campaigning, which centred on the far south, had an effect on local loyalties. Landowners in these regions were, like their ancestors during major Anglo-Scottish conflict, principally motivated by a desire to escape unscathed from the conflict between their two principal lords. The submissions to the king in March 1452 were the reactions of men who believed Earl William's murder spelt the end of Douglas dominance in the south. Earl James's counter-attack and the king's devastation of the south may have caused new shifts in local allegiances. The result of the struggle was confusion. Families had been divided. While Andrew Kerr had stood alongside Douglas, his brother, James, had been rewarded for backing the king. While John lord Haliburton abandoned Earl William in 1450, a number of his kin remained household servants of the Douglases. Such examples were probably not unusual, and with the reconciliation of their masters, which many must have feared was purely temporary, efforts were made to safeguard interests in future con-flicts. In June 1453 Andrew Kerr concluded an indenture with his Teviotdale neighbour, Robert Colville of Oxnam, a confirmed enemy of Douglas. They promised mutual loyalty, to settle 'discord' amongst their men and to give each other help in quarrels, 'thare legeans to the kyng and to the erle of Douglas alanerly outane'. Such a settlement probably ended a local conflict arising from the warfare of 1452, when Kerr's estates may have been attacked during his absence with Douglas. It also prevented new violence should their lords resort to arms once more, and sought to ensure the security of both men in a period of political unrest.[29]

Earl James appreciated the need to rebuild his own lordship in the south after the disruption of his brother's death and royal attack. He had, of necessity, to achieve this by working with men who, willingly or unwillingly, had deserted the Douglases in 1452. Unlike his brother, Earl James was in no position to terrorise these enemies. Even his brother's killer, Simon Glendinning, was restored to a place on the earl's council. Such a reconciliation was essential for Douglas's recovery of influence in Eskdale and for Glendinning's safety, but showed the extent to which the earl was forced to keep his promise not to persecute his tenants. Earl James also ended his brother's opposition to Douglas of Drumlanrig, who had invoked royal backing for his claim to Hawick in June 1452. In October, the earl recognised Drumlanrig's rights, ending a dispute which had rumbled on for two years. However, it was in Galloway that Douglas expended most effort. Between November 1452 and October 1453 he was in the province on three occasions in what was probably an attempt to repair his relationship with what had been the principal power-base of Archibald the Grim. Though Earl James proved unable, in two years, to repair the damage done to his family's affinity by thirty years of internal feuding and royal pressure, his efforts suggest that he was aware that his dependence on a few key adherents, the most valuable of whom was James Hamilton, made his position vulnerable.[30]

Douglas's actions beyond his own lordships suggest that he anticipated just such a renewal of conflict with his king. Given his brother's fate, he avoided the royal court. Instead during 1453 and 1454 his aim seems to have been to keep the king on the defensive by maintaining connections in Scotland and beyond which had hampered the earlier royal attacks. The most important of these alliances was with England. When Earl James and his brothers went south in May 1453, his contacts with the English king and court seemed a sure means of restraining the Scottish king. The threat of English garrisons in the Forest, Galloway and perhaps even Clydesdale naturally made James II cautious, while with the bloodless defeat of York and the successful renewal of the war with France in Gascony, Henry VI seemed to have ridden out the political storm which had beset his regime since 1450. Douglas was able to exploit his link with England to political advantage, negotiating the release of Malise Graham, earl of Menteith, in May 1453. The act was a favour to James Hamilton, Malise's brother-in-law, but was also aimed at James II. Malise had been in England since 1427 as a hostage for the unpaid ransom of James I. While Scottish kings left him to rot, Douglas secured his liberty and demonstrated the positive value of his links with England. However, in the summer of 1453 the earl's confidence in England was shaken by twin events, the destruction of the English army at Castillon and Henry VI's descent into madness. Though one of the first acts of the duke of York when he took up his formal powers as lieutenant in the following February was to send a herald to the marches with 'certain appointments' for Douglas, Earl James could no longer rely on the threat of English aid to keep the king at bay. Indeed, in May 1454 King James's help was sought by English magnates, the Percies and their allies, in their feud with York and the Nevilles. The recovery of Henry VI at the end of the year only intensified animosities. In early 1455 Douglas was to find himself without the prospect of English military backing.[31]

At the same time the earl would also find himself without active support from the magnates of the north and west of Scotland. In 1451 and 1452 the hostility of John earl of Ross, Alexander earl of Crawford and Douglas's brother, Archibald earl of Moray, towards the king prevented James II from concentrating solely on the south. The king's compromise with Douglas forced him to reverse the sentence of forfeiture passed on Crawford, to encourage Huntly to seek a local compromise with his enemies and to accept that Archibald, not James Crichton, held the earldom of Moray. In the struggle for dominance north of Tay, as in the south, determined opposition made the king's party abandon their gains. The tacit alliance of the Douglas earls with Ross and Crawford persisted, despite the king's efforts to break it, and safeguarded the Douglases' interests throughout the kingdom. However, this apparent security was shattered in the summer of 1453 by the sudden death of Alexander earl of Crawford. Crawford was described as 'ane felloun [who] held ane gret rowme in his tyme for he held all Angus in his bandoun and was richt inobedient to the king'. His heir was a child, and James II acted immediately to ensure royal rights of wardship and royal influence in the earl's 'rowme'. Within days of the news he was in Angus before moving on to Aberdeenshire, probably meeting his lieutenant, Huntly, at Kildrummy. Crawford's death removed Huntly's chief rival in the north-east and allowed him to concentrate his efforts on the lands to the west of Badenoch and Strathbogie, the Douglas lordships of Balvenie, Moray and Ormond.[32]

Crawford's death made the friendship of John earl of Ross and lord of the Isles vital for Earl James. The Douglases and Clan Donald were natural rivals in the north and the alliance between them, the product of local compromise, persisted because of a common mistrust of the king. To strengthen this bond Earl James travelled to Knapdale, at the southern extent of John's massive Hebridean realm, in May 1454. He 'spak thar with the erll of Ross and lord of Ilis and maid thaim all richt gret rewardis of wyne, clathis, silver, silk and English cloth and thai gaf thaim mantillis agane'. The exchange of gifts signalled the different cultural contacts of the two lords. Douglas gave his host goods from his connections with England and with continental trade, and received mantles, the most famous export of Gaelic Ireland. The discussions which followed marked a new phase in relations between the leading magnate dynasties of Anglicised and Gaelicised Scotland, in which Douglas sought to harness the military power of the Islesmen against his enemies. Central to his efforts was Donald

Balloch, lord of Dunivaig on Islay and the Glens of Antrim, Ross's cousin
and a man with a record of hostility to the Scottish crown. Douglas clearly
made an impression. Two months later Donald led a fleet of twenty-five
galleys on a raid up the Firth of Clyde which sacked Inverkip, Arran and
the Cumbraes, all lands of the king. With him was John Douglas, bastard
son of the fourth earl, whose presence suggests that Earl James supported
Donald's actions. An alliance with Donald Balloch did not guarantee the
continued support of Ross, and the latter may have used the raid to
strengthen his own position in negotiations with the king in late 1454. By
the following year a deal had been struck. Ross would withhold support
from Douglas in return for royal recognition of his possession of Urquhart
and a share of the northern lands of his erstwhile ally. In the spring of
1455 Ross was again in Knapdale. But his purpose this time was to restrain,
not encourage, the aggression of Donald Balloch. In 1455, no 'band or
ligue' would save Douglas, and Moray would take his place in the south
with his twin brother, leaving his earldom to the attentions of Huntly.[33]

Earl James's dealings with the Islesmen were a result of his isolation,
but, as Auchinleck reported, were 'demyt ill all' (disapproved of), and
provoked a hostile response from the 'comounis', the community. In actively
seeking allies amongst those perceived as enemies of the realm, Douglas
must have alienated sympathisers in the rest of the kingdom. However, his
support of the raid on the Clyde coast was not purely destructive, but
formed an attempt to expose and increase the king's difficulties during the
summer of 1454. These difficulties came not from his enemies, but from
his treatment of the family at the heart of his own 'secret council', the
Crichtons. Chancellor Crichton and his kin had influenced the timing and
direction of royal attacks on the Douglases, but the failure of these attacks
may have eroded the king's trust. The death of William Crichton in late
1453 ended this special relationship with the king. In May 1454 George
Crichton was made to name the king as his heir and resign many of his
estates. Once again, James II underestimated his opponents. Within a week,
George's disinherited son, James, had seized his father's castle of Blackness
on the Forth and was holding it against the king. The danger this represented was clear. Blackness lay close to the Douglas castles of Inveravon
and Abercorn, and Crichton may have used the threat of Douglas support
to force the king to buy him off. If so, the incident showed the continued
restrictions placed on James II by Douglas's existence and, followed by the
raid on royal lands in the west, served to maintain the king's anxiety.[34]

The sudden deaths of both James lord Crichton, the son of the chancellor,
and George Crichton in August 1454 did nothing to reduce this anxiety.
Though ending the family's influence, allowing the king to rely on a

broader-based council, the deaths also created new problems. Appropriately, the fall of the Crichtons would be a prelude to the fall of their rivals, the line of James the Gross. The deaths of the Crichtons had greatest impact in Annandale, where the family had sought to exercise influence since the Black Dinner. As recently as March 1453 George Crichton had secured a marriage alliance between his daughter and John, son and heir of Robert lord Maxwell, and with another ally, John Carruthers, in possession of Lochmaben, the Crichtons had considerable local influence which suddenly evaporated in August. Both the king and the Douglas earl would have had designs on this influence, while members of the local community, divided during the 1440s by a series of feuds, also sought new advantages. In what royal accounts described as 'the disorder of the country' of Annandale in 1454, 'the lard of Johnston's twa sonis tuk the castall of Lochmaben [from] carudderis ... throu treasson of the portar'. The Johnstones acted in the month after the Crichtons' deaths in a deliberate attempt by local men to overthrow the family's adherents. Any such effort would have Douglas's support and posed a danger to the king.

James II reacted with speed. He dispatched his justiciar, lord Abernethy, to Annandale to hold court. The exercise was no normal administrative task but a military and political expedition for which Abernethy later received £1,100. In particular, Abernethy secured access to Lochmaben by recognising the Johnstones' control, but it was only in early 1455, when Maxwell came to Edinburgh for confirmation of his stewardship, that James could be sure of his authority in Annandale. In the coming months the local support secured by Abernethy would prove vital.[35] The death of George Crichton had implications for the earl of Douglas well beyond Annandale. It ended the Crichtons' efforts since 1441 to obtain control over the estates of the Douglas lord of Dalkeith in the south-west and Lothian, the cause of conflict with James the Gross and his sons. However, George's death did not clear the field for Earl James's brother-in-law, Henry Douglas, who still controlled the barony of Dalkeith. Instead it is likely that during late 1454 the king sought direct control of the Dalkeith estates. In coming years the king actively backed the rights of the direct heir, James, to his whole inheritance, which his father had been deemed unfit to hold. In 1454 the heir was under age, allowing the king to act as his guardian. The attempt to claim a role which would dispossess Henry Douglas in Lothian and would justify the king's direct control of his ward's lands in Galloway, Eskdale and Nithsdale prepared the way for new conflict with Earl James.[36]

The opening shots of this new and decisive clash between the king and the Douglas earl were fired in Lothian. Probably in response to royal action, Earl James launched a series of attacks on the lord of Dalkeith's West

Lothian manors, Kingscavil, Bondington and Colden, burned Dalkeith itself and plundered the lands of lord Abernethy, the king's justiciar. The king countered by casting down the earl's castle of Inveravon. Given the simmering hostility of previous years, all involved must have understood that this dispute in Lothian, one of a number since 1440, would automatically grow into full-scale conflict. Earl James prepared the defences which had allowed his survival in 1452. He gathered his allies at Peebles, sent lord Hamilton to seek English aid and garrisoned and supplied his strongholds of Abercorn, Douglas, Strathaven and Threave. However, the king had learned from his earlier failure. Instead of levying the host for a single, massive attack, from March to July the king maintained constant pressure on his enemies, hounding them into defeat. He began in Clydesdale, Douglas's base in 1452. Moving to Glasgow, the king 'gaderit the westland men with part of the Ereschery', a force probably drawn from Ayrshire and Argyll, led by royal allies like Gilbert Kennedy, Duncan, lord Campbell and Stewart of Lorn, all of whom had extensive military retinues drawn from their Gaelic adherents. These western communities were those most threatened by an alliance between Douglas and Donald Balloch, and supported the king in revenge for the latter's raid. With this army the king moved down Clydesdale to Lanark before launching attacks which targeted Douglasdale and Strathaven and lord Hamilton's lands. Dispersing this force, James II moved to Edinburgh, where he levied 'ane host of lowland men'. At the head of this second force King James entered the middle march, making for the Forest where he plundered the lands of 'all that wald nocht cum till him'. By the end of March the king was back in Lothian, deploying his guns for the siege of Abercorn.[37]

This was the crisis point of the campaign. The capture of Abercorn would mark an irreversible royal victory, sealing the expulsion of Douglas from Lothian. But it was also the point at which the earl was at last free from royal attacks further south and able to retaliate. Though Abercorn held out, Douglas did not come to save it. The earl had not been inactive. He had gathered 'a great many, hostile, armed men in array of war' near Lanark to oppose the king's earlier campaign, and while he did not risk battle with the powerful royal army, it may have been at this point that his partisans in Galloway scored a major success by killing the king's sheriff, Andrew Agnew, and burning his lands. However, the English help on which Earl James had counted would not come. In March England was already moving towards political violence in which the northern magnates and their affinities would play a central role. Hamilton could get no help beyond vague promises of support if he and Douglas took an oath to become 'Inglis men'. Even before this the earl's party was crumbling away.

Andrew Kerr, who had been with Douglas in February, met Abernethy, Scott and Douglas of Cavers at Jedburgh to arrange his temporary withdrawal from the struggle, while James Tweedie, who had probably been in the earl's force at Lanark, submitted directly to the king in the burgh on 7 March, entering a bond with James II which recorded his surrender of himself and his tower of Drumelzier to the king. Force and the threat of force had brought about a collapse of Douglas support, which was completed when Hamilton appeared at the royal camp at Abercorn to 'put him lyf land and gudis in the kingis will'. Hamilton and the others recognised that time was running out to secure their survival in the event of royal victory, and switched sides when they could still count on royal gratitude. Meanwhile, untroubled by Douglas, the royal guns pounded Abercorn, bringing down many towers until, on St. George's Day, the castle was taken by storm. Despite their lord's failure to relieve them, the Douglas garrison had held out for a month while others surrendered. Their reward was the king's hatred, and in a letter to Charles VII, King James reported that he had hanged the castle's chief defenders.[38]

The loss of Hamilton 'left the erll of Douglas all begylit' and convinced him that, despite the political crisis in England, his only hope lay in seeking help from this last ally in person. Though his absence was seen as the desertion of his Scottish adherents, Douglas's efforts may have yielded results. During April the paid English garrisons of Carlisle and Roxburgh may have been put at Earl James's disposal. The former acted as a refuge for his wife and mother, while the latter may have given support to the family's continued resistance in the middle march. While Douglas went south, his brothers, Moray, Ormond, and Balvenie, sought to maintain an active presence and may once again have secured the help of Andrew Kerr in Teviotdale. In 1456 Kerr would be accused of 'in bryngyng of the Inglismen' into Jedburgh, Eckford, Crailing and Grahamslaw, harrying the 'kingis legis' and supplying and receiving 'commun traturis and haffand art and part' of their deeds. These acts may have occurred in April 1455. The lands attacked lay on the route between England and Roxburgh, and belonged to key enemies of the Douglases, Angus, Scott, and Cranstoun. Though it failed to save Abercorn, the raid showed the continued power of the family to hurt its enemies, and the brothers may have moved on to Eskdale to repeat their feat with the men of the lordship. Here their luck ran out. They were caught by a force of men from Annandale to the west and from Teviotdale to the east, led by their family's vassals, Lord Maxwell, the Johnstones and Walter Scott. These local enemies crushed the Douglas company at Erkinholme near Langholm on 1 May. Moray was killed, Ormond captured, and only Balvenie escaped with news of the defeat.[39]

The disaster at Erkinholme ended the independent resistance of the Douglases. With Moray's death the northern lordships of the family were also lost. Huntly moved in to occupy the castles of Darnaway and Lochindorb and the baronies of Balvenie and Boharm, and sought to secure his own rights to the earldom of Moray by marrying Douglas's recently widowed countess. To the west, Ormond's lands were seized by Gillespic, brother of John earl of Ross. The king too was thinking of the spoils of victory. In late April, after two months of conflict, James II sought legal sanction for his treatment of the Douglases, having them called before parliament to answer charges of treason. When the estates met in early June, a long list of charges, of conspiracy, rebellion and treason, of confederacy with England and of violent plundering, was read against the absent Douglas family and a judgement of forfeiture was passed against Earl James, Balvenie and their mother, Countess Beatrice. Moray, dead at Erkinholme, and Ormond, imprisoned only long enough for trial and execution, were not spared sentence. The verdict, given by six earls and fourteen lords, was a mark of the king's political as well as military success.[40]

Only Threave continued to defy the king. On its island in the Dee, equipped with its own guns and newly strengthened and at a distance from the royal artillery train, the castle could be expected to resist through much of the summer. In mid-June James II suspended parliament and began the process of moving the royal cannon south. To secure help from the new government of the duke of York, Douglas agreed to surrender Threave to the English, receiving £100 for the 'succour, victualling, relief and rescue' of the castle in mid-July. By then it was too late. Threave had fallen, not to royal guns but royal gold. Alarmed by the fate of Abercorn's defenders and sceptical of their lord's ability to save them, the garrison surrendered in return for generous payments. Douglas had lost his last Scottish stronghold. Already many more of his servants, even from his own household, had made their peace with the king. As James II turned from the capture of Threave to take the role of war leader against the English, Earl James, now an 'Inglis man', began a long career as an exile and enemy of Scotland.[41]

NOTES

1. *Buke of the Howlat*, 60, lines 378–79; *Calendar of Papal Registers*, x (1449–51), 96.
2. Douglas may have been lieutenant of the kingdom during late 1443 and early 1444, though there is no firm evidence of this, but he cannot have held the office once James II had been declared of age.
3. *Scotichronicon*, ed. Watt, viii, 262–63.
4. 'Auchinleck Chron.', in C. McGladdery, *James II*, 171–72; *R.M.S.*, ii, nos 291–92, 297–98, 301–305, 308–12, 316–19, 321–28, 331–99, 345, 508; *A.P.S.*, ii, 61, 62; J. Durkan, 'William

Turnbull, Bishop of Glasgow', in *Innes Review*, 2 (1951), 1–59, 24–25. Railston had been rector of Douglas and secretary of the fifth earl of Douglas while Spens was secretary of his son. Both were on the sixth earl's council in the year of the Black Dinner (Fraser, *Douglas*, iii, no. 303, 404; *H.M.C.*, Drumlanrig, i, no. 112).

5. *R.M.S.*, ii, nos 316–17, 357; 'Auchinleck', in C. McGladdery, *James II*, 172. Crichton was restored to the chancellorship in 1448 and, in the same year, carried letters pressing the claims of Douglas and Duchess Margaret to the duchy of Touraine (Bibliothèque Nationale, Ms Latin, no. 10187; *E.R.*, v, 336). Turnbull also served on the royal council throughout Douglas's dominance, suggesting his hostility to the earl only became open in 1450.

6. *R.M.S.*, ii, nos 292, 301, 308, 309, 315.

7. *E.R.*, v, 439; *Rot. Scot.*, ii, 346; Dunlop, *Bishop Kennedy*, 124 n. 3; *C. D. S*, iv, no. 1231; *R.M.S.*, ii, no. 360; S.R.O. GD 16/46/3; Griffiths, *The Reign of Henry the Sixth*, 515–22, 529–30; Beaucourt, *Charles VII*, v, 134. An account of events in Scotland in late 1450 and early 1451 is provided by the early sixteenth-century chronicle of John Law who had access to the Auchinleck chronicle and may have been drawing on a missing section of that work (C. McGladdery, *James II*, 128–30; *E.R.*, v, lxxxv–vi).

8. *E.R.*, v, lxxxv–vi; *R.M.S.*, ii, no. 404. Walter Scott and his brother Stephen remained allies of the Crichtons throughout the 1440s and both received royal charters and new lands in 1450–51. Cavers, for his part, had established links with Bishop Turnbull (*R.M.S.*, ii, nos 353, 356, 419, 423–24; Fraser, *Scotts of Buccleuch*, ii, nos 21, 33, 36, 41, 46, 47; Durkan, 'Bishop Turnbull', 13).

9. *E.R.*, v, lxxxv–vi, 521; *R.M.S.*, ii, nos 412–17. Duchess Margaret was still alive in early 1450 but dead by 1452. In early 1450 William Crichton granted lands in Annandale to the local man, Gilbert Corry, and his wife, the daughter of John Carruthers, keeper of Lochmaben castle since 1446, and in May he granted lands in the lordship to Douglas's companion, William Cranstoun (*R.M.S.*, nos 309, 319; *H.M.C.*, Drumlanrig, i, no. 80; *Laing*, no. 129).

10. *R.M.S.*, ii, nos 412–16, 436–37; S.R.O. GD 25/35, 37, 52. Somerville remained a member of the king's council which was involved in attacks on Douglas's position in 1450–51 (*R.M.S.*, ii, nos 403, 406, 417).

11. *A.P.S.*, vii, 143; *E.R.*, v, lxxxv–vi; *Rot. Scot.*, ii, 345–46; 'Auchinleck Chron.', in C. McGladdery, *James II*, 164–65; *H.M.C.*, xii, appendix 8, no. 201. At Jedburgh Earl William and James Douglas issued grants in favour of Alexander Hume. James Douglas cannot have gone to England before early May. Kerr's continued support of Douglas occurred despite a royal charter to him in late February, suggesting that he may have returned ahead of the earl (*R.M.S.*, ii, no. 422).

12. *R.M.S.*, ii, nos 447, 454–57; S.R.O., AD 1/53; 'Auchinleck Chron.', in C. McGladdery, *James II*, 164–65. In late June royal charters were issued to Alexander Hume and William Cranstoun, who had recently been with Douglas. In July Hume received another charter from the king which confirmed him in his lands in Stewarton, granted to him by Douglas in 1444 (Fraser, *Douglas*, no. 412; *R.M.S.*, ii, no. 484–85).

13. 'Auchinleck Chron.', in McGladdery, *James II*, 163, 169; *R.M.S.*, ii, nos 442, 447; A. Grant, 'The Revolt of the Lord of the Isles and the Death of the Earl of Douglas 1451–52', *S.H.R.*, 60 (1981), 169–74.

14. *R.M.S.*, ii, nos 463–82, 502–504. This refusal to give help was one of the king's complaints against Earl William in the following June (*A.P.S.*, ii, 73).

15. Pitscottie, *Historie*, i, 88–92; Godscroft, *Douglas and Angus*, i, 186; S.R.O., AD 1/53; *P.P.C.*, vi, 125–26. For analysis of stories concerning Earl William, see N. Macdougall, *James III* (Edinburgh, 1982), 18–20; McGladdery, *James II*, 137–44.

16. *R.M.S.*, ii, nos 490–92, 502, 507, 512, 514, 522–23; *Wigtownshire Chrs.*, no. 136; 'Auchinleck Chron.', in McGladdery, *James II*, 165.

17. *A.P.S.*, ii, 7, 73; M. Brown, *James I*, 24–31; R. Vaughan, *John the Fearless*, 263–81.

18. William Cranstoun accompanied Douglas to Rome in 1450 and his father remained

amongst the earl's councillors, while Glendinning, though not associated with Douglas in 1450–51, had been his regular companion up until 1449–50. Both men were among Douglas's local officials (*Rot. Scot.*, ii, 343; N.R.A.S., no. 832, Lauderdale Muniments, no. 77; British Library, Harleian MS, no. 6443, 19r–20v; Fraser, *Douglas*, iii, nos 306, 415, 419; *H.M.C.*, 12th report, appendix 8, no. 201; *Melrose Liber*, no. 564; Fraser, *Scotts of Buccleuch*, ii, no. 38).

19. *R.M.S.*, ii, nos 529–32; S.R.O., GD 89/10; GD 157/75; *Wigtownshire Chrs*, no. 138; Fraser, *Scotts of Buccleuch*, ii, no. 49. For a full itinerary, see A. Borthwick, 'The Council under James II: 1437–60' (unpublished Ph.D thesis, University of Edinburgh, 1989).

20. S.R.O., GD 25/55; Fraser, *Scotts of Buccleuch*, ii, no. 49; 'Auchinleck Chron.', in C. McGladdery, *James II*, 165–66; N. Macdougall, *James III*, 23–24.

21. *Bibliothèque Nationale*, Ms Latin, no. 10187; Stevenson, *Wars of the English in France*, ii, 315; *Accounts of the Lord High Treasurer of Scotland*, eds T. Dickson and J. B. Paul (Edinburgh 1877–1916), i, ccxvii; S.R.O. GD 10/14; *R.M.S.*, ii, nos 534–38, 540–43, 545–46, 549–52, 555, 557, 558, 583–84; Fraser, *Scotts of Buccleuch*, ii, no. 50; 'Auchinleck Chron.', in C. McGladdery, *James II*, 166.

22. 'Auchinleck Chron.', in C. McGladdery, *James II*, 166, 173; Godscroft, *Douglas and Angus*, 198; *A.P.S.*, ii, 73.

23. 'Auchinleck Chron.', in C. McGladdery, *James II*, 165–66; *Rot. Scot.*, ii, 357.

24. 'Auchinleck Chron.', in C. McGladdery, *James II*, 166; P. F. Tytler, *History*, ii, 386–87; *R.M.S.*, ii, no. 583; *Laing Chrs.*, no. 134. Kennedy had been given a portion of Stewarton by the king in June, giving him an additional interest in the success of the king's campaign (*R.M.S.*, ii, no. 583).

25. The 'appoyntement' of 1452 is printed in Tytler, *History*, ii, 386–87. In late July the king confirmed Andrew Agnew as sheriff of Wigtown, signifying his determination to retain the earldom at that point (*Wigtownshire Chrs.*, no. 140).

26. Tytler, *History*, ii, 386–87; Fraser, *Douglas*, i, 483–84; *Calendar of Papal Registers*, x, 130–31.

27. 'Auchinleck Chron.', in C. McGladdery, *James II*, 166, *Foedera*, xi, 324–27, 336; *Cal. Docs. Scot.*, iv, no. 1257.

28. S.R.O., GD 97/1/5; GD 150/109; GD 157/75; *R.M.S.*, ii, nos 383, 537–38, 594, 597–98; *Laing Chrs.*, no. 122; *Melrose Liber*, ii, no. 564; Fraser, *Scotts of Buccleuch*, ii, no. 53.

29. *R.M.S.*, ii, no. 535; *H.M.C.*, Roxburghe, no. 5. Colville's lands were not among those attacked with Kerr's help in 1455 (*H.M.C.*, Roxburghe, no. 7). Kerr made a bond with another local man, Thomas Robson, in 1454, perhaps for the same reason (*H.M.C.*, Roxburghe, no. 6).

30. *H.M.C.*, Hamilton, nos 14, 15; *Scotts of Buccleuch*, no. 51; *Scottish Greyfriars*, 102–103, no. 2; *Wigtownshire Chrs.*, no. 141. Ormond and Hamilton were also in Galloway in July 1453 (S.R.O., GD 10/13).

31. *Foedera*, xi, 336; *Cal. Docs. Scot.*, iv, no. 1266; B. Wolffe, *Henry VI* (London, 1981), 251–286; Griffiths, *The Reign of Henry the Sixth: The Exercise of Royal Authority* (London, 1981); R. Griffiths, 'Local Rivalries and National Politics: The Percies, the Nevilles and the Duke of Exeter', in *Speculum*, 43 (1968), 589–632, 605.

32. 'Auchinleck Chron.', in C. McGladdery, *James II*, 163; *E.R.*, v, 653; *Laing Chrs.*, no. 138; *Brechin Reg.*, no. 49; S.R.O., GD 1/220/66; GD 25/1/56; Borthwick, thesis, 446–47.

33. 'Auchinleck Chron.', in C. McGladdery, *James II*, 167–68; K. Nicholls, 'Gaelic Society and Economy', in A. Cosgrove (ed.), *A New History of Ireland, II, Medieval Ireland* (Oxford, 1987), 417; J. and R. W. Munro, *Acts of the Lords of the Isles*, Scottish History Society (Edinburgh, 1986), no. 58. The timing of Douglas's negotiations with Ross and the raid on Inverkip have been variously ascribed to 1452, 1453, 1454 and 1455. However, although the source for these incidents, the 'Auchinleck Chronicle', seems to date them to 1455, when Ross was supporting the king and Douglas was in England, the negotiations and raid are described either side of the king's siege of Blackness, which definitely took place in May–June 1454. As 'Auchinleck' specifically says this occurred 'that samyn moneth and yere' as Douglas's May meeting with Ross, 1454 is the obvious

date for this event and it would be perverse to assign Donald Balloch's raid in 'the saide yeire' to an earlier date. No salary was paid to the keeper of Brodick, destroyed in the raid, after July 1454, supporting this dating ('Auchinleck Chron.', in C. McGladdery, *James II*, 167–68; *E. R*, v, 610, 623, 649; S.R.O. GD 32/20/2; Fraser, *Maxwells of Pollock*, ii, no. 44). Ross's support for the king may also have been influenced by the restoration of his father-in-law, James Livingston.

34. 'Auchinleck Chron.', in C. McGladdery, *James II*, 167–68; *E.R.*, v, 611; S.R.O., GD 32/20/2; *H.M.C.*, vii, Atholl, no. 40.

35. 'Auchinleck Chron.', in C. McGladdery, *James II*, 163; Fraser, *Carlaverock*, ii, no. 40; *E.R.*, v, 632, 670; vi, 26, 63, 447; *H.M.C.*, 15th Report, appendix 9, Johnstone, no. 3.

36. Henry retained rights in Dalkeith in 1451. James II recognised the young James Douglas as lord of Dalkeith in 1456 and created him earl of Morton in 1457. George Crichton's widow, Janet Borthwick, renounced her claims to Morton within eight days of her husband's death, stating that he had compelled her to marry him. Her action probably had royal encouragement and allowed James II to press his ward's rights (*R.M.S.*, ii, no. 515; *A.P.S.*, ii, 78; *Morton Reg.*, ii, 333).

37. *A.P.S.*, ii, 76; 'Auchinleck Chron.', in C. McGladdery, *James II*, 166–67. James II was at Lanark on 8 March and at Peebles on 25 March, suggesting the exact timing of his campaigns (*H.M.C.*, Various Collections, 5, Tweedy, no. 14; S.R.O., GD 150/14; Borthwick, thesis, 446).

38. *A.P.S.*, ii, 76; 'Auchinleck Chron.', in C. McGladdery, *James II*, 167; *H.M.C.*, Various Collections, 5, Tweedy, no. 14; Roxburghe, no. 37; J. Pinkerton, *The History of Scotland* (London, 1797), i, 486–88.

39. 'Auchinleck Chron.', in C. McGladdery, *James II*, 167; J. Law, 'De Cronicis Scottorum Brevia', Edinburgh University Library, DC 763, f. 130; *H.M.C.*, Roxburghe, no. 7; *R.M.S.*, ii, no. 772; Fraser, *Scotts of Buccleuch*, ii, no. 57; *E.R.*, vi, 557.

40. *E.R.*, vi, 68, 265, 376, 466, 518; *Spalding Misc.*, iv, 128–30; *A.P.S.*, ii, 41–42, 76.

41. *E.R.*, vi, 119, 199, 200, 202, 203, 204, 208, 209; *Cal. Docs. Scot.*, iv, no. 1272.

CHAPTER FOURTEEN

Exile and End

'A SORE REBUKED MAN'

The disasters of the spring and summer of 1455 did more than condemn Earl James to life in exile, deprived of castles, lands and followers, sentenced to death as a traitor if taken by the king's men. They also marked the end of a Douglas dynasty able to claim predominance in the middle and western marches, claims which rested on, and were perpetuated by, achievements in war with England. This warleadership had given the family a European reputation which had been revived as recently as 1450 when Earl William progressed through the courts of the west. The king's destruction of the power of the Douglas earls was not simply an internal conflict of king and great magnate. It was the assertion of the Scottish crown's authority on the principal military frontier of the realm and its status in European politics. The dealings of the Douglas family with England, crucial to their defiance of the king, formed part of a charge of *lèse-majesté*, of usurping royal rights, in the earl's forfeiture, while in 1452 and 1455, James II actively sought the support of Charles of France, a king who was well acquainted with the house of Douglas.

With his victory in 1455 the Scottish king launched a war on England as a continuation of the conflict with Earl James. Even while his guns were trundled south to Threave, James II attempted to seize Berwick by surprise after the manner of the Bruce. Later in the year, the 'army of the lord king', perhaps part of the force sent to Galloway, was launched in a raid on the Isle of Man carried by Galwegian ships. Campaigns against two of the three remaining territories still lost to the Plantagenets were designed to demonstrate the king's ability to lead his realm in war against the English enemy. Over the next five years James II hammered the message home. He launched a series of raids, laid formal sieges against Berwick in 1457 and Roxburgh in 1460 and sought military alliances with Charles VII and the kings of Castile and Denmark. Such warfare in the marches was unsurpassed since the 1400s. Though outbreaks of political violence in England in 1455 and from 1459 encouraged James II's aggression, his actions were not simply motivated by English difficulties. For the first time since

Robert I, a king was providing effective personal leadership in war over a number of years. James II was showing the communities of the march, which for much of the previous century and a half had looked to the 'war wall' of the Douglases, that he could protect and lead them.[1]

Ironically King James's aggression was not without benefits for the man whose influence he sought to uproot in the south. The constant threat posed by the Scottish king in the late 1450s made James earl of Douglas a valuable client for successive English governments. Though Threave fell before it could provide Douglas's new lord with a third stronghold in the marches, the earl's services were still worth paying for. In August 1455 'our trusty and welbeloved th'erle of Douglas' was assigned an annuity of £500 sterling for life or until he recovered the bulk of his lands. Douglas received the kind of support given to earlier exiles, men branded as traitors by their Scottish enemies, like Edward Balliol and George Dunbar. In the 1450s and 1460s the courts of Europe were full of such exiles, from the Dauphin Louis, a refugee from his father, Charles VII, in Burgundy, and the Lancastrian royal family, soon to seek sanctuary in Scotland from their Yorkist opponents, to men like the Gascon, lord Duras, who had remained loyal to England after the conquest of his homeland by the French king. Like Duras and like Dunbar, Earl James had a value for the English which was both military and political. The Douglases knew the strengths and weaknesses of the Scots at war, they also retained connections and residual loyalties, especially in the region most open to attack, and the prospect of their restoration was a valuable diplomatic weapon for their hosts. While Earl James had little alternative to his new role, the precedents were not hopeless. The Dunbars, after leading English attacks on the marches, had returned from exile to recover their lands. With Scotland and England engaged in sporadic conflict from 1455, Douglas maintained hopes of restoration and demonstrated his value to his new lord by waging war in the marches in English service.[2]

In a period when the English were generally on the defensive, Douglas made an impact. In 1458 the Scots king agreed to recognise Earl James and his retainers as liegemen of Henry VI if the English council restrained Douglas from molesting the Scottish borders. Douglas's specific inclusion in the truce suggests James II's anxieties about his enemy's activities. The earl had clearly participated in English attacks into all three marches from late 1455, his presence designed to have, not just military, but political effect in reminding his former adherents of his survival. The wave of new forfeitures amongst kin and servants of Douglas in 1457 was a mark of continued royal mistrust of a group which included the earl's secretary, Mark Haliburton, and sister, Margaret, widow of Henry Douglas of Borgue who had not survived the defeat of 1455. Even James Lindsay of Covington, a royal councillor up to 1455, 'was excludit fra the counsall' and only escaped death through the payment of a fine. Others, like James Douglas of Railston and James Hog, both connected to Douglas before 1455, fled royal suspicion to join their old lord in exile.[3]

Such local successes could not obscure the earl's dependence on English support. From 1459 this help was disrupted by renewed internal conflict, and Douglas's priority was to ensure the continued favour of whatever party held power. In October 1459 he brought his retinue to support Henry VI and Queen Margaret of Anjou against York and his allies. After the bloodless royal victory at Ludford Bridge, Douglas followed the court to Coventry where he received a small portion of the earl of Salisbury's forfeited estate as a reward for 'his loyalty and great labours ... in resisting our rebels'. The return of the Yorkists from exile the following year, and the defeat of the royal army at Northampton in July left Douglas with the task of establishing his credit with the new government. James II made the transition possible. In late July 1460 he laid siege to Roxburgh, again exploiting English divisions. On 3 August the king was killed when one of his own guns exploded, but his host remained to force the surrender of Roxburgh five days later. The loss of one of the last two English-held castles in the Scottish marches was followed at the end of 1460 by the news that Queen Margaret and her son, Edward prince of Wales, were seeking active Scottish support; which gave Douglas, as an enemy of the Scottish government, hopes of employment by the Yorkists. In August Earl James was in London, where the Yorkist council was based, but any immediate hopes were disappointed. Faced with a growing military challenge from Queen Margaret in northern England, fighting against which York himself was killed in late 1460, the government had no resources to spare for Douglas. The earl, who had received no payments since 1459, was in financial difficulties. It was only in the summer of 1461, when York's son, Edward IV, had secured his place as king, that Douglas was sought out by the English government.[4]

Despite the recent past, Douglas had reason for optimism. Edward IV was committed to war with the Scots who were giving his Lancastrian enemies active support. This alliance had been agreed in April, with the Scots receiving Berwick in return for military aid in defence of Lancastrian-held castles in Northumberland. The situation was the reverse of that which Douglas had intended in 1455, and like James II, Edward IV gave priority to the defeat of the new allies. Of equal importance for Douglas, the death of James II in 1460 had removed his chief enemy. The council for the new king, James III, headed by Queen Marie of Gueldres, was hardly well-disposed towards the earl, but, as the events of the 1440s had shown, minority regimes were vulnerable to political and military pressure. It was during this kind of regime that the Dunbars had been restored, while, after 1437, even the heirs of the forfeited duke of Albany had recovered part of their inheritance. In the summer of 1461, with the sponsorship of the English king, Douglas travelled north to find allies within the Scottish kingdom.[5]

His objective was not the marches but the Western Isles. Repeating his efforts of the previous decade, Douglas sought out John earl of Ross and Donald Balloch. By 1461 John and Donald were in conflict with Colin Campbell earl of Argyll, the rising power in the western Highlands. The presence of Islesmen at parliament in early 1461, and a meeting on Bute in June between Ross and an embassy which included Bishop Kennedy, David earl of Crawford and lords Hamilton and Livingston, failed to settle the conflict, and Douglas was able to exploit these grievances. Accompanied once again by the bastard, John Douglas, Earl James met Ross at Ardtornish castle in Ardnamurchan, and by October had persuaded his host to enter a formal alliance with England. Douglas returned with ambassadors from Ardtornish, and in February 1462 an indenture was concluded in which Ross and Donald agreed to become liegemen of Edward IV. In return both received a pension, and should Scotland be conquered, a share of the spoils. Douglas was not forgotten. All agreed that if Scotland was conquered with his help, Earl James should recover his lands and hold them in future from the English king. For the Scottish magnates, the partition of Scotland was not the point of the agreement. English gold and soldiers and the renewal of the alliance of 1454 between Douglas and Clan Donald were intended as the means to put pressure on an unstable regency government, forcing the queen and council to make concessions to its enemies.[6]

Realistically, however, Douglas's hopes of such concessions depended not on his own efforts, but on the English king. In early 1462 Edward IV was ready to subsidise the Douglas cause heavily, restoring the earl's annuity and granting £100 a year to James's brother, John of Balvenie. Such sums suggest plans to unleash the earl, but with the departure of Queen Margaret to France, the Scots government agreed a truce in April to last through the summer. Douglas had the ground cut from under him. According to a contemporary letter, 'Erle Duglas ... a sorwefull and a sore rebuked man' was commanded to attend the talks as an Englishman, but if he were taken by the Scots, 'they to sle hym'. The return of Queen Margaret in October and renewed Lancastrian resistance backed, once again, by the Scots must have been welcomed by Douglas. While his kinsman, George earl of Angus, led an army to help Lancastrian garrisons in Northumberland, Douglas prepared a major offensive in war and politics. In November 1462 certain men in Roxburghshire were accused of 'treasonable communication and reception of James of Douglas, traitor and rebel', having been bought over by English money, according to one source. In the following January, Douglas received funds from the English king for 'special business' in the north, and in March led an attack on the west march and Galloway which defeated the force sent to oppose him under David earl of Crawford and lord Maxwell. These successes, accompanied by the death of Angus, the council's chief ally in the marches, were followed in July by a failed Scottish siege of Norham. Douglas took a leading part in a counter-attack, which with local help penetrated deep into the middle march and led to the defeat and capture of James lord of Rutherford. Events in the marches created an atmosphere of crisis in Scotland. The division between the queen, reputedly in correspondence with Douglas, and Bishop Kennedy was the kind of situation which the exiled earl could exploit. However, any such hopes were dashed, first by the capture and execution of John of Balvenie, another victim of the Scotts of Buccleuch, then by the death of the queen in December 1463. However, the real end of the earl's hopes came from his new master. After planning a major attack on Scotland through the summer, in December Edward IV agreed to a truce. In return for the end of Scottish support for Margaret of Anjou, Edward agreed to make Douglas observe a truce which, the following spring, was extended for fifteen years.[7]

The truce of 1464 ended Douglas's best hopes of restoration to a share of his lands and influence. These hopes had rested either on his participation in English military success or on Douglas's ability to extract a deal from the minority government by demonstrating his power to disrupt their hold on the marches. Both scenarios depended on the ability of the earl to win or compel support from within Scotland and, during the decade after his forfeiture, Douglas did continue to attract Scottish adherents. Evidence of James II's mistrust of a number of the earl's old followers in the late 1450s and the flight of several of these to England, and of the readiness of men in the middle march to aid and abet Douglas in the early 1460s, suggests that, though a condemned and proven traitor, he could still count on local connections. The most important of these was Andrew Kerr. After 1455 Kerr's support was vital to Douglas. He combined loyalty to his old lord with influence, both in the local community and in the kingdom. It was this influence which allowed Kerr to escape detailed charges of treason in 1456 and 1471. Both of these included his support of Douglas. On the latter occasion he was accused of the 'traitorous imbringing of James Douglas, traitor, from England within Scotland', and of assisting James and the English in 'the last battle', a charge which probably referred back to 1463.

However, Kerr was not just a border ally. In 1466 he backed the coup which gave control of James III and his council to the Boyd family. Any hopes that contact with the government would work in Douglas's favour were dashed by the fall of the Boyds in 1469. With Kerr's help, lord Boyd joined Douglas as an English pensioner.[8]

If the earl's continued links with Kerr and others in Scotland were proof of lingering attachments to his family, Douglas was never able to turn such support into a secure foothold in Scotland. His efforts to re-establish his place in Scotland were hampered by the sporadic nature of English support between 1455 and 1464, but even with more substantial backing, Douglas would have faced major difficulties. With the exception of Kerr, those who rallied to him in the 1450s and 1460s were minor men from Lothian, Annandale and Teviotdale. There is little sign of greater lords following their lesser neighbours. Lord Maxwell and the Johnstones in Annandale and the Scotts in the middle march had made their choice in 1455, and had little to gain and much to lose from the return of Douglas. In the years up to 1460, James II made sure that these sentiments were shared by the vast majority of landowners in the old Douglas heartlands. The forfeiture of the earl of Douglas had given James a territorial stake in the south greater than any ruler since the twelfth century, with Lauderdale, Wigtown, Galloway and Ettrick Forest added to Annandale and the earldom of March. The king was determined to use these lands as the base for royal lordship in the south. Their management was placed in the hands of local men, old Douglas officials or new royal adherents. The Pringles, Middlemasts and Scotts as rangers of the Forest, Glendinning in Eskdale and the Agnews in Wigtown all kept offices which their families had held from the Douglases, while the appointment of Donald McLellan as steward of Kirkcudbright and of Thomas Cranstoun as bailie of the Forest promoted lords who had supported the king in the early 1450s. After years of upheaval and with Douglas lurking just across the border, the king was concerned not to alienate local allies, but in the west march he did not place his whole trust in local men. Military command in the west, for over a century the preserve of the Black Douglases, was given to close royal adherents. Lochmaben was taken from Herbert Johnstone in 1455 and entrusted to Andrew Stewart, newly created lord Avondale, an intimate of the king and confirmed enemy of the Douglases. Threave was similarly placed in the custody of royal servants from outside the region, and the wardenship of the west march was also given to Avondale. James II was ensuring royal control over warfare in the west, and in 1456 he even withheld the lordship of Nithsdale from his own chancellor, William Sinclair earl of Orkney. Nithsdale's value was less in land than in rights of justice and war leadership.

The king was preventing such powers remaining the hereditary prerogative of a magnate in the west march.[9]

Yet elsewhere James II could not ignore the continued place of magnate leadership or traditions of Douglas lordship which stretched back a hundred and fifty years. Instead he sought to use both to the crown's advantage. Lord Hamilton, whose defection had been so crucial in 1455, had, as the husband of Countess Euphemia, widow of the fifth earl, a place in the wider Douglas family. He received a major share of the Clydesdale estates of Earl James and lands in the Forest, which formed his wife's portion of the Douglas lands. In both areas of traditional Douglas support, Hamilton could be expected to reconcile his tenants to the change of lord. It may have been through his efforts that his stepdaughter, Margaret of Galloway, returned to Scotland in 1459, obtaining a divorce from her husband, Earl James Douglas. Margaret, who may always have resented her marriages to the sons of James the Gross, could still claim to be heiress of Galloway. Her loss may have been accepted by the exiled earl, but the king quickly ensured that she was married to his own half-brother, John Stewart earl of Atholl, preventing her rights being raised in the south-west.[10]

Nowhere was the ambivalence of the king towards Douglas lordship clearer than in his relationship with George Douglas earl of Angus. By granting the wardenship of the middle march to the Red Douglas magnate, who was already warden of the east march, James II was promoting another great border dynasty. Though Angus's share of the border estates of his disgraced kinsman was limited to the minor lordship of Ewesdale, as lord of Jedforest and Liddesdale Earl George was now by far the greatest magnate in the march. However, James II did not regard Angus with mistrust. No family had greater reason to work for the continued exclusion of the Black Douglases and no other magnate possessed the power to secure the middle march for the crown. James needed Angus's military leadership in his southern war from 1455. He also needed the earl's political lordship in the middle march. It was Angus who presided over the trial of Kerr in 1456, and following the latter's acquittal, George concluded a bond with Kerr making him bailie of Jedforest. In return Kerr became the earl's retainer, a relationship which seems to have kept him free of treasonable contacts until the death of Angus in 1463, while a bond with George Turnbull of Bedrule attached a second Black Douglas tenant to Earl George. Like his ancestor, the first earl of Douglas, Angus was using private lordship to secure the loyalty of men in the marches. By 1460 he had built up a following which included both Kerr and Turnbull and the local enemies of the Black Douglases, Scott of Buccleuch and William Douglas of Cavers, Angus's cousin and the keeper of his castle of Hermitage. This achievement was vital for the king but it also rested on the traditional influence of the Douglases. Instead of being worried, James II actively sponsored such sentiments by granting Angus the lordship of Douglasdale. Although Douglasdale was a fitting reward for a trusted ally, the whole history of the house of Douglas was based on the exploitation of such royal generosity to form the basis of private lordship. Earl George, consistently styling himself lord Douglas, was a direct descendent of the first earl of Douglas, and since the 1390s his line had pressed their claim to headship of the family, and, with this, to the lands and power it held. With the death of James II, Angus played a miltary role in the marches even more reminiscent of earlier Douglas warlords, culminating in his leadership of armies into England in support of Henry VI. Like the fourth earl of Douglas, in return for this support Angus was offered a duchy by a king and faction desperate for support. Though Angus's death in the spring of 1463 cut short his ambitions in Scotland and northern England, the fall of the earl of Douglas did not bring an abrupt end to the methods or tradition they represented. The patronage of James II and the encouragement of Red Douglas ambitions would return to haunt the Scottish crown when Earl George's son, appropriately named Archibald, reached adulthood.[11]

The ambition of Archibald earl of Angus forms a backdrop to the last years of the Black Douglases. In the 1460s and 1470s James earl of Douglas remained an English pensioner. At the truce of December 1463, he had successfully petitioned to be made constable of Carrickfergus, the principal royal castle in Ulster. Douglas's interests suggest hopes of renewed contact with Donald Balloch, lord of the nearby Antrim glens, but these failed to yield any major advantages to the earl. Instead as the Anglo-Scottish truce turned into a formal peace in the 1470s, Douglas, though still used as a diplomatic bargaining piece, had no chance of recovering his lands. It was only in 1479, when Edward IV began planning a new war with the Scots, that Douglas was recalled to an active role. His rights were part of the English justification for war, and in early 1480 he was back in the marches. His servants, John Fraser and Richard Holland, the author of *The Howlat*, were sent into Scotland, while another, Patrick Haliburton, sought to renew the connection with John Mor, son of Donald Balloch. Their success was limited, with Haliburton handed over to the Scots by the earl of Ross, but in 1482 Douglas did get a major Scottish ally in Alexander duke of Albany, exiled brother of James III. Albany agreed to become Edward IV's man in return for support in the deposition of his brother. Douglas was to recover his lands from Albany as king of Scots. In the summer of 1482, Edward sent Albany north with an English army, but once in Scotland, the duke abandoned his backers and sought Scottish allies. Albany recognised that the English invasion, while it had forced Scotland into political crisis, was no basis for long-term recovery of power. Amongst other difficulties, the restoration of Douglas would hardly have been welcome to many southern landowners, amongst them one of Albany's close allies, Archibald earl of Angus. However, Angus had his own plans. When at the end of 1482 Albany sent Angus south to seek fresh English help, Red and Black Douglas came to a private arrangement. This 'convention' may have been Douglas's agreement to make Angus heir to his lost lands. Douglas was childless and may have been prepared to disinherit his nephew, Hugh, son of the earl of Ormond, in return for Angus's help in recovering his lands. Though the deal lapsed with the return of Angus to Scottish allegiance and the renewed exile of Albany, the Red Douglas earl did not abandon hopes of a Black Douglas inheritance.[12]

Earl James's exile was approaching its end. Though Edward IV had given his brother, Richard of Gloucester, rights to as much of south-west Scotland as he could conquer, when the latter seized the throne in 1483 he turned to Douglas to disrupt the Scottish marches. In the summer of 1484 Douglas, with Albany, led a force to Lochmaben in Annandale, where both men had claims on local loyalties. It was also a lordship split by violent feuds which the exiles may have hoped to exploit to advantage. However, it was a force of old enemies led by Crichton of Sanquhar and the Johnstones which confronted Douglas and Albany. The invaders were routed. While Albany escaped, this time Earl James fell into Scottish hands. To be spared the fate of his brothers and placed in prison was a mark of the decline of his standing in Scotland in the last years of his exile. However, even as a broken prisoner, Douglas was not forgotten. Four years later, when James III was killed by a coalition of rebel lords, Earl James's fortunes took a final turn. He was released into the less restrictive surroundings of Lindores abbey in Fife and given a pension of £400. The man behind this treatment was Angus, a leading figure in the new regime. His motive was to gain recognition of his rights to Earl James's old estates during a fresh period of minority politics. In early 1491, while Douglas, in his late sixties and worn out by three decades of exile, was on his deathbed, Angus was hovering nearby. Whether or not he received the bequest of the dying man to lands forfeited decades before, Angus persisted in his ambitions. Five years later the last of the Black Douglases, James's nephew, Hugh, now dean of Brechin, resigned any claims he had to the earldoms of Douglas, Avondale and Ormond to the earl of Angus. However, instead of clearing the way for the reconstruction of Douglas power in the marches, Angus's efforts only secured a further stage in its decline. In 1491, after Angus's approach to his dying kinsman and another period in which he sought

English help against Scots enemies, the earl was forced to surrender the lordship of Liddesdale. There would be no restoration of the Douglas earldom, and new families would dominate the marches in the coming century.[13]

THE FALL OF THE BLACK DOUGLASES

The fall of the Black Douglas earls in 1455 proved irreversible. The defeat of the family marked the end of a period of tensions and conflict, which stretched back to the return of James I and the almost simultaneous departure to France of the fourth earl of Douglas three decades before. In 1424 it had been the king who had been an exile, while the Douglas earl, the greatest lord in southern Scotland, 'great guardian' of the marches, 'eldest son of the pope' in the kingdom and, in France, duke of Touraine and lieutenant-general of King Charles's armies, was a prince of European power and ambition. The reversal of these fortunes, though it began in 1424, was only completed in the early 1450s. In only five years from 1450, the fourth earl's nephews, the sons of James the Gross, went from being the greatest magnate family in the kingdom to meet their respective fates: one stabbed by his king, one killed in battle, one beheaded and two exiled as traitors. Both the erosion of Douglas lordship since 1424 and the final series of clashes with the crown were dramatic shifts of power and fortune which had parallels across Europe. The fall of the Douglases can easily be seen as a European phenomenon, part of an age of tensions and conflicts between sovereigns and their greatest subjects, events traditionally regarded in western Europe as a major stage in the formation of modern states.

Closest to the fall of the Douglases in terms of time and place was the descent of English politics into a cycle of violence in the early 1450s. However, any comparison between these contemporary periods of tension reveals fundamental differences between the political societies of England and Scotland. In England the breakdown of political harmony resulted from the incompetence of king and council. The loss of France and the mishandling of patronage and politics in England sparked a series of popular revolts in 1450 and, in 1452 and 1455, attempts by Richard duke of York to seize power by force. Resulting tensions led to renewed wars from 1459 which saw the death of York and the establishment of his son as King Edward IV. From 1450, both magnates and popular rebels were concerned with the exercise of central power or with local grievances which reflected the crown's ability to interfere with and upset the structure of politics and justice in the shires. Consequently, political violence, even when based on local rivalries, centred on the control of royal government. In 1452 and

1455, Richard duke of York sought control of the royal council, and the transformation of the conflict into a dynastic civil war in 1460–61 occurred as a means to secure this control by creating a new king. Effective control of power was won and lost in short, sharp struggles, with even a street fight at St. Albans in 1455 delivering effective control of the whole kingdom.[14]

By contrast, conflict in Scotland after 1450, as since 1424, stemmed, not from the failure of kings to use their extensive powers to provide the strong and intrusive government expected by their subjects, but from the efforts of aggressive rulers to increase their authority in the face of resistance. The struggle between James II and the Douglases reflected the extent to which power remained diffused in Scotland. Although attempts were made to secure control of royal government by force, for example the murder of James I in 1437 and the arrest of his widowed queen two years later, and although such events could result in changes in the structure of regional power, the link between central and regional power was far weaker than in England.[15] The clashes between James II and the Black Douglases formed a very different kind of struggle. The conflict was about the extent of royal authority, not who exercised it. The readiness of a great magnate dynasty to resist royal demands was based on the family's perceived rights as private lords, rights which had no parallels in England. The existence of such perceptions was a product of a political development, both before and after 1300, which was very different to that of England.

But if the crown was less the lynchpin of Scottish political society, the limits on royal power hardly made kings safer or politics tamer than in England, as James II must have realised in the early spring of 1452 when he fled Stirling in the face of the advancing Douglases. Instead, the region-alised character of Scottish politics meant that, in contrast to England, political violence was less intensive and produced fewer pitched battles, but was more drawn out and covered greater areas of the kingdom. The Douglases were in open conflict with the king for six months of 1450–51, of 1452 and of 1455, while Ross and Crawford defied James II for parts of the same period. Such sustained resistance did not produce the kind of breakdown caused by conflict in a closely integrated and unified kingdom, like fifteenth-century England. Instead the existence of 'gret rowmes', the private lordships of magnates in the localities, provided the framework for regional politics and justice as they had done throughout the fourteenth and early fifteenth centuries. It was in defence of their rights to exercise effective lordship over local communities that the Douglases opposed James II. In aims, language and political expectations, the statements of the *Buke of the Howlat* and the Lincluden statutes were very diferent to those produced by Richard of York as the principal magnate opponent of Henry

VI's government. Although in the 1440s William earl of Douglas had used the authority of the crown to justify his search for political primacy, in rivalry with his royal lord after 1450 both he and his brother combined accusations of general misconduct against James II and his council with confident appeals to their rights as great regional magnates. The only similarity to this from fifteenth-century England was the rebellions of the Percies in 1403, 1405 and 1408, and even these challenges were still built round attempts to dethrone the king, Henry IV.[16] The absence of any similar dynastic challenge to James II by the Douglases was not a mark of weakness but of the family's ability to resist the king without such justification, relying instead on structures of lordship which had allowed previous earls to defy earlier kings and lieutenants.

The regionalised character of Scotland's politics and society was hardly exceptional. Many medieval realms possessed strong regional identities within the lordship of a single ruler. Like James I and James II in Scotland, the kings of mid-fifteenth century France, Charles VII and Louis XI, were forced to deal with entrenched structures of provincial authority which limited the power of the crown. The French princes possessed power of government which, like those granted by Charles VII to the fourth earl of Douglas in Touraine, extended over defined territorial principalities, many of which had historical integrity stretching back for centuries. These great regional lords, who at one extreme were virtually independent rulers like the dukes of Burgundy and Brittany, had dominated French politics during the crisis caused by Charles VI's incapacity from the 1390s to the 1420s. This period of princely regents was not dissimilar to the early years of the Stewart dynasty. These French princes regarded the efforts of Charles VII and Louis XI to extend their political reach as infringing rights which, although they included greater formal powers and claims than those of the Douglases, were also the basis of regional primacy. In France, as in Scotland, this friction produced conflict, which took the shape of major aristocratic rebellions against the king.[17]

In 1440 Charles VII was faced with a coalition which included his son, the Dauphin, and the dukes of Bourbon and Alençon in a revolt called the *Praguerie*. In 1465 Louis XI was opposed by the self-styled League of the Public Weal which, according to a contemporary, pitted the king and the city of Paris against seven dukes, twelve counts and 51,000 men-at-arms. Though dressed up in terms of the public good as complaints against royal misgovernment, such aristocratic coalitions were in practice the sum of individual princely grievances against the king. These *alliances* were not limited to moments of political crisis. Like Scottish 'bands and ligues', such private agreements between great lords were a recognised means of regulating regional politics. By excluding the crown from the equation, these agreements were also seen by kings as an implicit or explicit threat to royal authority. Even if there was nothing strictly equivalent to the League of the Public Weal against James II in early 1452, his knowledge that, of the lords who had agreed to the northern settlement of the late 1440s, Ross was attacking royal lands, Crawford was disaffected and the Douglases were intent on obstructing royal plans was sufficient to push the king into violence. In terms of relative power, the possible combination of the greatest magnate of the north and west, Ross, with the greatest from the south, Douglas, and several north-eastern lords, Crawford, Moray and Ormond, posed a threat as great as any faced by Charles VII and Louis XI from their princes. Like King Louis in 1465, in 1452 James II was forced to make concessions to his enemies against his will to end internal warfare. The objectives of Scottish and French kings in these periods of conflict were also comparable. In 1440 Charles VII's particular concerns were to extend the crown's authority within the rebels' principalities and to establish effective control over the kingdom's military resources. Though James II lacked the means to form the professional standing army created by Charles VII in the 1440s, he did seek to emphasise royal war leadership. Both Charles and James went directly from the suppression of revolt to war against the English enemy, stressing their national role in war. In both Scotland and France the royal government's encroachment on what were claimed as aristocratic prerogatives marked a major step towards the establishment of the monarchy as the only focus of national loyalty and political authority.[18]

In Scotland the fall of the Black Douglases was a vital step in this process. It removed one of the great magnate houses of the fourteenth century and broke up the last of the regional lordships which had dominated the political society of the kingdom in the early fifteenth century. The Douglases' forfeiture left the crown by far the greatest landowner and source of lordship in the English-speaking regions of the realm. The significance of the event was on a par with the French crown's recovery of the Angevin inheritance or even the annexations of Brittany and Burgundy as a political triumph and a move towards the unification of authority in the kingdom. Together with contemporary attempts to raise the status and power of monarchy in England and the Spanish kingdoms, historians have seen Stewart kings during the fifteenth century creating a 'new monarchy' as part of a political development spanning a number of European realms.

However, it is possible to find comparisons from the continent which suggest changes in Scottish politics were not as easy or automatic as this overview would indicate. In the same decade, across Europe there occurred a political conflict which shared many of the issues and even a parallel course to James II's campaign against the Black Douglases. The kingdom of Hungary had, like Scotland, seen a revival of royal authority in the first half of the fifteenth century. However, following the deaths of two kings in quick succession in 1437 and 1439, Hungary experienced a period of minority government, interregnum and civil war. It was only in the 1450s that the dynastic heir, Ladislas V Habsburg, returned to Hungary to assume power. He found his kingdom split by internal rivalries, especially between his close councillors and the family which had dominated minority politics and had built their reputation as the defenders of the kingdom against the threat of Turkish invasion, the Hunyadi dynasty. The family had been led by John Hunyadi, a warleader famed across Europe, whose death in 1456 precipitated a political crisis. The king responded by seeking to establish his own influence in the militarised borderlands of the kingdom. His efforts were blocked by Hunyadi's sons, Ladislas and Matthias, who killed the king's principal councillor, Ulrich von Cilli. This setback convinced the king of the need for drastic action. In early 1457 the Hunyadi brothers were summoned to court and once there were arrested. The elder, Ladislas, was executed, and Matthias was taken into exile by the king. Unlike the Douglases, though, the fortunes of the Hunyadi survived exile. Before the end of 1457 King Ladislas was dead and Matthias had returned to Hungary where he was not simply restored to his family lands but chosen to succeed as king.[19]

Like James II, Ladislas V was seeking to reduce the power of a great noble house, a dynasty with a European reputation as defenders of their kingdom, who had dominated recent minority politics. Both kings resorted to sudden strikes against their enemies which suggest doubts about their ability to win an open struggle. However, the death of King Ladislas without heirs in 1457 brought about a collapse of support in the kingdom. Though it was the elective element in Hungarian kingship which led to Matthias Hunyadi's accession to the throne, it is hard not to think that the death of James II at any point between 1450 and late 1455 would have brought a similar crisis for those who had backed the king and would have led to an immediate renewal of Black Douglas power. James, who was after all to meet a sudden, untimely end as the result of a military accident in 1460, could not feel secure of success at any point before the summer of 1455. His continued vulnerability hardly suggests that the steady or smooth extension of royal authority was the only possible path of Scottish political development.

Matthias Hunyadi, also known as Corvinus, ruled Hungary as a 'new monarch', but as the events which followed his death in 1490 showed, he had not significantly reduced the power of the Hungarian nobility. Hungary was far from unique in this respect. In the neighbouring kingdom of Bohemia, the nobility exploited religious and social unrest in the early fifteenth century to enlarge their powers at the expense of the crown, making Prague synonymous with aristocratic revolt across Europe. In the Scandinavian kingdoms, magnates supported the continued union of Denmark, Sweden and Norway precisely because the scale of this realm meant that royal power remained a weak and distant force. To an even greater extent, in the Empire the princes were in the process of turning their lands into effectively independent states. Closer to Scotland, during the 1450s and 1460s the great Anglo-Irish earls, who owed their regional power to the same period of crisis which produced the Black Douglases, were at the height of their power. Many of these aristocratic societies bore a closer similarity to Scotland than did the English and French realms, and to dismiss the continued domination of great magnates in realms across Europe as a mark of political underdevelopment is to equate strong royal government with good government and with some notion of progress.[20]

To regard the decline and fall of the Black Douglases in terms of such European comparisons is also to limit discussion of the factors involved in this to the very summit of political society. Just as the family's rise cannot be explained with sole reference to the Douglases relationship with the crown, in examining the eclipse of the Douglas earls it is important to look beyond the competition between royal and aristocratic ambitions. However, it is understandable that Black Douglas fortunes after 1424 have been set within a framework of royal politics. The return of James I, his son's minority and James II's assertion of authority in the early 1450s were events which had enormous significance for the house of Douglas. From 1424 James I brought a new energy to the exercise of royal authority and a new attitude to the place and purpose of the great magnates. His view of the kind of lordship exercised by a lord like Archibald fourth earl of Douglas was very different from that of his father, grandfather and uncle. Douglas's effective control of border war and politics, his influence in the church, and his virtual stranglehold on the management of local communities across the south appeared to James I as the usurpation of royal prerogatives. Instead the king sought to build an aristocracy which was bound to the crown by ties of service, which recognised the royal court as the kingdom's political focus and which regarded their local influence as subject to royal control. For the Douglas earl, as for other great magnates, recognition of James I's power was not the same as the recognition of a reduction in their own status and rights.

However, despite James's murder and the ensuing minority, this changed political atmosphere continued. From 1437, the power of the Douglas earls was linked to the politics of the council as well as to their regional lands and connections. The predominance which the family established after 1444 was a mark of exceptional power and status, but it also indicated the extent to which they stood alone in Scottish political society. As the son of a baron and royal councillor, Earl William Douglas had risen too fast to be wholly secure in this primacy, and it was probably not difficult for his enemies to convince James II that the Black Douglases represented a similar threat to that posed to his father by the Albany Stewarts. A great magnate with an elevated view of his dynasty's achievements and accustomed to holding sway on the royal council, Douglas was in a position to check the king's attempts to establish something like his father's personal primacy. The settlement of 1451, in which James II was forced into returning lands taken from Douglas, was, in this context, a humiliation which did huge damage to the prestige of the crown. James I had been killed after just such a humiliation, and his son's murder of Douglas was the act of a frightened, as well as an angry, man. Given the fate of James I and his own poor relations with his greatest subjects, James II could hardly have been confident of the protection afforded by his royal office. The confident and respected king who emerged in 1455 was only confident and respected because of his victory. From 1452 onwards James II recognised how much his authority rested on the defeat of the Black Douglases. Even more than James I's destruction of the Albany Stewarts, the fall of the Douglas earls was the event which secured the crown's political primacy.

However, the events which preceded this final conflict between the king and the Douglas earls can hardly be understood with reference to the ambitions and suspicions of crown and magnates working in isolation. In combination with the reigns of two active and aggressive kings with an elevated view of their status were factors which had determined the scale and shape of Douglas lordship throughout the century and half of the family's power: warfare, the internal relationships of the Douglas dynasty and, behind them, economic considerations. Working together, these were responsible for the growing vulnerability of the Black Douglases where it mattered most, in their ability to gain and retain the active support of lesser lords and communities across the south. The existence or renewal of earlier problems contributed greatly to the crown's ability to interfere with the structures of Douglas power. The pursuit of increased power through warfare had been a recurring theme of Douglas history since 1300. The fourth earl's defeat and death at Verneuil replicated earlier military disasters. The losses suffered by the earl's family and followers broke the continuity of Douglas lordship. What was different about Verneuil was the existence of a king able to exploit the defeat to alter the political balance in southern Scotland to his advantage. Similarly, the rivalry between descendants of the Good Sir James Douglas and those of his brother, the Tyneman, was nearly eighty years old by 1424, but the continued tensions between the Red and Black Douglases provided the crown with natural allies in the marches. William earl of Angus and his sons, James and George, were Douglas magnates acceptable to the crown both before and after the fall of their kinsmen. Fresh rivalry within the Black Douglas line itself owed its violent climax, if not existence, to bonds established between royal and Douglas politics in the 1420s and 1430s. The execution of the sixth earl following the Black Dinner was an act designed to secure the position of James the Gross both in the royal government and in control of the lands and influence of the Black Douglas family. Internal conflict of the kind which went back to the 1350s was, on this occasion, the means for the crown to widen its influence with families which had traditionally looked to the Douglases for lordship.

It is also possible to see increasing strains on Douglas lordship stemming from economic as well as political causes. Douglas finances probably rested most heavily on the profits of wool and other animal products. Frequent military activity and the nature of the land rendered the great border lordships of the Douglas earls best suited for pastoral farming. It was perhaps no coincidence that the period of the Douglases' greatest apparent security was in the 1370s and 1380s when the price of wool was at its height and provided magnates, as well as the crown, with the means to pay generously for service.

The decline in lucrative exports of wool to the Flemish cloth industry from the 1390s may have made it difficult for magnates like the Douglases to provide the gifts and fees which were part of their lordship. The fourth earl of Douglas's intervention in Stewart family politics, his usurpation of royal rights and expropriation of royal revenues, even the attraction of the rewards to be won in continental warfare, may have been influenced by his need to make up a growing shortfall in his income. James I's return and successful demonstration of political muscle not only shut off these sources of revenue, but also introduced the crown as a new competitor for the financial resources of the kingdom. The king's pursuit of his full financial rights and his attempts to gain new lands and rents must have put economic pressure on the great magnates. James II's similar efforts to enlarge the crown's landed resources as a means to provide for his new queen were amongst the issues which prompted the first attack on the Black Douglases. The situation in 1450 may have been rendered more serious by a new decline in the price of wool during the late 1440s and early 1450s which, perhaps, spurred the Douglases' renewed search for the rewards of royal government. Economic shifts since 1390 may have damaged the position of those, like the earl of Douglas, who held large areas of upland territory, rather than more compact but richer holdings. The greater political independence and cultural confidence of barons and lesser lords were not just the product of political changes. They may also have reflected the growing wealth of many such men relative to great magnates like the Douglases, a change reflected, for example, in the number of collegiate churches founded by barons after 1400. If still far richer than such men, the Douglas earls found their economic dominance gradually eroded from the 1390s. The usurpation of the earldom by a branch of the family which held rich but compact estates in Lothian and Clydesdale was a mark of economic as well as political change.[21]

However, behind the shifts of status, wealth and power within the political society of southern scotland was a further factor, the absence of war. After 1389 war on the Anglo-Scottish border gradually declined in scale, intensity and significance. The last communities in the Scottish west and middle marches in English allegiance were won back for Scotland by Douglas magnates in the 1380s, and with the restoration of the Dunbars in 1409, the last, sustained military threat to Lothian was ended. Any rise in the wealth of Lothian knights and barons was aided by their new freedom from attack. For the Douglases, however, the reduction of war with England to localised cross-border raiding and occasional larger efforts against English-held strongholds in the marches represented an alteration of the circumstances which had given the family its power and place in Scotland. The Good Sir James, the knight of Liddesdale, the first earl and Archibald the Grim all achieved regional power as leaders in war against England. They won support across the south through their ability to lead and protect local communities and their success in bringing men in the marches back into Scottish allegiance and into the following of the Douglas earls. Lesser men under military threat were bound to follow Douglas lords. With the end or reduction of this military threat these bonds were weakened. The signs of strain in Douglas predominance across the south were appearing even before James I's assumption of power. The fourth earl's French adventure was embarked on, in part, as the means to provide a new arena for Douglas military lordship, but ended in the disaster of Verneuil. With no military crisis on the marches, there was no need for southern lords to seek a new Douglas protector as there had been after Otterburn. To the lords of Lothian and even in the marches, King James and his allies represented an alternative source of effective lordship to that of the Black Douglases, and many were attracted from the earl's following to the rising star of the king. In similar fashion, after the Black Dinner broke the continuity of the Douglas line, many of the family's adherents looked beyond the earl in search of patronage and support.

In the fifteenth century barons, lords and freeholders across the south no longer needed the kind of military lordship provided by the Douglases. The Douglases' traditions of leadership and the ability to draw support from many families and communities had been vital in years of major conflict. But from 1409 the price of this lordship became too high. The proximity of the Douglas earls as lords, and the demands they made on lesser men, were probably a source of increasing resentment as the earls sought to maintain their influence against the claims of other lords. The efforts of the Douglases to re-assert their claims to special status on the marches and in the kingdom in the late 1440s were a response to political and social changes which undermined the family's power. However, despite the apparent resurgence of Douglas power during the minority of James II, it was clear to many of the family's key local adherents that Douglas lordship formed a check on their own local influence. The way in which landowners like Douglas of Cavers and Scott of Buccleuch, not to mention William Crichton, saw the opportunities of service to James I and returned as supporters of his son against the Douglases indicates major shifts in local political attitudes. Equally it was not surprising that Maxwell of Caerlaverock welcomed, and perhaps even anticipated, the Black Dinner. As steward of Annandale for the Douglases he was merely an important servant of the dominant regional magnate, a lord who was frequently in Annandale or its vicinity. As steward for the crown from 1440 Maxwell was, instead, in a position to exercise the rights of a more distant master, claiming effective leadership of the local community on his behalf, claims which led to local friction from the 1440s onwards. In the final conflicts between James II and the Douglas earl it was the decisions of such men, who had their own networks of kin and allies, to give their support to king or earl which decided the struggle.

As much as the crown, it was the baronial and minor noble vassals and neighbours of the Douglas earls who won most from the fall of their old lords. Anxious to avoid further disruption in the south, James II left these lesser lords in place. The politics of the 1460s were dominated by such men, not just those like the Kennedies and Flemings who had given James II crucial support, but also James Hamilton and Andrew Kerr, whose desertion of Earl James Douglas had come very late. In a kingdom now largely devoid of magnates like the earls of Douglas, these men and their peers, lords of parliament and newly created earls, formed the top rank of political society. The activities and ambitions of these families also dominated the local politics of the kingdom after 1455. In 1479, for example, the feuds in Teviotdale between Douglas of Cavers and his tenants the Cranstouns, and between the Rutherfords and Turnbulls, and in the west march between lord Maxwell and Douglas of Drumlanrig were debated in parliament as causes of dangerous local disorder. Just how dangerous became clear in the next decade when rivalries between the Maxwells and their neighbours, between Douglas of Cavers and his lord, the earl of Angus, and other disputes in the marches and across the kingdom found an outlet in the civil war between James III and his enemies in 1488.[22] In the opening decades of the century all these lords had been, willingly or unwillingly, in the shadow of the Douglas earl. As magnate, march warden and justiciar, the earl had the power to deal with local disputes, and it was as followers of Douglas that men like Maxwell would have participated in royal politics in 1401 and 1425. By the 1470s and 1480s the resolution of local conflicts and the exercise of political leadership in local communities fell to the crown, which in practice may have been a more distant and less demanding lord than the Black Douglases. What is most striking about the political society of the marches and Galloway after 1455 is not the extent of royal political control but the degree to which effective leadership passed from great lords, whose interests spanned the region, to lesser families, the Humes, Scotts, Kerrs and Douglases of Drumlanrig, whose main lands and influence were usually concentrated on a more limited area. The fall of the Douglases resulted less in the centralisation of Scottish politics than in the fragmentation of local lordship and a change in the character of the nobility.

A combination of factors, political, social, economic and international, placed new strains on the status and power of the Douglas earls and led, ultimately, to their defeat and expulsion from Scotland. It is possible to see the end of the family's century-and-a-half-long predominance in varying regions of southern Scotland as the natural consequence of new forces and different circumstances in play by the mid-fifteenth century. The prime function of fourteenth-century Douglas lordship was in binding the communities of the marches and even Lothian to the rest of the kingdom. The claims of the family to be the war leaders and defenders of those in Scottish allegiance was used as propaganda, but was based on the connection between Douglas landholding and influence and the maintenance of rights established in the Bruce settlement of the kingdom. Legal claims and the dynastic identity first established in defence of the Bruce monarchy continued to be at the core of the self-image and political power of lords and earls of Douglas.

It was, however, the military success achieved by these magnates that made the force of such claims less important to the Scottish crown and to local political society. Fear of a renewed political and military collapse of Scottish allegiance in the marches of the kind which followed Neville's Cross gradually diminished. With this, the fear that the removal or alienation of Douglas lords would lead to disaster was also diminished. However, even by the early 1450s, any weakening of the power held by the Black Douglases remained obscured by the continued predominance of the family in its border heartlands. In the conflict with the crown which ensued, the victory of the latter was neither easy nor automatic. For all the claims of the monarchy to the national leadership of the realm in war and politics, the appeal of James II's direct lordship was far from universal. While Scott, Maxwell and others gave him consistent support, lords of equal status, like Hamilton and Kerr, persisted in backing men accused as the king's enemies. The victory of James II was not simply one of royal ideology over the structures of magnate lordship, nor was it achieved by royal promises, many of them empty, to his supporters. Instead it was a hard-fought and far from one-sided civil war which the king won by the energy and aggression of his attacks on Douglas lands through early 1455. James II relied on methods of local warfare not far removed from those employed by Douglas lords since the 1300s, raising his armies from the retinues of local magnates, and coercing the communities of the south with small but effective forces.

During the opening stages of the 1455 campaign, the Tweeddale lord, James Tweedie of Drumelzier, who only weeks before had given active support to Earl James Douglas, submitted to the royal forces at Lanark. He recognised James II not just as king but as his personal lord, to whom he had given his manrent. Almost exactly a century earlier, during the later months of 1355, Tweedie's ancestor and namesake submitted to William lord of Douglas. He abandoned his allegiance to the English crown and 'was received to the loyalty and peace' of the Scottish king, David II, then an English prisoner. Like many others, though, the immediate significance of Tweedie's act was not his recognition of an absent and powerless king, but his formation of a personal bond with the lord of Douglas. A century apart, the submissions of the two lairds of Drumelzier symbolise the rise and fall of Douglas predominance in southern Scotland. But in their fall as well as their rise, Black Douglas fortunes rested on the continuing link between war and lordship in late medieval Scotland.[23]

NOTES

1. *A.P.S.*, ii, 75–77; *P.P.C.*, vi, 247–48; Stevenson, *Letters and Papers of the English in France*, ii, 317–31; Beaucourt, *Charles VII*, vi, 134–42; *E.R.*, vi, 204, 349; 'Auchinleck Chron.', in C. McGladdery, *James II*, 167; *Cal. Pat. Rolls*, vi, 346, 356.

2. Stevenson, *Letters and Papers of the English in France*, ii, 502, 503; *Cal. Docs. Scot.*, iv, nos 1272, 1277, 1278, 1279, 1283; *Foedera*, xi, 367, 381.

3. *A.P.S.*, ii, 42; M. Kelley, 'The Douglas Earls of Angus: A Study in the Social and Political Bases of Power of a Scottish Family from 1389 to 1557' (unpublished Ph.D thesis, University of Edinburgh, 1973), 61; S.R.O., SP6/20; GD 25/69, 72; *E.R.*, vi, 184, 196, 207, 347, 435.

4. *Foedera*, xi, 437; 'Auchinleck Chron.', in C. McGladdery, *James II*, 169–70; *Cal. Docs. Scot.*, iv, nos 1297, 1300, 1307, 1309. In August 1460 Douglas granted his goods and chattels in England to a London merchant probably in return for his debts (*Cal. Docs. Scot.* iv, no. 1309).

5. Waurin, *Chroniques* (Paris, 1859–63), ii, 302; *Three Fifteenth Century Chronicles*, Camden Society (London, 1880), 77–78; Dunlop, *Bishop Kennedy*, 221.

6. Macdougall, *James III*, 59; 'Auchinleck Chron.', in C. McGladdery, *James II*, 170; Munro, *Acts of the Lords of the Isles*, nos 72, 73, 74, 75; *Cal. Docs. Scot.*, iv, nos 1317, 1318, 1322, 1326, 1328.

7. *Cal. Docs. Scot.*, iv, nos 1322, 1323, 1324; N. Davis (ed.), *The Paston Letters: Papers of the Fifteenth Century* (Oxford, 1976), 2 vols, ii, 285; *E.R.*, vii, 152; Macdougall, *James III*, 60–63; C. D. Ross, *Edward IV* (London, 1974), 49–63; Fraser, *Douglas*, nos 95, 96; Waurin, *Chroniques*, iii, 169–73; *Three Fifteenth Century Chronicles*, 159; *E.R.*, vii, 285, 289; Dunlop, *Bishop Kennedy*, 236–37; Fraser, *Scotts of Buccleuch*, ii, 63–64; *R.M.S.*, ii, no. 786.

8. *H.M.C.*, Roxburghe, nos 7, 54; Macdougall, *James III*, 72, 74, 82.

9. *E.R.*, vi, 26, 63, 189, 201, 205, 208, 225–26, 352, 556; Fraser, *Douglas*, iii, no. 85. Ettrick and Galloway were among the estates permanently entailed to the crown in 1455 (*A.P.S.*, ii, 42). As when the Dunbars were forfeited in 1400, a number of Douglas tenants gained the superiority of their lands when their lords were dispossessed (Kelley, thesis, 58–59).

10. Fraser, *Douglas*, iii, nos 429, 430; *H.M.C.*, Hamilton, 16, 17, 18; *E.R.*, vi, 101, 226, 227, 228, 498, 571, 646.

11. Fraser, *Douglas*, iii, nos 82, 86, 88, 89, 90, 94, 95, 96, 97, 434, 435, 437, 439; *H.M.C.*, Roxburghe, nos 7, 34; S.R.O. GD 1/479/23; Boardman, thesis, 111–32. Significantly, Angus also entered an indenture with lord Hamilton, heir to much Douglas influence in Lanarkshire (Fraser, *Douglas*, iii, no. 436).

12. *Foedera*, xi, 510; *Cal. Docs. Scot.*, iv, nos 1339, 1466, 1469, 1489, page 412–13. In 1475 Douglas provided a contingent of four men-at-arms and forty archers for Edward IV's invasion of France, and at some point between 1461 and 1484 he was married to Anne Holland, widow of John Neville earl of Westmoreland. A niece of his, Margaret, probably a daughter of his sister Margaret and Henry Douglas of Dalkeith, was married into the Percy family. Such connections with northern English lords were probably sponsored by the crown to give Douglas support in the marches (*Cal. Docs. Scot.*, iv, nos 1511, 1526).

13. Macdougall, *James III*, 211–12, 241–42; N. Macdougall, *James IV* (Edinburgh, 1989), 89–90, 152; A. Grant, 'Richard III and Scotland', in A. J. Pollard (ed.), *The North in the Age of Richard III* (Stroud, 1996); *E.R.*, x, 116–17, 183, 253–54; British Library, Add. Ms. no. 6443, 24v.

14. R. L. Storey, *End of the House of Lancaster*; J. L. Watts, 'Polemic and Politics in the 1450s' in, M. L. Kekewich, C. Richmond, A. F. Sutton, L. Visser Fuchs, J. L. Watts (eds), *The Politics of Fifteenth Century England: John Vale's Book* (Woodbridge, 1995), 3–42; Griffiths, *The Reign of Henry the Sixth*, 610–771; A. J. Pollard (ed.), *The Wars of the Roses* (London, 1995).

15. For a discussion of this which may take the argument too far see A. Grant, *Independence and Nationhood*, 196–99 and A. Grant, 'Crown and Nobility in Late Medieval Britain', in Mason (ed.), *Scotland and England*, 34–59.

16. Kekewich *et al*, *The Politics of Fifteenth Century England*, 187–89, 194–202; J. M. W. Bean, 'Henry IV and the Percies', in *History*, 44 (1959), 212–27. Like the Douglases, in 1403

the Percies sent 'mortal *diffidatio*' to Henry IV, formally dissolving bonds of allegiance to the king (B. Wilkinson, *A Constitutional History of England in the Fifteenth Century* (London, 1964), 45).

17. P. S. Lewis, *Late Medieval France: The Polity* (London, 1968), 31–110, 190–201; P. S. Lewis, 'France in the Fifteenth Century: Society and Sovereignty', in P. S. Lewis, *Essays in Later Medieval French History* (London, 1985), 3–28; B. Guenée, 'Espace en Etat en France médiévale', in *Annales*, 32 (1968), 744–58.

18. R. Vaughan, *Philip the Good*, 379–91; M. Vale, *Charles VII*, 70–86; P. S. Lewis, 'Decayed and Non-Feudalism in Later Medieval France' and 'Of Breton Alliances and other matters', in *Essays in Later Medieval France*, 41–90.

19. J. Held, *Hunyadi: Legend and Reality* (New York, 1985), 1–16, 170–73.

20. J. M. Klasen, *The Nobility and the Making of the Hussite Revolution* (New York, 1978); M. Roberts, *The Early Vasas* (Cambridge, 1968), 1–44; F. R. H. du Boulay, *Germany in the Later Middle Ages* (London, 1988), 91–114; S. G. Ellis, *Reform and Revival: English Government in Ireland, 1470–1534* (Woodbridge, 1986); A. Cosgrove, 'The Emergence of the Pale', in A. Cosgrove (ed.), *A New History of Ireland, II, Medieval Ireland* (Oxford, 1987), 557–68.

21. S. G. E. Lythe, 'Economic Life', in J. M. Brown (ed.), *Scottish Society in the Fifteenth Century* (London, 1977), 66–183; A. Grant, *Independence and Nationhood*, 77–82, 95–96; R. H. Britnell, 'The Economic Context', in A. J. Pollard (ed.), *The Wars of the Roses*, 41–65. For a contemporary English comparison, see J. M. W. Bean, 'The financial position of Richard duke of York', in J. Gillingham and J. C. Holt (eds), *War and Government in the Middle Ages* (Cambridge, 1984), 182–98.

22. *A.P.S.*, ii, 122; Macdougall, *James III*, 133, 237, 241, 254; Boardman thesis, 159–260.

23. *R.R.S.*, vi, no. 137; *H.M.C.*, Various Collections, 5, Tweedie, no. 14.

Bibliography

A. PRIMARY MANUSCRIPT SOURCES

Archives Nationales, Paris:
J677, no. 15
J677, no. 20
J678, nos 21–25
J680, no. 71
K57, no. 9/12
X1a no. 8604

Bibliothèque Nationale, Fonds Français:
no. 7858
no. 32510
no. 32511
Fonds Clairambault, 41, no. 144
Fonds Latins, no. 10187

British Library:
Harleian Manuscripts, nos 6434–6443

Edinburgh University Library:
Manuscript DC7.63 (Law, J., 'De Cronicis Scotorum Brevia')

National Library of Scotland:
MS 72. Morton Chartulary
Advocates Manuscripts
  22.2.4
  80.4.15

National Register of Archives, Scotland:
no. 776: Earl of Mansfield Muniments
no. 832: Lauderdale Muniments

Scottish Record Office:
GD 1 Miscellaneous Collections
GD 8 Kilmarnock Charters
GD 10 Broughton and Cally Muniments
GD 11 Bruce of Kennet Charters
GD 12 Calendar of Swinton Charters

GD 15 Cardross Writs
GD 16 Airlie Muniments
GD 20 Crawford Priory Collection
GD 25 Ailsa Muniments
GD 30 Shairp of Houston Muniments
GD 32 Elibank Papers
GD 39 Glencairn Muniments
GD 40 Lothian Muniments
GD 44 Gordon Castle Muniments
GD 45 Dalhousie Muniments
GD 48 Rossie Priory Papers
GD 52 Forbes Papers
GD 65 Boswell of Balmuto Muniments
GD 70 Scott of Brotherton Muniments
GD 72 Hay of Park Papers
GD 75 Dundas of Dundas Papers
GD 78 Hunter of Barjarg Muniments
GD 82 MacGill Charters
GD 89 Skirling Writs
GD 90 Yule Collection
GD 97 Dunbeath Muniments
GD 98 Inventory of the Douglas Collection
GD 103 Antiquaries Charters
GD 109 Bargany Muniments
GD 111 Curle Collection
GD 119 Torphichen Writs
GD 121 Murthly Castle Muniments
GD 124 Mar and Kellie Muniments
GD 134 Cranstoun of Corhouse Papers
GD 148 Craigans Writs
GD 150 Morton Papers
GD 154 Agnew of Lochnaw Muniments
GD 157 Scott of Harden Muniments
GD 158 Hume of Marchmont Papers
GD 188 Guthrie of Guthrie Manuscripts
GD 212 List of Dr Maitland Thomson Note Books
GD 246 Hope, Todd and Kirk W. S.
GD 254 Lindsay of Dowhill Muniments
GD 350 Borthwick Muniments
RH 2, 6 Register House Charters
AD1 Crown Office Writs
SP6 State Papers
B30 Burgh Records

B. PRINTED PRIMARY SOURCES AND WORKS OF REFERENCE

i) Record Sources

A.B. Ill. Illustrations of the Topography of Antiquities of the Shires of Aberdeen and Banff, eds J. Robertson and G. Grut, 4 vols, Spalding Club (Aberdeen, 1847–69)

Aberdeen Burgh extracts. Extracts from the Council Register of the Burgh of Aberdeen, Spalding Club (Aberdeen, 1844)

The Acts of the Parliaments of Scotland [A.P.S.], ed. T. Thomson and C. Innes, record commission, 12 vols (1814–75)

Benedict XIII, Letters. Calendar of Papal Letters to Pope Benedict XIII, ed. F. McGuirk, Scottish History Society (Edinburgh, 1976)

Brechin Register. Registrum Episcopatus Brechinensis, Bannatyne Club (Edinburgh, 1856)

Bryce, W. M. (ed.), *The Scottish Greyfriars*, 2 vols (Edinburgh and London, 1909)

Calendar of Close Rolls (London, 1892–)

Calendar of Documents relating to Scotland, preserved in H.M. Public Record Office, ed. J. Bain and others, 5 vols (Edinburgh, 1881–88)

Calendar of Entries in the Papal Registers relating to Great Britain and Ireland: Papal Letters, eds W. H. Bliss and others, 16 vols (London, 1896–)

Calendar of Entries in the Papal Registers relating to Great Britain and Ireland: Petitions to the Pope, vol. i (London, 1896)

Clement VII, Letters. Calendar of Papal Letters to Pope Clement VII of Avignon 1378–94, ed. C. Burns, Scottish History Society (Edinburgh, 1976)

Calendar of Scottish Supplications to Rome [C.S.S.R.], vol. i, eds A. I. Cameron and E. R. Lindsay, Scottish History Society (Edinburgh, 1934)

C.S.S.R., vol. ii, ed. A. I. Dunlop, Scottish History Society (Edinburgh, 1956)

C.S.S.R., vol. iii, ed. A. I. Dunlop and I. B. Cowan, Scottish History Society (Edinburgh, 1970)

C.S.S.R., vol. iv, ed. A. I. Dunlop and D. MacLauchlan (Glasgow, 1983)

Coldingham Correspondance. The Correspondence, Inventories, Account Rolls and Law Proceedings of the Priory of Coldingham, ed. J. Raine, Surtees Society (London, 1841)

Connolly, M. (ed.), 'The Dethe of the Kynge of Scotis: A new edition', *Scottish Historical Review*, 71 (1992), 46–69

Copiale Prioratus Sancti Andree, ed. J. H. Baxter (St. Andrews, 1930)

The Exchequer Rolls of Scotland [E.R.], ed. J. Stuart and others, 23 vols (Edinburgh, 1878–1908)

Foedera, Conventiones, Litterae et Cuiuscunque Generis Acta Publica, ed. T. Rymer, 20 vols (London, 1704–35)

Fraser, A. (ed.), *The Frasers of Philorth*, 3 vols (Edinburgh, 1878)

Fraser, W. (ed.), *The Maxwells of Pollok*, 2 vols (Edinburgh, 1863)

Fraser, W. (ed.), *The History of the Carnegies, Earls of Southesk and their kindred*, 2 vols (Edinburgh, 1867)

Fraser, W. (ed.), *The Red Book of Grandtully*, 2 vols (Edinburgh, 1868)

Fraser, W. (ed.), *The Book of Carlaverock*, 2 vols (Edinburgh, 1873)

Fraser, W. (ed.), *The Scotts of Buccleuch*, 2 vols (Edinburgh, 1878)

Fraser, W. (ed.), *The Red Book of Menteith*, 2 vols (Edinburgh, 1880)

Fraser, W. (ed.), *The Chiefs of Grant* (Edinburgh, 1883)

Fraser, W. (ed.), *The Douglas Books,,* 4 vols (Edinburgh, 1885)

Fraser, W. (ed.), *Memorials of the Earls of Haddington*, 2 vols (Edinburgh, 1889)

Fraser, W. (ed.), *The Melvilles Earls of Melville and the Leslies Earls of Leven*, 3 vols (Edinburgh, 1890)

Fraser, W. (ed.), *The Elphinstone Family Book*, 2 vols (Edinburgh, 1897)

Hay, R. (ed.), *Genealogie of the Sainteclaires of Roslyn* (Edinburgh, 1835)

Henry IV, Letters. The Royal and Historical Letters of Henry IV, ed. F. C. Hingeston, 2 vols (London, 1860)

Calendar of Signet Letters of Henry IV and Henry V (London, 1978)

Highland Papers, vol. i, ed. J. R. N. MacPhail, Scottish History Society (Edinburgh, 1914)

Historical Manuscripts Commission: Reports of the Royal Commission on Historical Manuscripts [*H.M.C.*] (London, 1870)

John of Gaunt's Register, 1379–83, ed. E. C. Lodge and R. Somerville, Camden Society (London, 1937)

Kelso Liber, Liber S. Marie de Calchou, Bannatyne Club, 2 vols (Edinburgh, 1846)

Laborde, L. de, *Les Ducs de Bourgogne. Etudes sur les Lettres etc*, 3 vols (Paris, 1849–52)

The Lag Charters, Scottish Records Society (Edinburgh, 1958)

Laing Charters. Calendar of the Laing Charters, 854–1837, ed. J. Anderson (Edinburgh, 1899)

Melrose Liber. Liber Sancte Marie de Melros, Bannatyne Club, 2 vols (Edinburgh, 1837)

Moray Register. Registrum Episcopatus Moraviensis, Bannatyne Club (Edinburgh, 1837)

Morton Register. Registrum Honoris de Morton, Bannatyne Club, 2 vols (Edinburgh, 1837)

Munro, J. and R. W. (eds), *The Acts of the Lords of the Isles*, Scottish History Society (Edinburgh, 1986)

Newbattle Register. Registrum S. Marie de Newbattle, Bannatyne Club (Edinburgh, 1837)

North Berwick Chartulary, Carte Monialium de Northberwic, Bannatyne Club (Edinburgh, 1847)

Paisley Register. Registrum Monasterii de Passelet, Maitland Club (Glasgow, 1832)

Proceedings of the Privy Council, ed. H. Nicholas, Records Commission, 7 vols (London, 1834–37)

Regesta Regum Scotorum [*R.R.S.*], v, ed. A. A. M. Duncan (Edinburgh, 1988)

Regesta Regum Scotorum [*R.R.S.*], vi, ed. A. B. Webster (Edinburgh, 1982)

Registrum Magni Sigilli Regum Scotorum [*R.M.S.*], ed. J. M. Thomson and others, 11 vols (Edinburgh, 1882–1914)

Rotuli Scotiae in Turri Londonensi et in Domo Capitulari Westmonasteriensi, ed. D. MacPherson, 2 vols (London, 1814–19)

Saint Giles Chartulary. Registrum Cartarum Ecclesie Sancti Egidie de Edinburgh, Bannatyne Club (Edinburgh, 1859)

S.H.S. Miscellany. Miscellany of the Scottish History Society (Edinburgh, 1893–)

Spalding Miscellany, v. *Miscellany of the Spalding Club*, v, Spalding Club (Aberdeen, 1852)

Stevenson, J. (ed.), *Letters and papers illustrative of the Wars of the English in France during the reign of Henry the Sixth*, Rolls Series, 2 vols (London, 1861–64)

Stevenson, J. (ed.), *Documents Illustrative of the History of Scotland 1286–1306*, 2 vols (London, 1870)

Stones, E. L. G. (ed.), *Anglo-Scottish Relations, 1174–1328. Some Selected Documents* (Oxford, 1965)

Teulet, A., *Inventaire Chronologique des Documents relatifs à l'histoire d'Ecosse*, Abbotsford Club (Edinburgh, 1839)

Treasurers Accounts. Accounts of the Lord High Treasurer of Scotland, eds T. Dickson and J. B. Paul, vol. i (Edinburgh, 1877)

Wigtown Charter Chest. Charter Chest of the Earls of Wigtown, Scottish Record Society (Edinburgh, 1910)

Wigtownshire Charters, ed. R. C. Reid, Scottish History Society (Edinburgh, 1960)

Yester Writs, Calendar of Writs preserved at Yester House, 1166–1503, Scottish Record Society (Edinburgh, 1930)

ii) Literary and Narrative Sources

Barbour, John, *The Bruce*, ed. W. M. Mackenzie (London, 1909)

Barbour's Bruce, ed. M. P. McDiarmid and J. A. C. Stevenson, Scottish Text Society (Edinburgh, 1980–84)

Basin, Thomas, *Histoire de Charles VII*, 2 vols (Paris, 1933)

Chronique de Jean le Bel, ed. J. Viard and J. Déprez, 2 vols (Paris, 1904–1905)

Boece, Hector, *Scotorum Historia* (Paris, 1526)

Bower, Walter, *Scotichronicon*, ed. W. Goodall, 2 vols (Edinburgh, 1759)

Bower, Walter, *Scotichronicon*, ed. D. E. R. Watt, 9 vols (Aberdeen, 1987–)

Chartier, Jean, *Chronique de Charles VII Roi de France*, ed. Vallet de Viriville (Paris, 1858)

Chronique des Quatres Premiers Valois, ed. S. Luce (Paris, 1862)

Chronique de Mathieu d'Escouchy, ed. G. du Fresne de Beaucourt, 2 vols (Paris, 1863–64)

Fordun, Johannis de, *Chronicis de Gentis Scotorum*, ed W. F. Skene, 2 vols (Edinburgh, 1871–72)

Froissart, Jean, *Chronicles of England, France and Spain*, ed. and trans. T. Johnes, 5 vols (London, 1805–10)

Froissart, *Chronicles*, ed. and trans. G. Brereton (London, 1968)

The Scalachronica of Thomas Gray of Heton Knight, Maitland Club (Glasgow, 1837)

The Chronicle of Walter of Guisborough, ed. H. Rothwell, Camden Third Series, 2 vols (London, 1957)

Holland, Richard, 'The Buke of the Howlat', P. Bawcutt and F. Riddy (eds), *Longer Scottish Poems*, vol. i (Edinburgh, 1987)

Hume, David of Godscroft, *The History of the House and Race of Douglas and Angus* (Edinburgh, 1748)

The Chronicle of Henry Knighton, Rolls Series, 2 vols (London, 1889)

Chronicon de Lanercost, ed. J. Stevenson, Bannatyne Club (Edinburgh, 1839)

Lindsay, Robert of Pitscottie, *The Historie and Cronicles of Scotland*, Scottish Text Society (Edinburgh, 1899–1911)

The Paston Letters: Papers of the Fifteenth Century, ed. N. Davis, 2 vols (Oxford, 1976)
Legends of Saint Ninian and Saint Machar, ed. W. M. Metcalfe (Paisley, 1904)
Three Fifteenth Century Chronicles, Camden Society (London, 1880)
Walsingham, Thomas, *Historia Anglicana*, 2 vols, Rolls Series (London, 1863)
Waurin, Jean de, *Chroniques et Anchiennes Istoires de la Grant Bretagne*, 5 vols (London, 1864–91)
The Westminster Chronicle, 1381–94, eds L. C. Hector and B. Harvey (Oxford, 1982)
Wyntoun. The Original Chronicle of Andrew of Wyntoun, ed. F. J. Amours, Scottish Text Society, 6 vols (Edinburgh, 1908)
Wyntoun, *Chronicle*, ed. Laing. Wyntoun, Andrew of, *The Orygynale Cronykil of Scotland*, ed. D. Laing, 3 vols (Edinburgh, 1872–79)

C. SECONDARY SOURCES

Agnew, A., *The Hereditary Sheriffs of Galloway*, 2 vols (Edinburgh, 1893)
Allmand, C. T., *Henry V* (London, 1992)
Balfour-Melville, E. W. M., *James I King of Scots* (London, 1936)
Barbé, L. A., *Margaret of Scotland and the Dauphin Louis* (London, 1917)
Barrow, G. W. S., *Robert Bruce* (Edinburgh, 1976)
Barrow, G. W. S., *The Anglo-Norman Era in Scottish History* (Oxford, 1980)
Barrow, G. W. S., *Kingship and Unity, 1000–1306* (Edinburgh, 1981)
Beaucourt, G. du Fresne de, *Histoire de Charles VII*, 6 vols (Paris, 1881–91)
Boardman, S., *The Early Stewart Kings, Robert II and Robert III* (East Linton, 1996)
Brooke, D., *Wild Men and Holy Places* (Edinburgh, 1994)
Brown, M., *James I* (Edinburgh, 1994)
Burne, A. H., *The Agincourt War* (London, 1956)
Cowan, I. B., *The Parishes of Medieval Scotland*, Scottish Record Society (Edinburgh, 1967)
Dowden, J., *The Bishops of Scotland* (Glasgow, 1912)
Du Boulay, F. R. H., *Germany in the Later Middle Ages* (London, 1988)
Duncan, A. A. M., *Scotland: The Making of the Kingdom* (Edinburgh, 1975)
Dunlop, A. I., *The Life and Times of James Kennedy, Bishop of St. Andrews* (Edinburgh, 1950)
Elliot, G. F. S., *The Border Elliots* (Edinburgh, 1897)
Ellis, S. G., *Reform and Revival: English Government in Ireland, 1470–1534* (Woodbridge, 1986)
Forbes Leith, W., *The Scots Men at Arms and Life Guards in France*, 2 vols (Edinburgh, 1882)
Frame, R., *English Lordship in Ireland, 1318–61* (Oxford, 1982)
Frame, R., *The Political Development of the British Isles, 1100–1400* (Oxford, 1990)
Francisque-Michel, *Les Ecossais en France. Les Francais en Ecosse*, 2 vols (London, 1862)
Given-Wilson, C., *The English Nobility in the Late Middle Ages* (London, 1987)
Goodman, A., *John of Gaunt. The Exercise of Princely Power in Fourteenth Century Europe* (London, 1992)

Grant, A., *Independence and Nationhood* (Edinburgh, 1984)

Grant, F. J., *The Court of the Lord Lyon*, Scottish History Society (Edinburgh, 1946)

Griffiths, R. A., *The Reign of Henry the Sixth: The Exercise of Royal Authority* (London, 1981)

Held, J., *Hunyadi: Legend and Reality* (New York, 1985)

Hicks, M., *Bastard Feudalism* (London, 1995)

Johnston, G. H., *The Heraldry of the Douglases* (Edinburgh, 1907)

Kekewich, M. L., Richmond, C., Sutton, A. F., Visser Fuchs, L., Watts, J. L. (eds), *The Politics of Fifteenth Century England: John Vale's Book* (Woodbridge, 1995)

Klasen, J. M., *The Nobility and the Making of the Hussite Revolution* (New York, 1978)

Laing, H., *Impressions from Ancient Scottish Seals*, Bannatyne Club (Edinburgh, 1850)

Lewis, P. S., *Late Medieval France: The Polity* (London, 1968)

Macdougall, N. A. T., *James III* (Edinburgh, 1982)

Macdougall, N. A. T. *James IV* (Edinburgh, 1989; reprinted East Linton, 1997)

McFarlane, K. B., *The English Nobility in the Later Middle Ages* (Oxford, 1973)

McFarlane, K. B., *England in the Fifteenth Century* (London, 1981)

McGladdery, C. A., *James II* (Edinburgh, 1990)

McKerlie, P. H., *History of the Lands and their Owners in Galloway*, 2 vols (London, 1906)

McNamee, C., *The Wars of the Bruces: Scotland, England and Ireland, 1306–1328* (East Linton, 1997)

Macquarrie, A., *Scotland and the Crusades, 1095–1560* (Edinburgh, 1985)

Newhall, R. A., *The English Conquest of Normandy, 1416–24* (Yale, 1924)

Nicholson, R., *Edward III and the Scots* (Oxford, 1965)

Nicholson, R., *Scotland, the Later Middle Ages* (Edinburgh, 1974)

Oram, R. and Stell, G., *Galloway, Land and Lordship* (Edinburgh, 1991)

Paul, J. B. (ed.), *The Scots Peerage*, 9 vols (Edinburgh, 1904–14)

Pinkerton, J., *The History of Scotland* (London, 1797)

Prestwich, M., *The Three Edwards* (London, 1980)

Rae, T. I., *The Administration of the Scottish Frontier, 1513–1603* (Edinburgh, 1966)

Raine, J., *North Durham* (London, 1852)

Rawcliffe, C., *The Staffords, Earls of Stafford and Dukes of Buckingham* (Cambridge, 1978)

Richardson, J. S., Wood, M. and Tabraham, C. J., *Melrose Abbey*, H.M.S.O. (Edinburgh, 1981)

Roberts, M., *The Early Vasas* (Cambridge, 1968)

Robson, R., *The Rise and Fall of the English Highland Clans* (Edinburgh, 1989)

Ross, C., *Edward IV* (London, 1974)

Royal Commission on the Ancient and Historical Monuments of Scotland: Galloway, 2 vols (Edinburgh, 1914)

Royal Commission on the Ancient and Historical Monuments of Scotland: East Lothian (Edinburgh, 1924)

Royal Commission on the Ancient and Historical Monuments of Scotland: Roxburghshire, 2 vols (Edinburgh, 1956)

Royal Commission on the Ancient and Historical Monuments of Scotland: Selkirk (Edinburgh, 1957).

Simpson, W. D., Breeze, D. J. and Hume, J. R., *Bothwell Castle*, H.M.S.O. (Edinburgh, 1985)

Smith, J. H., *The Great Schism* (London, 1970)

Storey, R. L., *The End of the House of Lancaster* (London, 1966)

Sumption, J., *The Hundred Years War: Trial by Battle* (London, 1990)

Tuck, A., *Crown and Nobility* (London, 1985)

Tytler, P. F., *A History of Scotland*, 9 vols (Edinburgh, 1828–43)

Vale, M. G. A., *Charles VII* (London, 1974)

Vaughan, R., *John the Fearless* (London, 1966)

Vaughan, R., *Philip the Good* (London, 1970)

Walker, S., *The Lancastrian Affinity, 1361–99* (Oxford, 1990)

Watt, D. E. R., *A Biographical Dictionary of Scottish Graduates to A.D. 1410* (Oxford, 1977)

Wilkinson, B., *A Constitutional History of England in the Fifteenth Century* (London, 1964)

Wolffe, B., *Henry VI* (London, 1981)

Wormald, J. M., *Lords and Men in Scotland: Bonds of Manrent, 1442–1603* (Edinburgh, 1985)

Wylie, J. H. and Waugh, W. T., *The Reign of Henry the Fifth*, 3 vols (Cambridge, 1914–29)

Yeoman, P., *Medieval Scotland, an archaeological perspective* (Edinburgh, 1995)

D. ARTICLES AND BOOK CHAPTERS

Anderson, J. M., 'The beginnings of St. Andrews University, 1410–18', *Scottish Historical Review*, 8 (1911), 225–48

Barrow, G. W. S., 'Lothian in the First War of Independence', *Scottish Historical Review*, 55 (1976), 151–71

Barrow, G. W. S., 'The Aftermath of War', *Transactions of the Royal Historical Society*, Fifth Series, 28 (1978), 103–25

Bean, J. M. W., 'Henry IV and the Percies', *History*, 44 (1959), 212–27

Bean, J. M. W., 'The Financial Position of Richard duke of York', in J. Gillingham and J. C. Holt (eds), *War and Government in the Middle Ages* (Cambridge, 1984), 182–98

W. N. M. Beckett, 'The Perth Charterhouse before 1500', *Analecta Cartusiana*, 127 (1988), 1–74

Boardman, S., 'Lordship in the North-East: The Badenoch Stewarts I. Alexander Stewart Earl of Buchan', *Northern Scotland*, 16 (1996), 1–30

Boardman, S., 'Chronicle Propaganda in Fourteenth Century Scotland: Robert the Steward, John of Fordun and the 'Anonymous Chronicle'', *Scottish Historical Review*, 76 (1997), 23–43

Britnell, R. H., 'The Economic Context', in A. J. Pollard (ed.), *The Wars of the Roses* (London, 1995), 41–65

Brown, A. L., 'The English Campaign in Scotland, 1400', in H. Hearder and H. R. Loyn (eds), *British Government and Administration. Studies presented to S. B. Chrimes* (Cardiff, 1974), 40–54

Brown, J. M., 'The Exercise of Power', in J. M. Brown (ed.), *Scottish Society in the Fifteenth Century* (London, 1977), 270–80

Brown, M., '"That old serpent and ancient of evil days": Walter earl of Atholl and the death of James I', *Scottish Historical Review*, 71 (1992), 23–45

Brown, M., 'Scotland Tamed: Kings and Magnates in Late Medieval Scotland: A review of recent work', *Innes Review*, 45 (1994), 120–46

Brown, M., 'Regional Lordship in North-East Scotland: The Badenoch Stewarts II. Alexander Stewart Earl of Mar', *Northern Scotland*, 16 (1996), 31–54

Brown, M., '"Rejoice to Hear of Douglas": The House of Douglas and the Presentation of Magnate Power in late Medieval Scotland', in *Scottish Historical Review*, forthcoming

Burns, J. H., 'Scottish Churchmen and the Council of Basle', *Innes Review*, 13 (1963), 1–53, 157–89

I. Campbell, 'A Romanesque Revival and the Early Renaissance in Scotland c. 1380–1513', *Journal of the Society of Architectural Historians*, 54 (1995), 302–25

J. Campbell, 'England, Scotland and the Hundred Years War in the Fourteenth Century', in J. R. Hale, J. Highfield and B. Smalley (eds), *Europe in the Late Middle Ages* (London, 1965), 184–216

Cherry, M., 'The Courtenay Earls of Devon: the formation and disintegration of a late medieval aristocratic affinity', *Southern History*, 1 (1979), 71–97

Chevalier, B., 'Les Ecossais dans l'armée de Charles VII jusqu'à la bataille de Verneuil', *Jeanne d'Arc: une époque, un rayonnement* (Paris, 1982), 85–94

Contamine, P., 'Froissart and Scotland', G. Simpson (ed.), *Scotland and the Low Countries, 1124–1994* (East Linton, 1996), 43–58

Cosgrove, A., 'The Emergence of the Pale', in A. Cosgrove, ed., *A New History of Ireland, II, Medieval Ireland* (Oxford, 1987), 557–68

Ditcham, B. L. G., ''Mutton Guzzlers and Wine Bags': Foreign Soldiers and Native Reactions in Fifteenth Century France', in C. T. Allmand, *Power, Culture and Religion in France* (Woodbridge, 1989), 1–13

Duncan, A. A. M., '*Honi soit qui mal y pense*: David II and Edward III, 1346–52', *Scottish Historical Review*, 67 (1988), 113–41

Duncan, A. A. M., 'The War of the Scots, 1306–23', *Transactions of the Royal Historical Society* (1992), 125–51

Duncan, A. A. M., 'The Laws of Malcolm MacKenneth', in A. Grant and K. Stringer (eds), *Medieval Scotland: Crown, Lordship and Community* (Edinburgh, 1993) 239–73

Durkan, J., 'William Turnbull, Bishop of Glasgow', *Innes Review*, 2 (1953), 1–59

Ford, C. J., 'Piracy or Policy: the Crisis in the Channel, 1400–1403', *Transactions of the Royal Historical Society*, 29 (1979), 63–78

Frame, R., 'Power and Society in the Lordship of Ireland, 1272–1377', *Past and Present*, 76 (1977), 3–33

Frame, R., 'The Defence of the English Lordship', in T. Bartlett and K. Jeffery, *A Military History of Ireland* (Cambridge, 1996), 76–98

Goodman, A., 'A Letter from an Earl of Douglas to a King of Castile', *Scottish Historical Review*, 64 (1985), 68–78

Goodman, A., 'The Anglo-Scottish Marches in the Fifteenth Century: A Frontier Society?', in R. A. Mason (ed.), *Scotland and England, 1286–1815* (Edinburgh, 1987), 18–33

Goodman, A., 'Religion and Warfare in the Anglo-Scottish Marches', in R. Bartlett and A. MacKay (eds), *Medieval Frontier Societies* (Oxford, 1989), 245–66

Goodman, A., 'Introduction', in A. Tuck and A. Goodman (eds), *War and Border Societies in the Middle Ages* (London, 1992), 1–29

Grant, A., 'Earls and Earldoms in late Medieval Scotland 1310–1460', J. Bossy and P. Jupp (eds), *Essays Presented to Michael Roberts* (Belfast, 1976)

Grant, A., 'The Development of the Scottish Peerage', *Scottish Historical Review*, 57 (1978), 1–27

Grant, A., 'The Revolt of the Lord of the Isles and the Death of the Earl of Douglas', *Scottish Historical Review*, 60 (1981), 169–74

Grant, A., 'Crown and Nobility in Late Medieval Britain', in R. A. Mason (ed.), *Scotland and England 1286–1815* (Edinburgh, 1987), 34–59

Grant, A., 'The Otterburn War from the Scottish point of view', in A. Tuck and A. Goodman (eds), *War and Border Societies in the Middle Ages* (London, 1992), 30–64

Grant, A., 'Aspects of National Consciousness in Late Medieval Scotland', in C. Bjørn, A. Grant and K. Stringer (eds), *Nations, Nationalism and Patriotism in the European Past* (Copenhagen, 1994), 68–95

Grant, A., 'Richard III and Scotland', in *The North in the Age of Richard III* (Stroud, 1996), 115–48

Griffiths, R. A., 'Local Rivalries and National Politics: The Percies, the Nevilles and the Duke of Exeter', *Speculum*, 43 (1968), 589–632

Guenée, B., 'Espace and Etat en France mediévale', *Annales*, 32 (1968), 744–58

Hannay, R. K., 'A Chapter Election at St. Andrews in 1417', *Scottish Historical Review*, 13 (1916–17), 327

M. A. Hicks, 'Bastard Feudalism: Society and Politics in Fifteenth Century England', in M. A. Hicks, *Richard III and his Rivals* (London, 1991), 1–40

R. G. Inglis, 'Ancient Border Highways', *Proceedings of the Society of Scottish Antiquaries*, 58 (1924), 203–23

Lewis, P. S., 'France and the Fifteenth Century: Society and Sovereignty', in P. S. Lewis, *Essays in Later Medieval French History* (London, 1985) 3–28

Lewis, P. S., 'Decayed and Non-Feudalism in Later Medieval France', in P. S. Lewis, *Essays in Later Medieval French History* (London, 1985) 41–69

Lewis, P. S. 'Of Breton Alliances and other matters', in P. S. Lewis, *Essays in Later Medieval French History* (London, 1985), 70–90

Lyall, R. J., 'The Lost Literature of Medieval Scotland', in J. D. McClure and M. R. G. Spiller (eds), *Bryght Lanternis: Essays on the Literature of Medieval and Renaissance Scotland* (Aberdeen, 1989)

Lythe, S. G. E., 'Economic Life', in J. M. Brown (ed.), *Scottish Society in the Fifteenth Century* (London, 1977), 66–83

McDiarmid, M. P., 'Richard Holland's *Buke of the Howlat*: an interpretation', *Medium Aevum*, 38 (1969), 277–90

McNiven, P., 'The Scottish Policy of the Percies and the Strategy of the Rebellion of 1403', *Bulletin of the John Rylands Library*, 62 (1979), 498–530

MacQueen, H. L., 'The Laws of Galloway, A preliminary survey', in R. Oram and G. Stell, *Galloway, Land and Lordship* (Edinburgh, 1991), 131–43

MacQueen, H. L., 'The Kin of Kennedy, 'Kenkynnol' and the Common Law', in A. Grant and K. Stringer, *Medieval Scotland: Crown, Lordship and Community* (Edinburgh, 1993), 274–96

Macrae, 'The English Council and Scotland in 1430', *English Historical Review*, 54 (1939), 415–26

Mason, E., 'Legends of the Beauchamps' ancestors: The Use of Baronial Propaganda in Medieval England', *Journal of Medieval History*, x (1984), 25–40

Nicholls, K., 'Gaelic Society and Economy', in A. Cosgrove (ed.), *A New History of Ireland, ii, Medieval Ireland* (Oxford, 1987), 397–438

Reed, J., 'The Ballad and the Source: Some literary reflections on *The Battle of Otterburn*', in A. Tuck and A. Goodman (eds), *War and Border Societies in the Middle Ages* (London, 1992), 94–123

Reid, R. C., 'Edward de Balliol', *Transactions of the Dumfriesshire and Galloway Antiquarian and Natural History Society*, 35 (1956–57), 38–63

Reid, R. R., 'The Wardens of the Marches; its origins and early history', *English Historical Review*, 35 (1917), 479–96

Reid, W. S., 'The Douglases at the court of James I, *Juridical Review*, 56 (1944), 77–88

Sayles, G. O., 'The Rebellious First Earl of Desmond', in J. A. Watt, J. B. Morrall and F. X. Martin (eds), *Medieval Studies presented to Aubrey Gwynn* (Dublin, 1961), 203–29

Scammell, J., 'Robert I and the North of England', *English Historical Review*, 73 (1958), 385–403

Simpson, M. A., 'The Campaign of Verneuil', *English Historical Review*, 49 (1934), 93–100

Stewart, M., 'Holland of the Howlat', *Innes Review*, 23 (1972), 3–15

Stewart, M., 'Holland's *Howlat* and the Fall of the Livingstones', *Innes Review*, 23 (1975), 67–79

Storey, R. L., 'The Wardens of the Marches of England towards Scotland' *English Historical Review*, 74 (1957), 593–615

Stringer, K., 'Periphery and Core in Thirteenth Century Scotland: Alan son of Roland, Lord of Galloway and Constable of Scotland', in A. Grant and K. Stringer, *Medieval Scotland: Crown, Lordship and Community* (Edinburgh, 1993), 82–115

Swanson, R., 'The University of St. Andrews and the Great Schism, 1410–19', *Journal of Ecclesiastical History*, 26 (1975), 223–45

Swinton, G. S. C., 'John of Swinton: a border fighter in the middle ages', *Scottish Historical Review*, 16 (1919), 261–79

Thompson, J. M., 'A Roll of the Scottish Parliament, 1344', *Scottish Historical Review*, 35 (1912), 235–40

Tuck, A., 'Richard II and the Border Magnates', *Northern History*, 3 (1968), 27–52

Tuck, A., 'Northumbrian Society in the Fourteenth Century', *Northern History*, 6 (1971), 22–39

Tuck, A., 'War and Society in the Medieval North', *Northern History*, 21 (1985), 31–52

Tuck, A., 'The Emergence of a Northern Nobility', *Northern History*, 23 (1986), 1–17

Tuck, A., 'The Percies and the Community of Northumberland in the Fourteenth Century', in A. Tuck and A. Goodman (eds), *War and Border Societies in the Middle Ages* (London, 1992), 178–95

Vale, M., 'Seigneurial Fortification and Private War in Later Medieval Gascony', in M. C. E. Jones (ed.), *Gentry and Lesser Nobility in Later Medieval Europe* (Gloucester, 1986), 133–48

Webster, A. B., 'English Occupations of Dumfriesshire', *Transactions of the Dumfries-shire and Galloway Antiquarian and Natural History Society*, 35 (1956–57), 64–80

Webster, A. B., 'Scotland without a King', in A. Grant and K. Stringer, *Medieval Scotland: Crown, Lordship and Community* (Edinburgh, 1993), 228–38

Wormald, J. M., 'Taming the Magnates?', in K. J. Stringer (ed.), *Essays on the Nobility of Medieval Scotland* (Edinburgh, 1985), 270–80

Young, A., 'Noble Families and Political Factions in the Reign of Alexander III', in N. H. Reid (ed.), *Scotland in the Reign of Alexander III* (Edinburgh, 1990), 1–30

E. THESES

Boardman, S. I., 'Politics and the Feud in Late Medieval Scotland' (unpublished Ph.D. thesis, University of St. Andrews, 1989)

Borthwick, A., 'The Council under James II: 1437–60' (unpublished Ph.D. thesis, University of Edinburgh, 1989).

Brown, M. H., 'Crown-Magnate Relations in the Personal Rule of James I, 1424–1437' (unpublished Ph.D. thesis, University of St. Andrews, 1991)

Ditcham, B. G. H., 'The Employment of Foreign Mercenary Troops in the French Royal Armies, 1415–70' (unpublished Ph.D. thesis, University of Edinburgh, 1978)

Grant, A., 'The Higher Nobility of Scotland and their Estates, 1371–1424' (unpublished D.Phil thesis, University of Oxford, 1975)

Kelley, M. G., 'The Douglas Earls of Angus: A study of the social and political bases of power of a Scottish family from 1389 to 1557' (unpublished Ph.D. thesis, University of Edinburgh, 1973)

MacDonald, A., 'Crossing the Border: A Study of the Scottish Military Offensives against England c. 1369–1403' (unpublished Ph.D. thesis, University of Aberdeen, 1995)

O'Brien, I., 'The Scottish Parliament in the Fifteenth and Sixteenth Centuries' (unpublished Ph.D. thesis, University of Glasgow, 1980)

Väthjunker, S., 'A Study of the Career of Sir James Douglas: The Historical Record versus Barbour's *Bruce*' (unpublished Ph.D. thesis, University of Aberdeen, 1992)

Index